Foliage Penetration Radar

Foliage Penetration Radar

Detection and Characterization of Objects Under Trees

Mark E. Davis

SciTech Publishing, Inc.
Raleigh, NC
scitechpub.com

SciTech Publishing, Inc.

Published by SciTech Publishing, Inc.
911 Paverstone Drive, Suite B
Raleigh, NC 27615
(919) 847-2434, fax (919) 847-2568
scitechpublishing.com

Editor: Dudley R. Kay
Editorial Assistant: Katie Janelle
Production Manager: Robert Lawless
Typesetting: MPS Limited, a Macmillan Company
Cover Design: Brent Beckley
Printer: Sheridan Books, Inc., Chelsea, MI

Printed in the United States of America
10 9 8 7 6 5 4 3 2 1

ISBN: 978-1-891121-00-5

Library of Congress Cataloging-in-Publication Data

Davis, Mark E. (Mark Edward), 1945-
Foliage penetration radar : detection and characterization of objects under trees / Mark E. Davis.
p. cm.
Includes bibliographical references and index.
ISBN 978-1-891121-00-5 (hardcover : alk. paper) 1. Ground penetrating radar. 2. Forest canopies.
3. Earth–Surface–Remote sensing. 4. Aerial observation (Military science) I. Title.
TK6592.G7D38 2011
621.3848'5–dc22 2011005731

Dedicated to my late father,

Capt. Jack E. Davis, USN

Contents

Preface

The story of foliage penetration RADAR has had many authors over its almost half century of development. This attempt at reconstructing the early developments owes a great debt to Mr James Rodems, formerly of Syracuse University Research Corporation who lead the research, development and early deployment of one of the two systems in the 1960s. The majority of the material in Chapter 1 came from his archives and personal descriptions of the motivation and trials that led to both ground based and airborne testbed.

There were many pioneers in the second phase of FOPEN development during the late 1980s to mid 1990s. But without the continuous support and technical leadership of Dr Serpil Ayasli of MIT Lincoln Laboratory, the breadth of innovation in phenomenology, waveforms, and image understanding would not have matured into today's solid foundation of science. Two testbeds were developed as independent efforts, each under a strong leader: Stanford Research Institute's FOLPEN under Roger Vickers, and Swedish Defence Research Estableshment's CARABAS under Hans Hellsten. Several other testbeds were constructed during this period to provide complementary geoscience or military research objectives. Each of the airborne testbeds that collected and refined the ultra wide band synthetic aperture RADAR signals is covered in Chapter 2. They were conceived to implement an important set of innovations, leading to understanding of the importance of frequency choice, polarization, radio frequency interference removal, and target and clutter characterization for efficient detection of objects under dense forests. Much of this development and test was funded by the Defense Advanced Research Projects Agency under the program management of a sequence of leaders that included Dom Giglio (1988–1995), Mark Davis (1995–1998) and Lee Moyer (1999–2005).

Modern foliage penetration RADAR continues to advance with the continuous improvement in high speed digital signal processing. The single most impediment to its general use is the proliferation of personal and wideband communications into the radio frequency spectrum. Frequency spectrum allocation and protection of specific frequencies for safety of life and emergency communications requires careful attention to the choice of waveform. It will continue to be important to develop cognitive processing to avoid interference to or from other users of this spectrum.

The author would like to acknowledge all of the pioneers who preceeded and succeeded his involvement in foliage penetration radar development. The past 15 years has been a very enjoyable journey into the scientific and geopolitical evolution of ultra wideband radar. He would also like to thank his parents, Jack and Mary Lou Davis for encouraging his scientific development. And most importantly he would like to thank his wife Diane Rogers Davis, and two sons Colin and Shelby for the patience and encouragement in a long journey into RADAR development, test and operation.

Mark E. Davis
medavis@ieee.org
March 2011

CHAPTER 1

History of Battlefield Surveillance

The military has long known of the importance of surveillance on the battlefield. The development of the balloon prior to the 1861 American Civil War, as illustrated in Figure 1–1, gave the commander on the ground the ability to see longer distance and with more safety than any forward ground observer [1]. Since the military had more access to balloons and the early airplane in First World War, the opposing side obviously sought a counter to this battlefield surveillance capability.

Often, airborne surveillance was opposed simply by hiding in woods to deny the other side an accurate knowledge of one's tactical intent. However, when these surveillance platforms were given weapons, the ensuing counter was to shoot at the platforms from the ground position. An alternative to offensive retaliation was to exploit a natural obscurant such as maneuvering in the fog or rain or creating a smoke screen to deny long-range surveillance. Until RADAR was developed, armies of the world successfully employed tactics of concealment and deception to deny their adversaries the current knowledge of their position and maneuver tactics.

In 1903, Christian Hülsmeyer conducted the first experiments using RADAR by scattering radio waves off ships; and his patent followed in 1904. However, at that time the German military did not support his work because it considered radio wave components to be still underdeveloped. As Skolnik points out, RADAR did not generate much scientific interest until the 1920s and 1930s [2]. Marconi's research as well as developments at the Naval Research Lab demonstrated the ability to detect ships on the surface and aircraft in the air. Both long-wave and microwave RADAR were being developed in Europe and the United States. By the start of the Second World War, RADAR technology had progressed sufficiently to detect aircraft and ships at long ranges from ground installations. Soon aircraft were being provided RADARs

FIGURE 1–1

Battlefield surveillance in 1861 [1]

for all weather detection of air and ground targets. However, ground clutter (especially forests) was a significant problem to early airborne RADARs, since it competed with detection of targets, or concealed those objects hidden under the clutter. No real attempts were made to image or penetrate this obscuration for detecting objects, primarily due to the lack of stable waveform and signal processing technology.

In the early 1960s, the US Army developed the first battlefield surveillance RADAR—the OV-1 APS-94 side-looking array RADAR (SLAR), which was for detecting military encampments and large groups of artillery and mechanized vehicles on the battlefield [3]. In the early 1970s, the army determined that there was also a need for detecting large numbers of moving vehicles, at a significant range from the forward edge of the battle area (FEBA). The first ground moving target indication (GMTI) system for battlefield surveillance was developed as the standoff target acquisition system (SOTAS). It was constructed using the APS-94 RADAR with a moving target mode and operated from a UH-1 helicopter. The helicopter was necessary to minimize the platform motion and to provide sufficiently low minimum discernable velocity (MDV) detections over a wide area. The SOTAS prototype was tested in the United States and in Germany under the return of forces to Germany

FIGURE 1–2

SOTAS RADAR during 1980 REFORGER deployment [4]

(REFORGER) exercises, as shown in Figure 1–2, and was accepted by military leaders as a capable operational capability. Because of the demonstrated importance of detecting slowly moving troops at long distance, the army started to develop an operational system to be installed on a Black Hawk UH-60 helicopter. The Blackhawk had a larger capacity payload and a longer endurance than the UH-1. But the system development was stopped in 1978 by the Secretary of Defense because even longer endurance and more survivability of a manned platform were needed [5].

The battlefield surveillance capabilities of SLAR and SOTAS soon led to the development of the Joint Surveillance and Target Acquisition system (JOINT STARS) for use by the army and air force [5]. The JOINT STARS standoff battlefield surveillance capabilities could be integrated on a high-altitude, multiengine aircraft for longer endurance and significantly longer standoff for survivability. The benefits of JOINT STARS combining SAR and GMTI on the battlefield are extensively documented and have been reproduced and fielded on many international platforms. All these early battlefield surveillance RADAR systems were developed in the microwave frequency band. Microwave frequencies were important to provide all-weather, long-range, high-probability detection of vehicles and structures and to allow systems to be small enough that they could be carried on tactical aircraft [6].

However, there was one important operational issue—the opposing combatants understood X-band RADAR's limitations to see through forest cover. Tactics had been developed to deny microwave RADARs the ability to image movement and locate ground forces. Hiding in tree lines and using other forms of camouflage and concealment quickly countered operational RADAR. This concealment tactic had become highly effective against RADAR, as it had for early optical surveillance. Thus, there was an evolving need to detect fixed and moving targets under foliage, as a complement to the very capable microwave battlefield surveillance RADAR systems.

The first development of foliage penetration (FOPEN) RADAR occurred during the Vietnam conflict, where early systems were needed to detect and recognize ground-moving targets [7]. Specifically, there was a compelling need to detect and locate insurgent soldiers walking through the dense tropical forests. Two innovations were needed: (1) coherent waveforms and the associated signal processing; and (2) RADAR installations on major hills and masts. These two innovations increased the target signal-to-noise and minimized the clutter spread that masked the small returns from personnel and vehicles. However, it did not provide any ability to detect stationary, man-made objects. A parallel development of FOPEN synthetic aperture RADAR (SAR) was needed to detect man-made objects under the trees. The required technology innovations for foliage penetration SAR were wideband image processing and coherent discrimination of man-made objects from the background clutter.

FOPEN RADAR has continued to be a developing technology to provide geospatial and military users with detection and characterization of objects under dense foliage. Many areas of the earth are remote and inhospitable for characterization, as well as monitoring the effects of weather, atmosphere and geological changes on the region. Similarly, military commanders want to know about recent construction or tactical maneuvers in an area covered by dense foliage. RADAR has the inherent ability to characterize a wide area, to assess changes in fixed objects, and to detect and track moving objects. Early RADARs were limited to detection and tracking of objects by the attenuation and scattering of clutter between the RADAR and the features being observed. Forests have been particularly difficult environments due to the scattering of the waveforms and severe attenuation at microwave frequencies.

Over the past 40 years, the advances in waveform synthesis and digital signal processing have given the RADAR community the ability to observe the behavior of fixed and moving objects under foliage. The most significant advance was in synthesis and reception of ultra-wideband waveforms, where the signals can achieve over 50% fractional bandwidth. Digital signal processing enables the RADAR to compensate for scattering of the signal by the foliage and to discriminate the man-made objects from the surrounding clutter. However, before the details of these innovations are explored, it is beneficial to trace the history and early motivation for foliage penetration RADAR.

1.1 Early FOPEN MTI RADAR

In the mid 1960s, the US Army Scientific Advisory Board explored the technical feasibility of detecting insurgents in deep forests. Coincidentally, two University of Rochester graduate students were conducting experiments to understand the modulation effect of personnel walking between a radio transmission tower and their radio receiver [7]. They conducted a series of successful experiments using a 175 MHz CW signal to characterize the source of signal modulation. Based on their results, the university researchers submitted a 1965 proposal for "ORCRIST, An Anti-Guerrilla Detection System" to the Army Land Warfare Laboratory (LWL) in Aberdeen, Maryland. This is the first documentation of a foliage penetration RADAR system development. The army soon hired one of the researchers, Louis V. Surgent Jr., who would lead the development of FOPEN technology at LWL for the next 10 years.

Early analysis of propagation loss through forests was very pessimistic about the ability to detect personnel in dense woods. The losses quoted for 100 MHz propagation varied from 0.02 dB to 1.0 dB per meter depending on the source of radio wave, the density of the forest, and the receiver characteristics. However, the DARPA SEACORE project made extensive measurements of high frequency (HF) and very high frequency (VHF) propagation in order to characterize communications links in tropical jungle environments. Their hypothesis was that the most efficient radio propagation was not via a straight line through the forest but over the tops of the trees with eventual diffraction into the forest. When communications, and eventually RADAR propagation was from a tower, the foliage loss was dramatically reduced [8].

As a result of these observations on communications, the Army LWL conducted a series of tests in a dense Georgia forest with light undergrowth, using a 140 MHz CW signal to quantify the potential RADAR propagation losses. Both horizontal and vertical polarizations were tested, and horizontal polarization exhibited the lower propagation loss. Figure 1–3 shows the propagation losses of horizontal signals between two Yagi antennae, one transmitting and the other receiving. Two different transmitter heights were used (2 meters and 13 meters) to determine the effects of propagation via a direct path and over the tops of the 10 to 15 meter high trees.

The solid line on Figure 1–3 represents the free-space, one-way propagation loss, which would be proportional to R^{-2}. The two dotted lines correspond to a loss proportional to R^{-4}, which is representative of multipath propagation, between the direct path and ground bounce to the receiver. The returns from the 2 meter high antenna follow the R^{-4} loss very closely. However, the 13 meter high antenna appears to exhibit lower losses as a function of range.

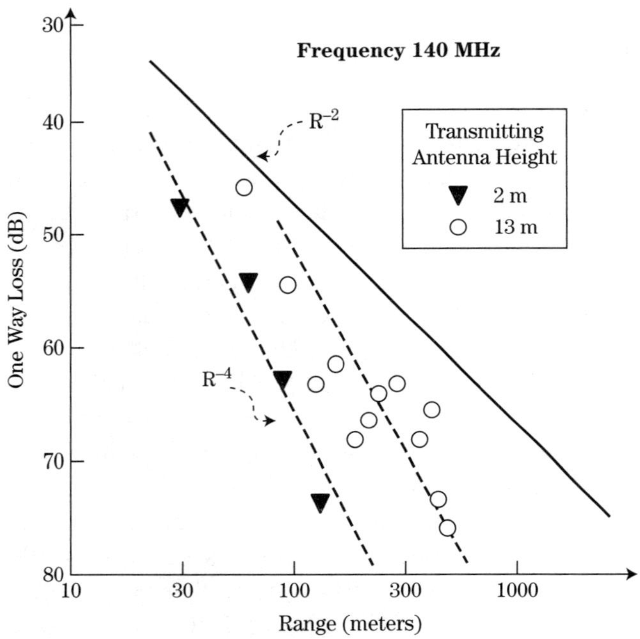

FIGURE 1–3

Propagation of CW signals between two 140 MHz Yagi antennae [7]

To better illustrate the difference between free-space propagation and the measured losses, Figure 1–4 plots the differential loss between R^{-2}propagation and the measured propagation loss. The 2 meter high antenna exhibits an additional R^{-2}loss in signal, over the full range of measurements. However, the 13 meter high antenna exhibits a nearly constant 20 dB attenuation, independent of range. This was interpreted to verify the SEACORE observations for propagation over the forest, but more quantitative tests were required before developing an operational FOPEN system [9].

The first FOPEN GMTI system was constructed in 1967 and served as a brassboard for data collections in Florida, Puerto Rico, and Panama. Initial trials were with a noncoherent system, as coherent system components were still relatively immature. This brassboard was taken to Panama and Hawaii in 1968, and was deployed on a moderate height 13 meter mast for evaluations verifying detection performance in heavy tropical and jungle environments.

Field observations from this initial system verified that horizontal polarization was preferable for minimizing the foliage loss and detecting personnel with weapons. The wavelength was near resonance with the weapons and provided a better detection mechanism than just the body return of the personnel walking under foliage (referred to as dismounts in military terms, and generally in this work). Measured detection probability was over 90% out to 200 meters, as long as the weather was dry and winds were low.

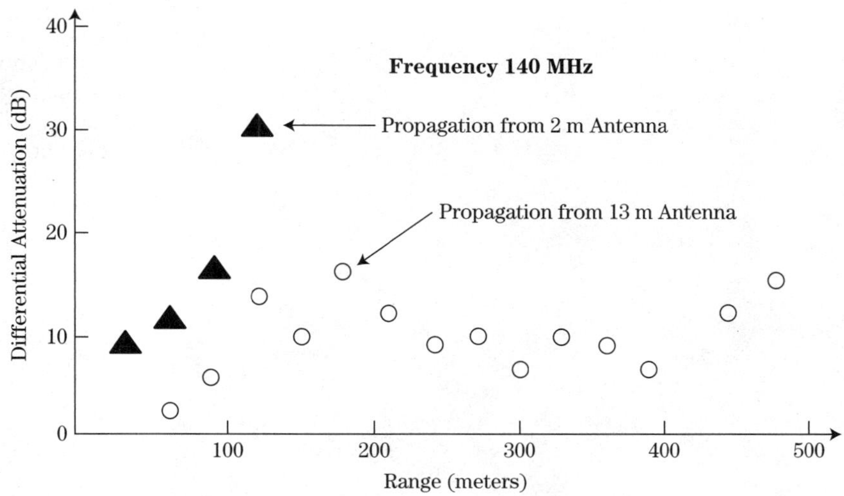

FIGURE 1–4

Differential propagation loss compared with free space [7]

However for a noncoherent system, the proximity of moving trees near the antenna caused amplitude modulation of the transmitted signal. This clutter modulation made the receiver-processor lose sensitivity, reducing detection range and increasing false alarms. In addition to these sensitivity issues, the humid jungle environment made the antenna mast and the electronics unreliable. The Army decided that a new system approach was needed to meet the needs of jungle warfare. The recommendations were to develop a coherent system, to go to a higher mast to get above the tree line, and to conduct more environmental design and testing.

1.1.1 Camp Sentinel RADAR

Under the Camp Sentinel program, DARPA and MIT Lincoln Laboratory undertook the development of an operational area surveillance RADAR to detect insurgents moving under thick jungle canopy. The RADAR system shown in Figure 1–5 operated at a higher frequency (435 MHz) than the earlier brassboard systems. The motivation was to obtain more accurate location of personnel at much longer ranges. Lincoln Laboratory developed an initial Camp Sentinel II service test system, using a solid-state transmitter and a mechanically scanned antenna. This early system had a moderate detection range to 200 meters. But it was signal sensitivity limited by the low solid-state power amplification was and target discrimination limited by clutter modulation with the mechanically scanned antenna [10].

FIGURE 1–5

Camp Sentinel II RADAR installed on hill in Vietnam *Source*: MIT Lincoln Laboratory [10]

To provide longer detection range and higher losses at ultra high frequencies (UHF), the system was upgraded to Camp Sentinel III with a high power-aperture design. A 3.5 meter diameter, 1 meter high cylindrical antenna was developed through the Army Harry Diamond Laboratory. This antenna was mounted on a 33 meter high tower along with a 2 kilowatt tube transmitter. The high tower was chosen primarily to extend the propagation range out to 2 kilometers. Stepped surveillance with 32 beam positions were provided by the cylindrical antenna shown in Figure 1–6, thus eliminating the mechanically scanned antenna masking of low Doppler targets by clutter modulation. The waveform was a coherent, range-gated pulse Doppler MTI dwell, and provided relatively accurate location of the intruders. An automatic alarm processor was developed and installed in the remote operation shelter to alert the operators when a person came within detection range. The recent innovation by the Army Harry Diamond Laboratory of the Kalmus filter, provided a reduction of false alarms from the foliage by using a balanced Doppler processor. This accomplishment was obtained by integrating out the oscillatory, swaying foliage motion competing with the linear motion of a walking person. Finally, an audio output was provided to the operator to give aural indications and discrimination of the detected motion [7].

The Camp Sentinel RADAR was a very large system, weighing over 8,000 pounds. For transportation to theater, it was packaged on a tractor trailer and carried in a C-130 or under a Chinook helicopter. This provided for defense around larger fixed encampments, where there was sufficient height above the

FIGURE 1–6

Camp Sentinel RADAR Antenna [11]

surrounding terrain. However, it required mounting on a hill and atop the 30 meter tower, and thus it presented a very distinctive visual landmark.

1.1.2 M-FOPEN RADAR

The Army LWL developed a more mobile system for use in forward deployed encampments. Called the multipurpose-FOPEN (M-FOPEN) system, it could be carried by a single person as shown in Figure 1–7, and be set up in the jungle in an hour [7]. The objectives were to have a system supporting remote operations, where a large fixed installation was not feasible.

The characteristics of the M-FOPEN RADAR are given in Table 1–1. In contrast to the UHF Camp Sentinel RADAR, this system operated at VHF (140 MHz) to greatly reduce the foliage loss and to minimize the need for a heavy high power-aperture design. The lower frequency and reduced foliage loss also enabled mounting a 50 watt solid-state transmitter directly onto the antenna. Figure 1–8 shows the RADAR antenna installed on the 13 meter tower as tested in Florida light tropical forest environment.

A homodyne receiver was developed to enable exploitation of phase in the detection of moving targets in strong clutter. By making the system coherent, the target Doppler could be easily differentiated from the background clutter. Even in windblown clutter, the distinct characteristics of a person walking in the forest could be detected by the automatic alarm processor and discriminated by the trained operator.

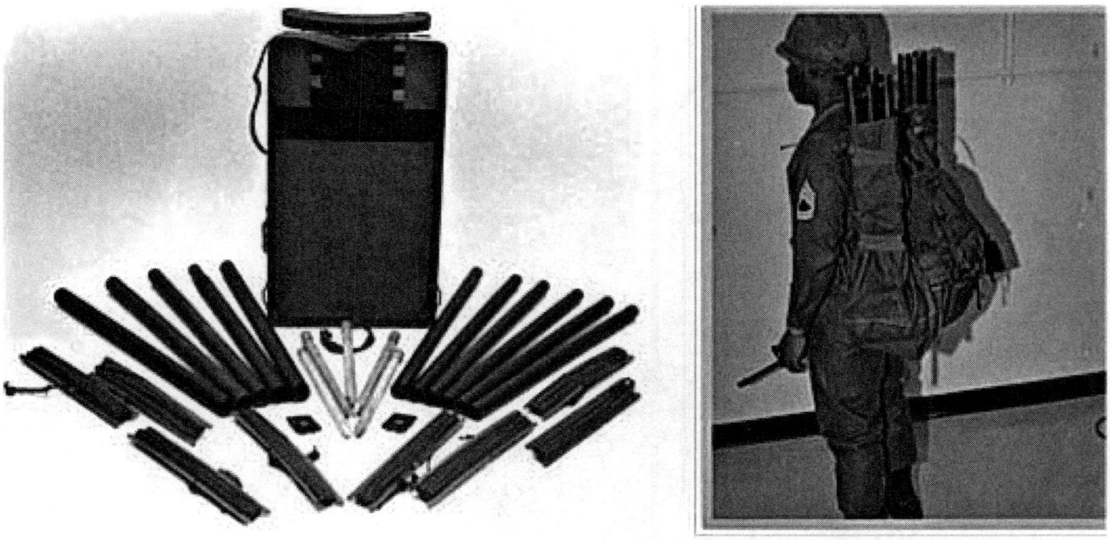

a. M-FOPEN Components b. M-FOPEN Backpack

FIGURE 1–7

M-FOPEN man-transportable RADAR [12]

The most basic operation was a single 15 meter mast, which was an additional 60-pound assembly. This system was designed to operate at 140 MHz frequency center frequency with a 50 Watt peak power solid-state amplification for output directly to the transmit antenna. This closely spaced integration reduced the losses between a remote power amplification subsystem and aided in more power aperture at reduced overall system weight.

The waveform was an unmodulated 0.1 microsecond pulse with a 15 kilohertz pulse repetition frequency yielding an average power of 0.75 watts. Again, the use of simple pulsed, coherent waveform was chosen to maintain the very low system weight and power for transportability. Based on the earlier tests in Florida, Panama, and Puerto Rico, a horizontally polarized antenna

Table 1–1 Characteristics of M-FOPEN RADAR [7]

Frequency	[MHz]	140	Antenna Gain	[dBi]	9
Transmit Power (peak)	[Watt]	50	Antenna Beamwidth	[deg]	45
Pulse Width	[μsec]	0.1	Polarization		Horizontal
PRF	[KHz]	15	Height	[m]	15
Receiver		Homodyne	Detection		Kalmus filter

FIGURE 1–8

M-FOPEN for tests in Florida environment [12]

was constructed to minimize the clutter backscatter from trees. With coherent processing and horizontal polarization, the detection of slowly moving insurgents carrying weapons were greatly enhanced by the Doppler separation of the motion compared with the jungle foliage backscatter.

Eight of these man-portable systems were sent to theater for operation at forward deployed operational sites. The employment of the systems required training of the operators on the potential siting and effects of near-range foliage, hills in the area, and other geographic effects. Figure 1–9 shows a page from one of the M-FOPEN operator's manuals. It clearly indicates the effects of tree line clutter and height of the transmitter on propagation.

During operation in the dense jungle environment, the propagation losses during heavy rain and humidity were found to limit the operational range. So for larger established camps, an additional two base-station systems were developed. These latter had a higher power-aperture design by combining several Yagi antennas and integration with a 20 kilowatt peak power amplifier. This extended the detection range for triple canopy to over 500 meters. The

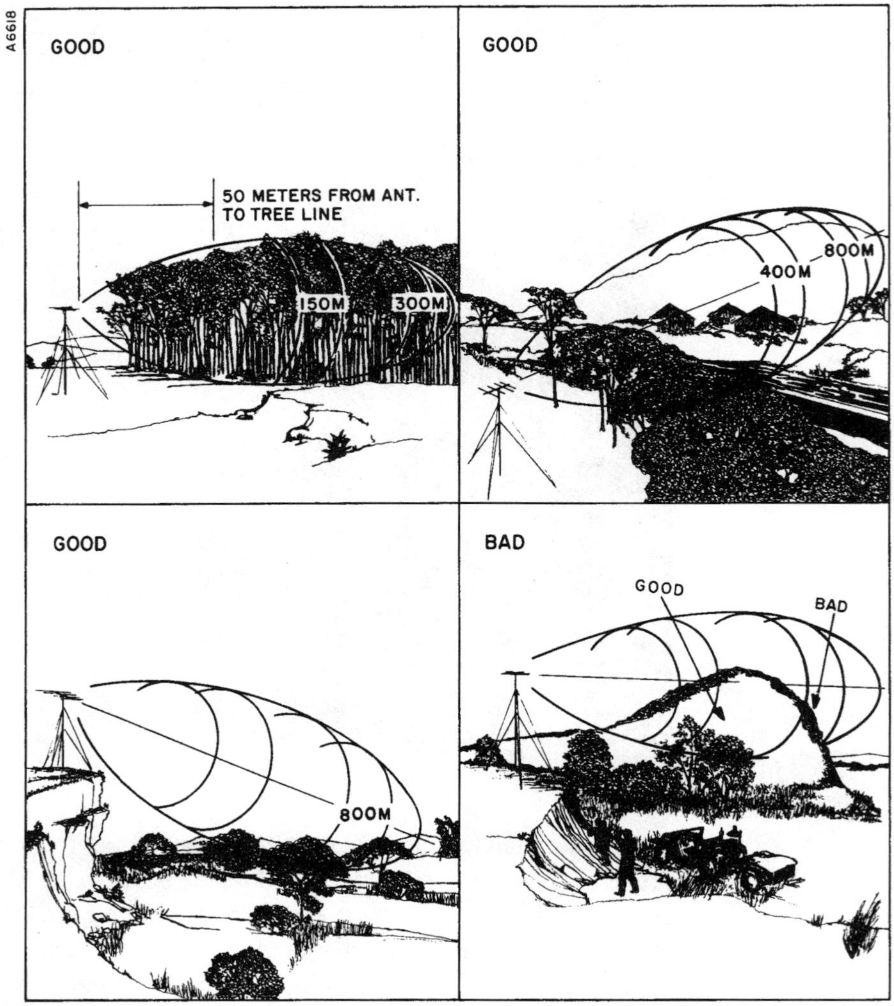

FIGURE 1–9

M-FOPEN siting instructions for operation in jungle environment [12]

displays and controls had multiple outputs for characterizing the low signal-to-clutter returns from the background and a large A-scope display for coarsely localizing the threat.

These two early FOPEN MTI systems were extensively tested in double- and triple-canopy forests in Central America and Vietnam. They clearly established the capabilities of VHF and UHF coherent RADAR for providing effective detection of dismounts and vehicles in forests out to modest tactical ranges.

These state-of-art (early 1970s) FOPEN systems provide new insight into MTI RADAR for detecting slowly moving targets in foliage. Coherent systems were essential for detecting and differentiating target Doppler in dense foliage and for windblown clutter near tree lines. There was also strong interference by man-made signals in the region, and adaptive processing techniques to remove these interfering sources were needed. On the positive side, the earlier assessment of the propagation effects on foliage scattering and signal losses were not generally well known. By having an elevated antenna, the losses were an order of magnitude less than for point-to-point propagation. When the siting of the systems was chosen correctly, detection performance was at significantly longer ranges than early estimates.

Based on the testing in dense tropical jungle, several operational results demanded further developmental testing. Foliage clutter that was either near the transmitter or subject to wind-induced motion affected the adaptive processing. These moving clutter or clutter motion effects produced false detections that were similar to approaching targets. The Doppler frequencies of dismount targets were generally larger than the foliage internal clutter motion, and acoustic processing techniques could be used to differentiate human motion. The short-term solution was to provide an audio channel and adequate training for experienced operators. These techniques were well known in the sonar world and quickly were accepted in the RADAR community. Finally, the volumetric scattering during heavy rains and high winds affected the RADAR sensitivity, even at VHF.

1.1.3 Doppler Frequency of Dismount Targets

Both the operational Camp Sentinel and M-FOPEN RADARs were designed and built as coherent RADAR systems to use the dismount Doppler signature for discrimination against clutter. M-FOPEN used a homodyne receiver to mix the return signal down to baseband as shown in Figure 1–10. The Kalmus tracker (target identification processor) could filter the baseband signal, providing a direct indication of positive and negative Doppler [14]. Two velocity integrators were provided, one for slowly moving persons and one for faster-moving vehicles. The automatic alarm provided the user with a rough bearing, range indication, a characterization of the threat, and an audio output for potential reduction of false alarms.

The output of the phase comparator was processed through a comb filter to produce the spectrogram shown in Figure 1–11 and displayed the major spectral characteristics. In addition to the digitally processed output, the signal could be amplified and sent to the audio channel for an experienced operator to “recognize” the sounds of people approaching, animals in the area, or

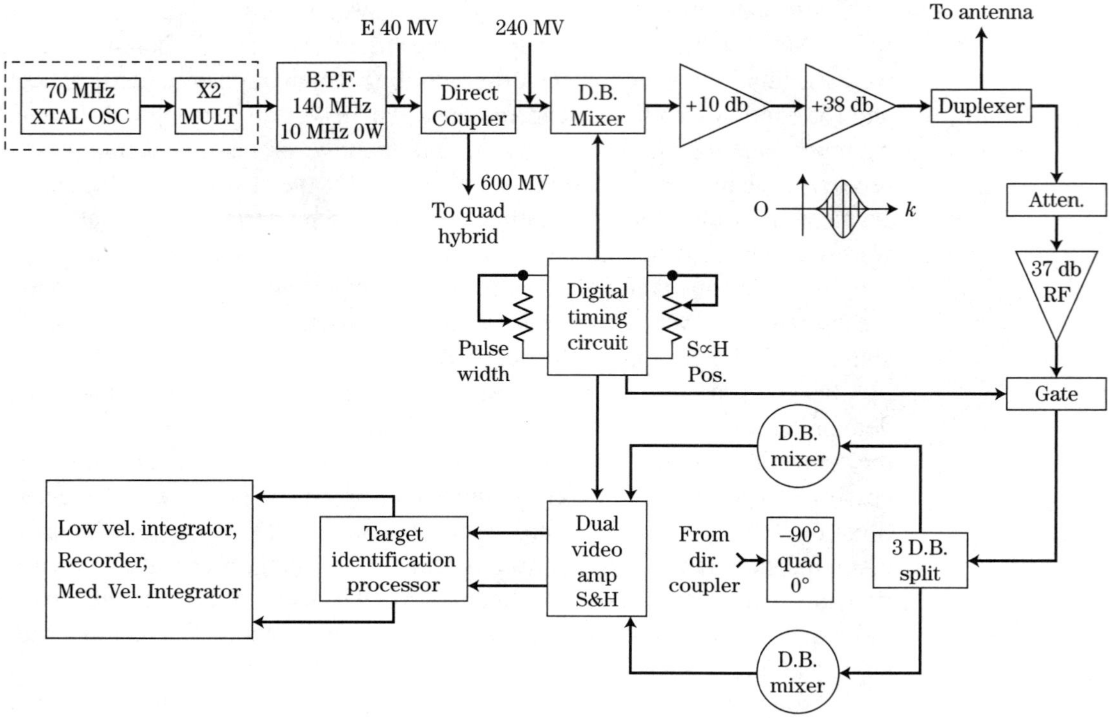

FIGURE 1–10

Block diagram M-FOPEN receiver [7]

windblown foliage clutter. Extensive analyses of the data from several collections of dismounts in foliage clutter were made. The objective was to be able to discriminate a dismount from clutter, and additionally a dismount running, walking, and carrying a weapon. The distinguishing characteristics of the clutter turned out to be the Doppler frequency, amplitude, and time in an isolated range cell [13]. A dismount produced the amplitude versus Doppler signature in a detection cell as shown in Figure 1–12, from the motion of the body toward the RADAR. The background clutter produces a wide variety of features occurring at lower frequencies, primarily below 0.1 Hz. However, the clutter return for a given cell fluctuates over a wider range in high winds, with an amplitude standard deviation on the order of 5 dB. Even in low winds the amplitude standard deviation was seldom less than 1 dB. Hence, it was clearly recognized that the temporal characteristics of a spectrogram was needed to discriminate the motion of dismounts in the jungle environment.

The dismount returns were generally smaller in scattering than the background clutter but could be discriminated by a difference in Doppler

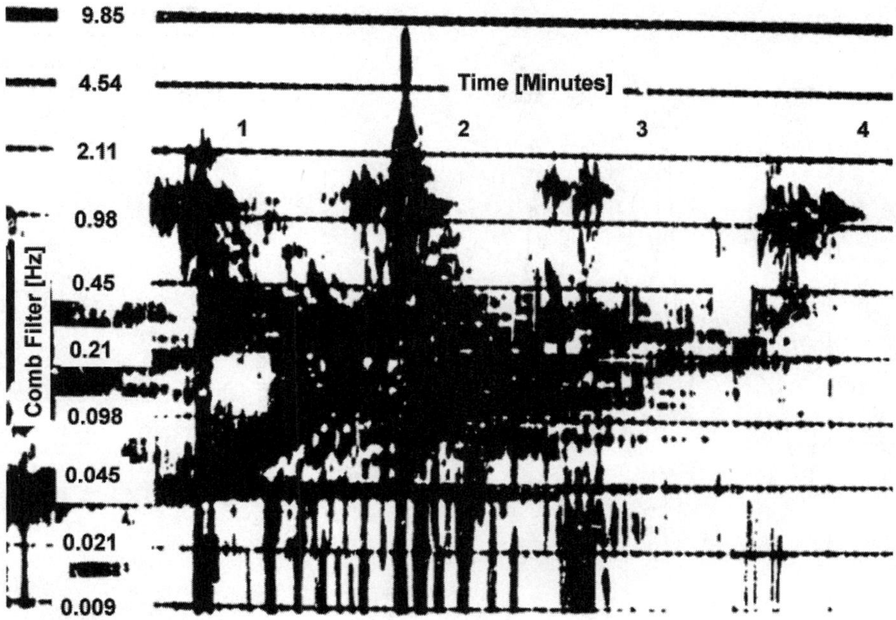

FIGURE 1–11

Dismount target Doppler versus time [7]

frequencies. At the M-FOPEN center frequency of 140 MHz, the Doppler frequency in Hertz is approximately the same value as the target radial motion in m/sec. Most importantly, the dismount motion will exhibit a wide variation in the Doppler spectrum, depending on the motion of personnel toward the RADAR location. Figure 1–12 shows the variation of the target in a single range cell, consisting of both a dismount and foliage clutter.

The ground clutter motion is expected to occur below 0.045 Hz for calm winds but may also include the motion of persons' feet as they contact the ground. The body of the dismount is anticipated to be within a range of radial velocities between 0.3 and 0.6 m/sec, consistent with moving through a jungle trail. Note that a second contribution at 0.96 Hz was thought to be either the peak of the leg motion or arms of the dismount. The analog filters used in M-FOPEN presented an "exponential" build-up of the return from transient responses due to the charging of the filters. These effects can be clearly seen in the 0.96 Hz transient responses. Due to the poor subclutter visibility of the sensor and limitation of the signal processing technology, the desirable capability to discriminate soldiers marching with and without weapons was never realized due to the limited signal-to-clutter ratio of the systems.

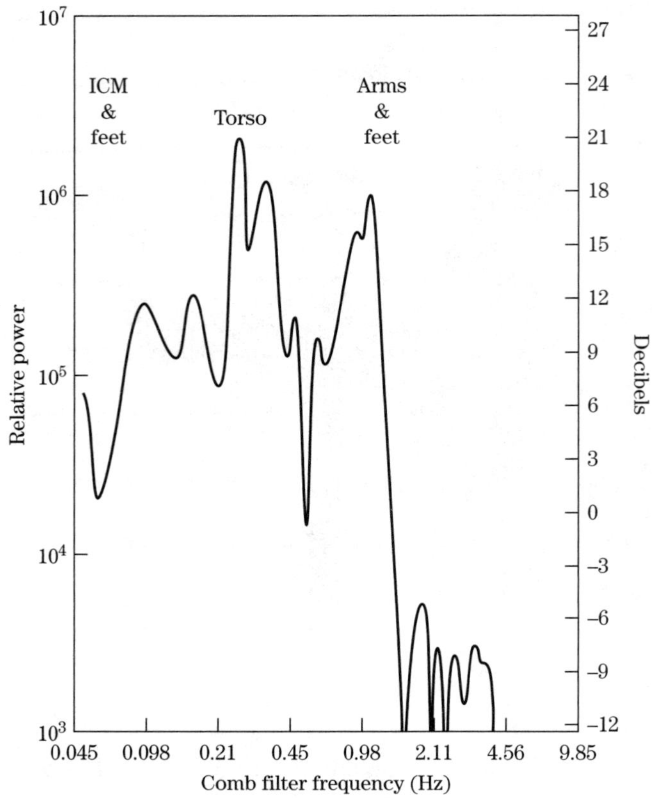

FIGURE 1–12

Amplitude characteristics of dismount from M-FOPEN [7]

In summary, the use of coherent processing with a comb filter was a breakthrough in detecting moving dismounts in clutter. The magnitude and spectral content of the clutter is dependent on wind speed, the proximity of the tree lines to the RADAR, and the height of the antenna. The limitations could be as small as 1 dB for low winds and distant separation and as high as 10 dB for a widely spread clutter Doppler in proximity. Because the spectrum of foliage clutter for a 140 MHz RADAR is below 0.4 Hz, a high-pass filter could be used in the signal processing to distinguish the return of dismounts from the effects of the windblown tree clutter.

1.2 Synthetic Aperture Dual Frequency RADAR

It was clear from operation of the foliage penetration GMTI systems that if the targets were not moving it was impractical to detect the important tactical objects—vehicles and structures. A need existed to develop a method of detecting stationary objects, and an evolving technology was the synthetic aperture RADAR (SAR). Excellent cross-range resolution could be obtained with coherent processing of long collections of RADAR data. With the operational experience in Vietnam, an operational need was established for a tactical SAR system that could provide real-time images over a wide area. Since this system was to be flown in a manned aircraft, the ranges needed to long enough for both the pilot and aircraft surviving against countering threats during the RADAR collection.

The Air Force Cambridge Research Laboratory (AFCRL) initiated an aggressive project to understand detection of stationary, man-made objects under dense foliage and to provide a standoff surveillance system. It had been observed that simultaneous illumination of objects with multiple frequencies would provide a phase discriminator not available in single-frequency SAR. This phase discrimination was strongest when the objects were near one wavelength (i.e., near resonance) and when the wavelength was long compared with the natural clutter. It will be shown in Chapter 2 that natural clutter exhibits a constant differential phase, whereas a man-made structure, such as a dihedral reflector, will show a predictable phase difference versus frequency. The synthetic aperture dual frequency RADAR (SADFRAD) development exploited this difference to easily detect man-made objects under dense foliage [15].

The block diagram in Figure 1–13 shows the method of phase comparison between the two signals. A common crystal oscillator provides long-term coherence between two frequencies, f_1 and f_2, where $f_2 = 2f_1$. By multiplying the base oscillator, the reference in both SAR images was to a common phase. The transmitted signals T_1 and T_2 are related to the frequencies $f_1 = 2\pi\omega_1$ and $f_2 = 2\pi\omega_2$ by the relations [16]

$$T_1 = A_1 \cos \omega_1 t \tag{1.1}$$

$$T_2 = A_2 \cos 2\omega_1 t \tag{1.2}$$

The signal returns are proportional to the round-trip range delay $(2R_0/c)$ and the Doppler shift ω_d.

$$S_1 = A_1 \cos\left[(\omega_1 \pm \omega_d)t - \frac{2\omega_1 R_0}{c} \mp \omega_d t_0 + \varphi_1\right] \tag{1.3}$$

$$S_2 = A_2 \cos\left[2(\omega_1 \pm \omega_d)t - \frac{4\omega_1 R_0}{c} \mp 2\omega_d t_0 + \varphi_2\right] \tag{1.4}$$

The characteristic phase shifts of the two targets are φ_1 and φ_2, respectively. Since S_2 is double the frequency of S_1, the returned signals are at the same frequency, and the difference of the two results in a differential phase shift of

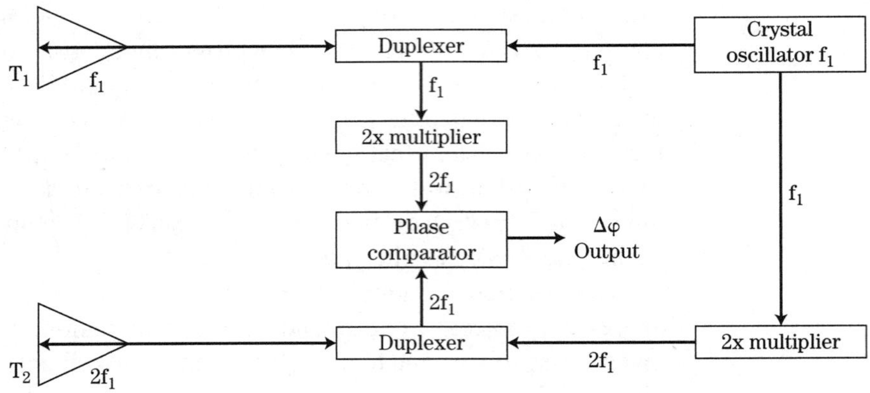

FIGURE 1–13

Dual frequency phase comparison processing © 1974 IEEE [17]

the target of $\Delta\varphi$ are given by

$$\Delta\varphi = 2\varphi_1 - \varphi_2 \tag{1.5}$$

This differential phase shift of the target will be independent of target range and relative motion between the RADAR platform and the target [17].

Early tests verified the ability to phase discriminate manmade objects from the cultural background. AFCRL conducted scaled measurements of spheres and dipoles in an anechoic chamber using S-band (3.0 GHz) and X-band (9.0 GHz) signals. The phase multiplier would be 3× in this case. Data was obtained for various sizes of targets with respect to wavelength, and a comparison of the phase from these objects, as shown in Figure 1–14, verified the theory [16].

Based on these successful chamber tests AFCRL developed a Synthetic Aperture Dual Frequency RADAR (SADFRAD) in the HF/VHF bands. The characteristics of the system are summarized in Table 1–2. The dual-band antenna consisted of two closely interlaced, asymmetrical folded dipoles, as seen on the side of the C-121 test aircraft in Figure 1–15. The lower frequency of 30 MHz was transmitted on the outer loop dipole, and the higher frequency of 60 MHz was from the inside dipole. With the coincident phase center of the two antennas, an improved performance was obtained with required radiation pattern, sidelobe levels, complementary front-to-back ratios, and impedance match over the bandwidth. This integration ensured alignment of pixels in the SAR image and enabled real-time imaging of wide areas.

SADFRAD provided significant advantages over existing SAR systems because of the reduced foliage penetration loss, resonant target detection, and real-time strip-map imaging processor. The RADAR block diagram is shown

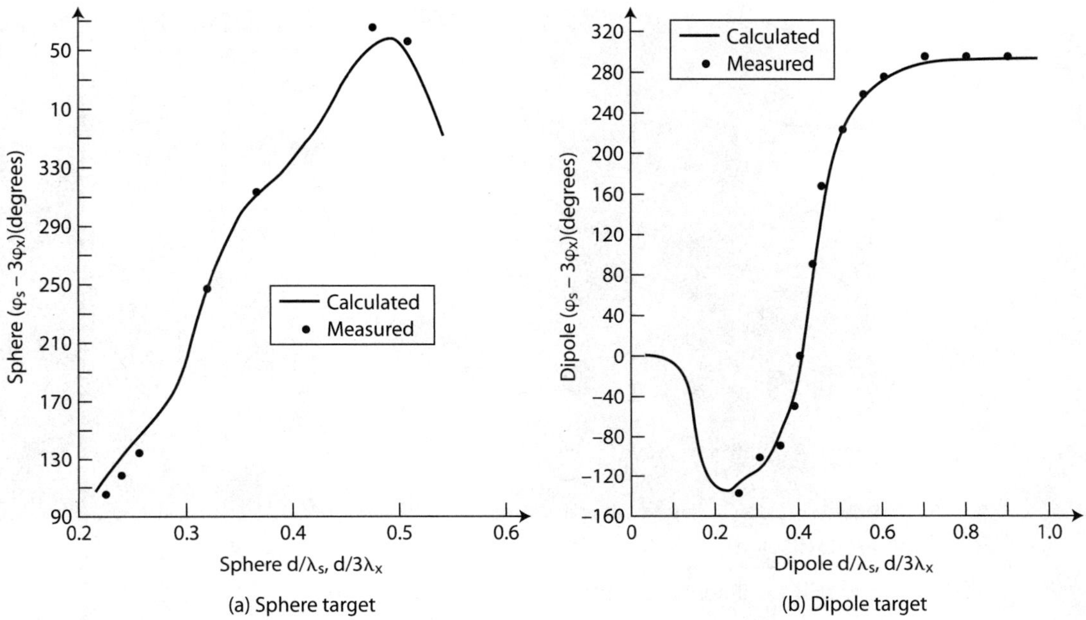

FIGURE 1–14

Measurement of standard targets with phase coherent RADAR © 1974 IEEE [17]

in Figure 1–16. The dual-channel receiver consists of a phase stable limiting channel to generate bipolar video data for the SAR processor and a linear channel for stretched bipolar video. These two separate receive systems are necessary because the RADAR system provides both instantaneous target data and background clutter reference data. The latter required the development of a data stretcher to provide a 128:1 slowdown of the incoming data to match the bandwidth of the onboard tape recorder.

Table 1–2 SADFRAD RADAR Characteristics [15]

Altitude	[Km]	2.5	Pulse Width	[μsec]	3.0
Velocity	[m/s]	90	PRF	[Hz]	750
Frequency	[MHz]	30.25, 60.5	ADC	[MHz]	5, 3 bits
Bandwidth	[MHz]	5	Range Resolution	[m]	30
Antenna Type		Folded Dipole	Cross Range Resolution	[m]	30
Polarization		Horizontal	Slant Range	[Km]	6.5
Waveform		15:1 Carley Code	Swath Width	[Km]	3.2
Average Power	[Watt]	5–10	Developed		AF Cambridge Research Lab

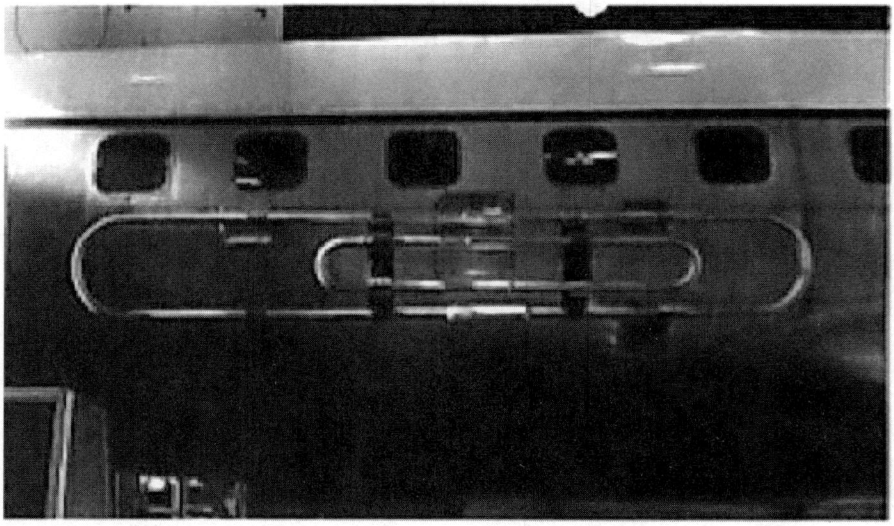

FIGURE 1–15

SADFRAD antenna mounted on C-121 aircraft [15]

The major development for AFCRL was the real-time SAR processor, which included digital pulse compression and focused azimuth compression. It provided simultaneous digital signals proportional to the amplitude of the target at the two frequencies in addition to the dual harmonic target differential phase signature. These outputs provided RADAR strip-map imagery (i.e., range versus cross-range) to drive the display.

The display consisted of three monitors. Real-time target amplitude data at the two frequencies was provided in a black-and-white monitor. In addition, the target differential phase signature data were presented as 1 of 16 colors on a color monitor.

1.3 Summary

These early RADARs developed for foliage penetration were in response to military needs to find and locate insurgents in a severe tropical environment. Little quantitative data existed to characterize the clutter and propagation losses in this environment. Based on a series of data collections in tropical regions, the decision was made to rapidly develop experimental systems and to get them into operational tests in a remote operational environment. Although some limited testing had been carried out prior to deployment, extensive system design and performance verification did not follow. These systems

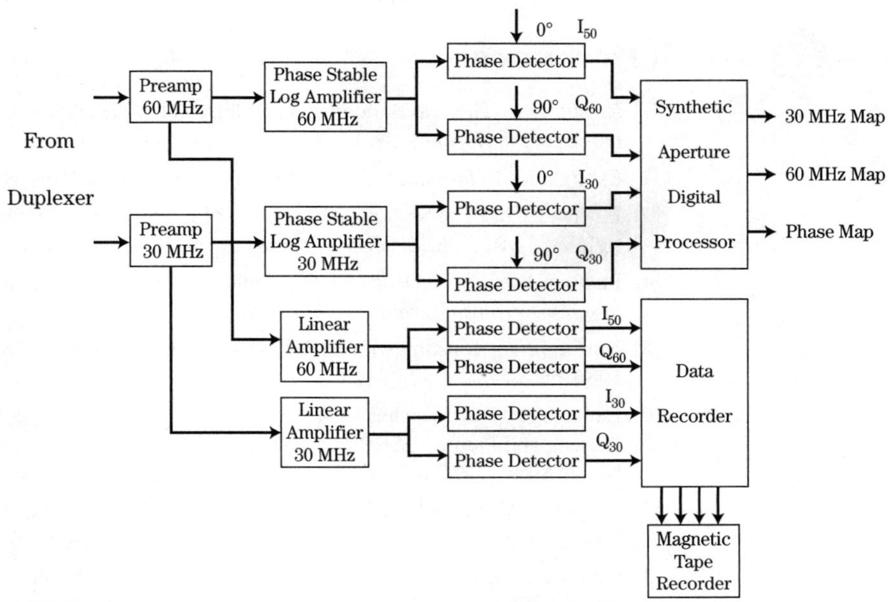

FIGURE 1–16

Block diagram of SADFRAD RADAR characteristics processing [15]

performed well, and both the technical and military communities learned from the experience.

Only the ground-moving target indication RADARs were taken to the military operations in Southeast Asia. The development of SAR capabilities was attempted, but the military planners could not justify the development due to several factors. First, the resolution of FOPEN SAR was limited to tens of meters. Operational SAR systems were significantly better than this and were not accepted due to the unreliable image recognition results. Second, the SAR systems were large and could not be carried on aircraft that would survive in a military environment. Finally, the state of the art in real-time signal processing was not mature enough to meet the needs of the military users.

No documented developments for peacetime use of foliage penetration RADAR systems are found for the period from 1975 to 1985. Because of the lack of mature coherent subsystems in the FOPEN RADAR band, it required a major push by the military to reenergize the development of this new capability. It would take significant breakthroughs in understanding the phenomenology of foliage penetration, the concept of operations in a crowded electromagnetic spectrum, and breakthroughs in signal and image processing before FOPEN would obtain the support and funding to become a viable system.

1.4 References

[1] Illustration, "The War Balloon at Gen McDowell's Encampment Preparing for a Reconnaissance,"", *Harper's Weekly*, October 26, 1861 p. 279.

[2] Skolnik, M. I., *Introduction to RADAR Systems*, McGraw Hill, New York, 1962, p. 8.

[3] Fowler, C. A. "The Standoff Observation of Enemy Ground Forces; from Project PEEK to JointSTARS," *IEEE Systems Magazine* Vol. 12, No. 6, June 1997, pp. 3–17.

[4] Photo of SOTAS RADAR on UH-1 Helicopter in Germany from US Army in Germany, *http://www.usarmygermany.com/Sont.htm*

[5] Northrup, T., "Jousting with JOINT STARS", *US Army Field Artillery Journal,* January 1986, pp. 24–25.

[6] Entzminger, J. N., Fowler, C. A., and Kenneally, W. J., "Joint STARS and GMTI: Past, Present, and Future," *IEEE Transactions on Aerospace and Electronic Systems* Vol. 35, No. 2, April 1999, pp. 748–761.

[7] Surgent, L.V. Jr., "Foliage Penetration RADAR: History and Developed Technology," US Land Warfare Laboratory Report AD/A000805, July 1974. (Publically released)

[8] Gordon, G. A. and Holt, E., "An Estimate of the HF/VHF Surface-Wave Communications Wave Reaches in the West German Forest Environment," Defense Nuclear Agency Report DNA-TR-82-07, January 1982, pp. 12–17.

[9] Johnson, J. R., et al., "Analysis of Tactical Intelligence Experience in South East Asia," General Research Corporation, McLean VA, DTIC ADC0050509, February 1976, pp. D-5–D-9. (Publically released December 14, 2000)

[10] Bryant, T. G., Morse, G. B, Bovak, L. M, and Henry, J. C., "Tactical RADARs for Ground Surveillance," *Lincoln Laboratory Journal*, Vol. 12, No. 2, 2000, p. 342.

[11] Picture of Camp Sentinel RADAR Antenna courtesy of Lee Moyer Technology Systems Company, Bethesda, MD.

[12] Photographs and details of M-FOPEN RADAR provided by Mr. James Rodems, former Syracuse University Research Company division manager.

[13] Larson, R., "Preliminary Target Detection of a Man Portable RADAR System," Honeywell, St. Paul, MN, Technical Report LWL-CR-06P66, April 1968, p. 27. (Declassified December 31, 1974)

[14] Kalmus, H. P., "Direction Sensing Doppler Device," *Proc. IRE,* June 1955, p. 698.

[15] Centofanti, J. J., "Synthetic Aperture Dual Frequency RADAR (SADFRAD—A Unique Airborne Sensor," AFCRL-70-0676, Air Force Cambridge Research Laboratories, Hanscom Field, Bedford, MA, AD515249, December 9, 1970. (Publically released December 31, 1982)

[16] "Report on AFCRL Research 1967–1970," AFCRL-71-0022, April 1971, pp. 245–249.

[17] Goggins, W. B., Blacksmith, P., and Sletten, C. J., "Phase Signature RADARs," *IEEE Transactions on Antennas and Propagation* Vol. AP-22, No. 5, September 1974, pp. 774–780.

CHAPTER 2

Foliage Penetration SAR Collection Systems

Both the military and scientific imaging communities learned from the early foliage penetration (FOPEN) developmental RADAR systems operated in the late-1960 to mid-1970 time frame. Two important system realities affected the growth of the technology: (1) foliage attenuation limited the systems to short-to-medium-range operation; and (2) manned aircraft could not be adequately protected at these ranges. Remotely piloted vehicles (RPV; also known as unmanned air systems, or UAS, in today's vocabulary) were just starting to be developed. They would address the ability to collect data in in hospitable environments. More importantly, the development of wideband data links would enable significant processing and image interpretation on the ground.

By the late 1980s, the image collection community had determined that SAR could provide acceptable and useful detection and characterization of forested regions. These SAR systems required small antennas and modest power; which was acceptable for experiments and might be possible on RPV installations. In 1988, the NASA Jet Propulsion Laboratory started the AIR-SAR program and flew a multiple-frequency SAR platform until 2004 [1]. At approximately the same time, several research groups started experimental FOPEN SAR systems, notably Stanford Research Institute (SRI) [2] and Sweden's Defence Research Agency (FOA) [3].

Airborne ground-moving target indication (GMTI) FOPEN RADAR systems were significantly more difficult to implement, especially on airborne moving platforms. The size of the antenna for both detection and localization of moving targets prohibited installation on a fixed-wing aircraft. As presented in Section 1.1, the X-band SOTAS development verified the benefit of stationary rotary wing operation for GMTI RADAR. But the size of the antenna at UHF and the lack of unmanned helicopters would not give rise to airborne FOPEN GMTI RADAR for more than 2 decades, when the

FOPEN reconnaissance, surveillance, tracking, and engagement RADAR (FORESTER) system would be developed for the remotely piloted A-160 Hummingbird [4].

With the advances in critical RADAR technologies of wideband waveform generation and digital image formation, the community could start the task of understanding the capabilities and limitations of FOPEN SAR. SAR systems were just starting to gain acceptance in the surveillance community, which had relied on high-resolution optical pictures for decades. Figure 2–1 presents the motivation for the need for tactical FOPEN SAR and an advanced look at what it will provide the operational user—whether it is the military or commercial customer of the image products [5]. All four panes in the figure are of the same scene; a forested region with several vehicles parked under the foliage and in the tree lines, but collected with different imaging technologies. On the left is a moderate to high-resolution optical picture, but the vehicles cannot be observed until the sensor is nadir looking.

The next image to the right is a typical 1 meter resolution X-band image of the scene taken on the same day. Sporadic detections were obtained, but only when the glint of targets could be captured in the image. Neither of these two image products would satisfy the user, especially when high area coverage rate is needed. The next two images to the right, which are UHF and VHF SAR images, show a more optimistic ability to detect the fixed targets. The UHF

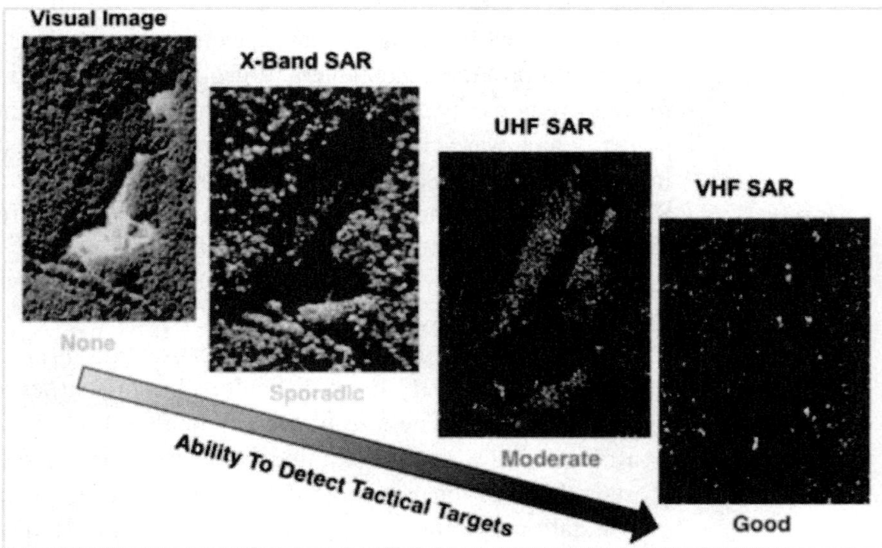

FIGURE 2–1

Comparison of optical and several RADAR image sources
Source: MIT Lincoln Laboratory [5]

panel shows images of many of the man-made targets but high false alarms with the foliage clutter in the scene. The detection at VHF is higher where the foliage attenuation is significantly lower and the target cross sections are larger than the clutter. However, there is limited resolution (i.e., pixels on target) to characterize the objects in the image.

This realization of reliable imaging capabilities for FOPEN SAR was important. It started a 5-year campaign to recharacterize the foliage clutter so that better SAR system engineering could be made possible. It was also realized at that time that a better understanding of the foliage scattering phenomenology would derive civilian uses for the systems. There was a definite dual-use message in the development objectives in the early 1990s.

Every new FOPEN RADAR system developed needed to answer the question of "why VHF or UHF?" This question is easy to answer. Optical photographs and microwave RADARs cannot reliably detect man-made objects that have been hidden in the dense forest cover. Two emerging technologies were being developed that could reduce the unreliable detection of targets under foliage. The first technology was ultra wideband (UWB) waveforms that would enable high-resolution SAR images at both VHF and UHF frequencies. The second technology was use of polarization of the RADAR signal in the FOPEN SAR processing.

High-resolution imagery serves two purposes: (1) provide a better separation of the object scattering from the background clutter; and (2) provide more detail of scattering of objects for characterization. In applications for foliage or terrain characterization, this factor is not as strong a motivation. However, to find a small vehicle or a buried land mine, image resolution is a major consideration.

Polarization diversity has been evolving as a significant capability for both target detection and characterization of terrain and man-made objects. If characterization is an important system objective, then polarization must be factored into the system waveform and processing approach from the start.

The system engineering task was for not only the FOPEN SAR design but also the concept of operations (CONOPS), as illustrated in Figure 2–2.The relevant questions were how much of the system:

- Could be installed on either a small manned or unmanned vehicle?
- Could be processed in real time and onboard the UAS?
- Needed motion measurement and compensation for collection geometries?

The global positioning system (GPS) was not generally available at that time; so inertial measurement and guidance systems were stressed for the long data runs while obtaining a strip map.Tactical data links did not have the bandwidth

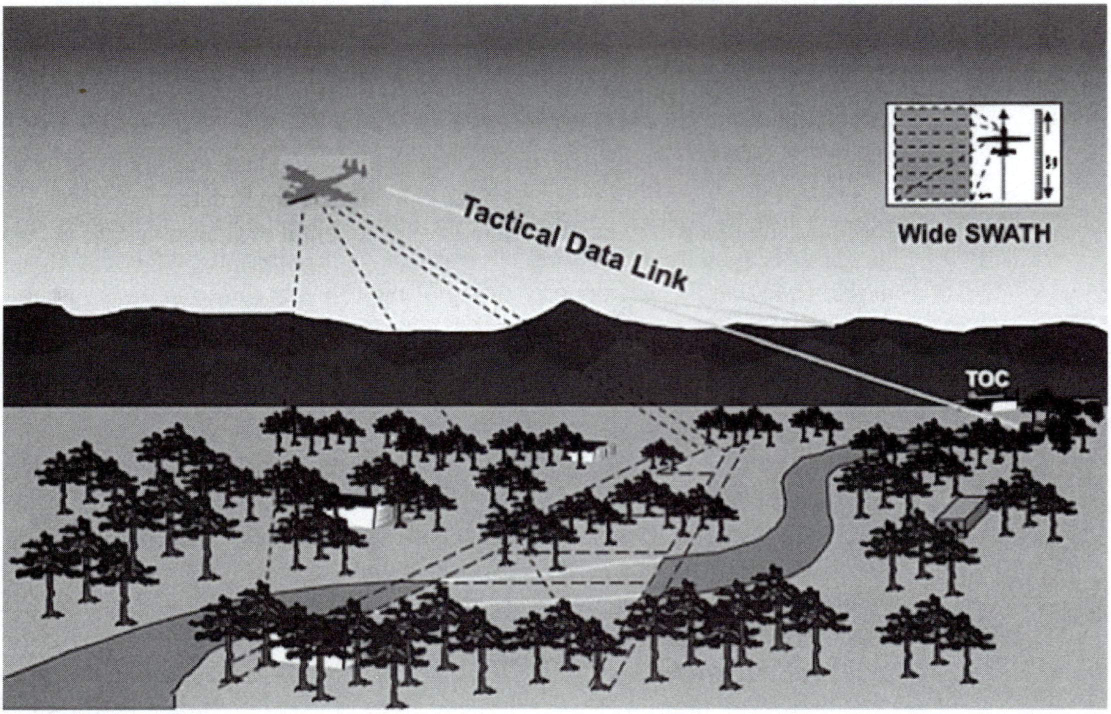

FIGURE 2–2

Future requirements for FOPEN RADARs on data link control

to send down all of the data for image processing on the ground, and the ground stations needed to be close to the flight path for real-time operation.

These initiatives and several other military and space science programs were addressing significant CONOPs issues. However, the first task was to gain a significant assessment of the foliage characteristics—scattering and losses. This would enable sizing the RADAR systems and computers that could be built. But remember we had the GPS coming into reality and Moore's Law in our favor. The first addressed the motion measurement and navigation problem that plagued real-time SAR systems. The latter gave the potential for higher processing through put on small vehicles. So there was soon to be a rebirth of foliage penetration RADAR—albeit focused on SAR systems and not GMTI RADAR.

The first FOPEN SAR system (SADFRAD), summarized in Chapter 1, exploited the coherence of man-made objects when illuminated with dual frequencies. However, the HF did not provide adequate resolution on the objects to characterize the type. It was important to push FOPEN SAR into higher frequencies to improve the range and cross-range resolution.

Both applications provided strong existence proof of the utility of VHF and UHF propagation through forests and the detection mechanism.However, the development of efficient signal processors and the ability to counter the effects of moving clutter and radio frequency interference (RFI) needed significant development. These capabilities were more than a decade in the future.

This chapter will give details on early FOPEN SAR data collection systems built for both civilian and military experimental evaluation. We will first revisit the merits of VHF and UHF for foliage penetration operation. Both are effective for part of the detection and characterization of foliage and of man-made objects under foliage. Understanding the relative merits was important in choice of frequency, bandwidth, and polarization for the several prototype systems.

2.1 SAR Resolution

Synthetic aperture RADAR (SAR) obtains fine resolution for ground images through two effects. The range resolution δ_R, similar to conventional RADARs, is obtained primarily by the bandwidth of the waveform B. Cross-range resolution is obtained by a physical antenna angular pattern and the ability to coarsely resolve objects within the real beam. However, for fine cross-range resolution, it is necessary to form a synthetic aperture length by flying a length L and coherently integrating the returns to obtain the resolution δ_{CR}. This is especially true for imaging from VHF and UHF RADARs, where real beam apertures with any reasonable angular pattern would be impractical on airborne platforms. This section will treat the basic factors for obtaining resolution in range and cross-range with a SAR system. The extension to an UWB SAR will be developed in more detail in Chapter 3 for UWB phenomenology and in Chapter 4 for UWB SAR image formation.

The range resolution δ_R of a pulse in the slant plane is directly related to the bandwidth of the RADAR and any weighting to reduce the range sidelobe levels by [6]

$$\delta_R = \frac{k_R c}{2B \cos \gamma_g} \tag{2.1}$$

where:

- c Speed of light;
- B Bandwidth of the waveform;
- k_R Range broadening factor due to aperture weighting;
- γ_g Grazing angle with respect to the local terrain.

Normally, a SAR system will illuminate the ground at small grazing angles, and the range resolution is determined primarily by the waveform bandwidth.

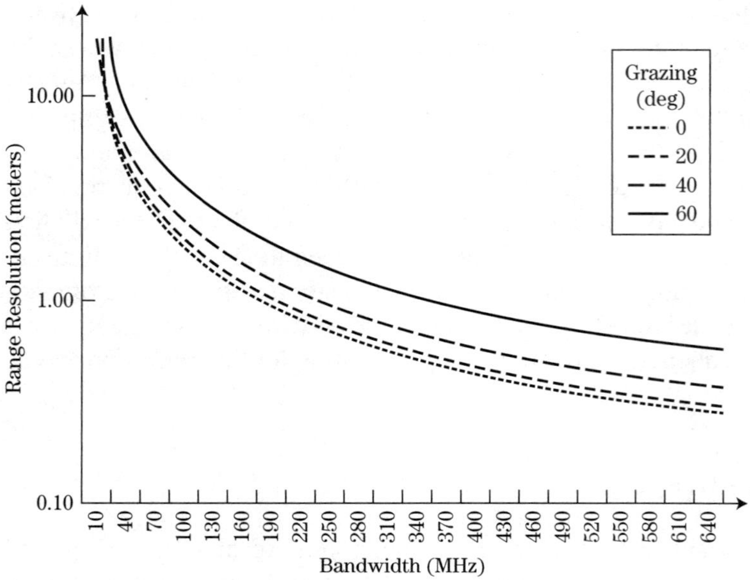

FIGURE 2–3

Ground range resolution versus signal bandwidth and grazing angle

However, for foliage penetration RADAR higher grazing angles is important for providing less foliage loss and better signal-to-clutter ratio. So at higher grazing angles the target signal return has the potential to be enhanced relative to the background clutter but with a reduction in ground plane range resolution. Figure 2–3 illustrates the ground plane resolution for VHF and UHF waveforms as a function of bandwidth and grazing angle.

Figure 2–3 illustrates the importance of bandwidth when compared with the carrier frequency as well as the importance to range resolution. For resolutions under a meter, the required bandwidth is above 150 MHz, independent of any range sidelobe weighting. Since the FOPEN SAR must operate at center frequencies comparable to the signal bandwidth, it was necessary to consider the impact of fractional bandwidth on the system design to achieve fine-range resolution.

The bandwidth of the signal waveform extends from the low-frequency component f_L to the high frequency f_H. If a uniform distribution of the signal spectral density is assumed, the bandwidth B is the difference between f_H and f_L. This spectral density is determined by the transmit waveform generation and any spectral effects that are provided by the antenna system dispersion.The fractional bandwidth ΔB of the system is calculated by the ratio of the bandwidth B to the center frequency

$$\Delta B = \frac{2(f_H - f_L)}{(f_H + f_L)} \tag{2.2}$$

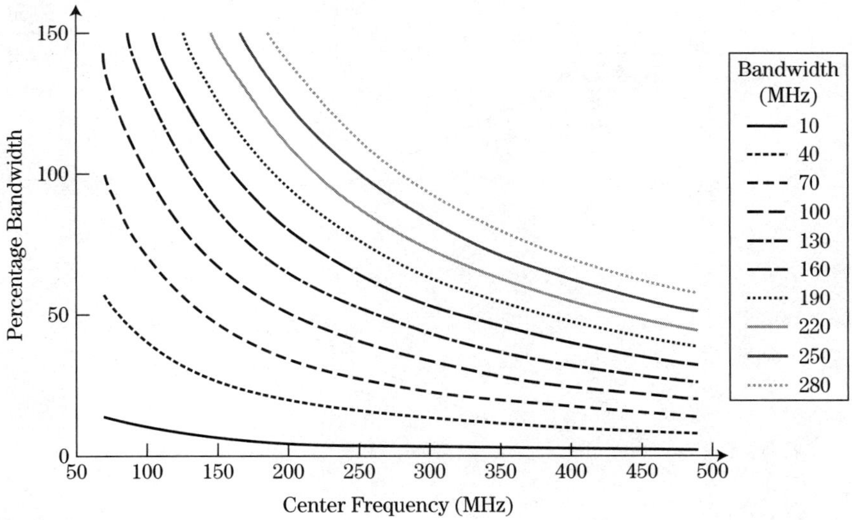

FIGURE 2–4

Waveform percentage bandwidth for VHF/UHF SAR

The IEEE convention is that a system is considered to be UWB if the fractional bandwidth ΔB is greater than 25% [7]. For most systems in the VHF and UHF RADAR bands, this fractional bandwidth is significantly above 25%, as shown in Figure 2–4. It is also apparent that almost all cases of bandwidth and center frequency for FOPEN SAR exceed the definition of ultra wideband. A UWB system affects all aspects of the RADAR design (waveform, antenna, signal processing, and phenomenology), as will be shown in Chapter 4.This creates a technical challenge that needed to be addressed in developing FOPEN SAR capabilities. But just as important was the impact of the system characterization as UWB, which restricts where and when the system can be operated. Any operational system must be in compliance with the National Telecommunications and Information Administration (NTIA) in the United States and its counterpart in most of the world. A UWB RADAR needs to operate outside of the conventional RADAR bands, which has caused a significant political challenge. Chapter 5 will address the design complications to meet this requirement for frequency allocation.

The advantage of SAR systems is the improved cross-range resolution over that of a real-beam antenna obtained by flying a long synthetic aperture as shown in Figure 2–5. The cross-range resolution δ_{CR} for a broadside SAR operation and integration through an angle of θ_I is given by [6]

$$\delta_{CR} = \frac{k_{CR}\lambda_c}{4\sin(\theta_I/2)} \tag{2.3}$$

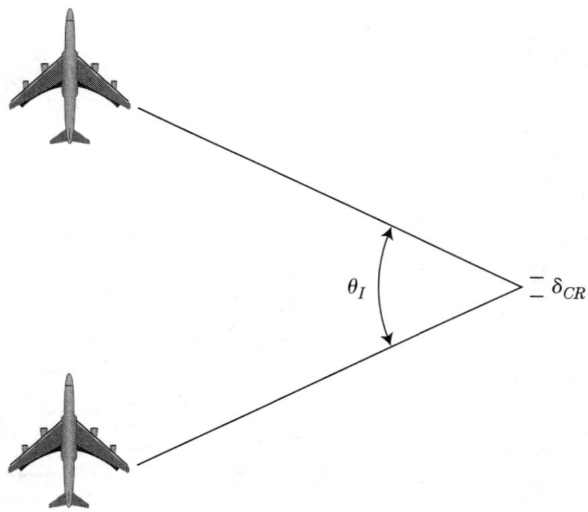

FIGURE 2–5

SAR geometry for cross-range resolution

where:

k_{CR} Cross-range broadening factor due to aperture weighting;
λ_c Wavelength of the RADAR's center frequency;
θ_I Azimuthal integration angle during SAR image formation.

For VHF and UHF frequencies, the angles needed to get significant resolution are very large. As a result, both the fractional bandwidth and the integration angle are UWB compared with conventional microwave frequency SAR systems. The achievable cross-range resolution as a function of the frequency and integration angles is detailed in Figure 2–6. For VHF it is necessary to have

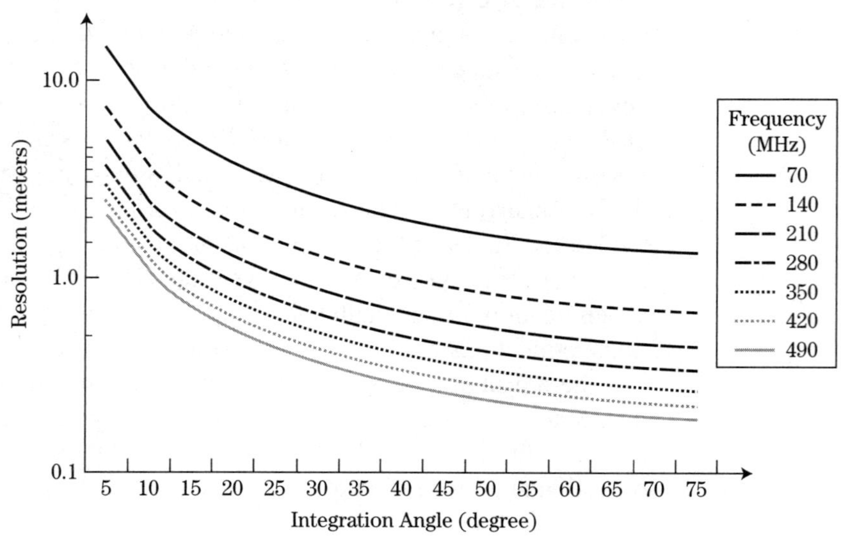

FIGURE 2–6

SAR resolution with variable integration angle and frequency an integration angle over 45 degrees to obtain better than 5 meters of cross-range resolution. An integration angle this large posed a major development in system requirements for the integration times, motion measurement, and motion compensation, as well as achieving the comparable range resolution.

The FOPEN SAR data collection systems built in the 1990s had to factor these issues into many aspects of the RADAR design. The remaining parts of this chapter will summarize the differences in design chosen by the airborne RADARs used to demonstrate the capabilities for detecting man-made objects under dense foliage and buried in shallow ground.

2.2 FOPEN SAR Systems

The early FOPEN SAR systems were developed for detecting and characterizing objects under both foliage and through ground penetration. The latter capability was important, as demining operations were required after military actions in war-torn areas. In addition, for finding objects that have been hidden, the systems'long wavelengths and polarimetric sensing found usefulness in characterizing land use, land cover, and terrain elevation in many geographic areas.

Significant progress was made in the design of antennas and transmitters for FOPEN SAR. The antennas needed to have wide azimuth coverage to enable the requisite illumination angle for achieving the desired cross-range resolution. They also needed to have an efficient match to the transmit waveform over a very large bandwidth to support the range resolution. As will be shown in the following two chapters, polarization has found an important place in FOPEN SAR for characterization of the clutter and objects. Providing UWB polarimetric antennas was an early challenge. The design of the transmit waveform and match to the antenna was also important to limit the spectral transmission as controlled by the need for frequency allocation constraints. These early systems addressed the technical obstacles that were important in the design and use of operational FOPEN SAR systems.

Several early FOPEN SAR systems were developed and flown in the early 1990s. They represent significantly different approaches for image formation processing, the details of which will be presented in Chapter 4. They also provided extensive data on the characterization of both the clutter and the detection of objects under and near the clutter. The five RADARs illustrated in the following sections present a wide variation in frequency, waveform design, image processing, and the use of polarization. Four of the systems have been installed on fixed wing aircraft, which vary widely in size and speed. Two were multiengine planes that allowed onboard signal processing and real-time observation of the data during flight, and two were flown on small

Table 2–1 Comparison of technology from FOPEN SAR experimental systems

	FOLPEN II	Carabas II	P3 UWB	GeoSAR	BoomSAR
Waveform	Impulse	Freq Jump	Notched LFM	Notched LFM	Impulse
Frequency MHZ	200–400	20–80	225–740	280–460	50–1100
Polarization	HH	HH	HH, HV, VV	HH, HV or VV	Full Pol
Transmit RFI	N/A	Freq Sequence	Notch	Notch	N/A
Image Formation	Back Projection	Back Projection	RMA	INSAR	Back Projection

tactical aircraft with onboard data recording and subsequent image formation processing and analysis after the flight.

Much had to be learned to refine the eventual design objective of installation on unmanned air vehicles for operation over remote, and often hostile, environments. The fifth RADAR system to be examined was an instrumentation RADAR installed on a computer controlled "cart" that would provided performance verification of target characterization by allowing high dynamic range collection of both foliage and ground penetration data.

Each of these experimental FOPEN SAR systems embodied a new technology that had the potential for enabling operational system design. Table 2–1 summarizes the critical technologies employed in the design and development of each of the systems. The critical RADAR designs and applications, along with the sections that cover their design, include:

- *Impulse waveform:* (Sections 2.2.4 and 4.2) A very narrow pulse that has wide spectral content
- *Frequency jump burst (FJB):* (Section 5.1.2) A waveform that covers the required bandwidth by incremental transmission of narrowband pulses, combined with coherent reconstruction
- *Notched linear frequency modulation (LFM):* (Section 5.1.3) Use of an LFM waveform over a wide bandwidth, with narrow regions of exclusion of critical frequencies
- *Polarization:* (Sections 3.4 and 6.1.1) Transmitting one linear polarization and receiving one or more polarizations. The first letter is the transmit polarization, and the second is the receive polarization (e.g., HV is transmit horizontal and receive vertical)
- *Back projection algorithm (BPA):* (Section 4.3.1) An image formation process that directly, coherently adds the contribution of each pulse to the

appropriate image resolution cell, after appropriate motion compensation for the imaging platform position and orientation.

- *Range migration algorithm (RMA):* (Section 4.3.2) The image formation process that uses a two-dimensional mapping of range and Doppler curvature during the imaging process to provide efficient image formation processing
- *Interferometric SAR (InSAR):* (Sections 2.2.3 and 8.4) The technique of forming two SAR images, displaced by a distance to enhance differences in either the terrain height or surface target motion for image processing

Details of these technologies and their experimental results for clutter and target characterization will be covered in subsequent chapters.

Significant publication of the design details and results occurred in that decade. As a result, the lessons have been shared and formed the designs of more recent systems for the next decade. Several SAR data collections provided comparison on the same terrain and objects. These collections will be covered in Chapter 3 along with the characterization of foliage clutter and attenuation. One of those collections included both an X-band and a UWB UHF collection over Camp Roberts in California to evaluate both foliage penetration and digital elevation model (DEM) generation. The image is of a small segment of the wooded area shown in Figure 2–7.

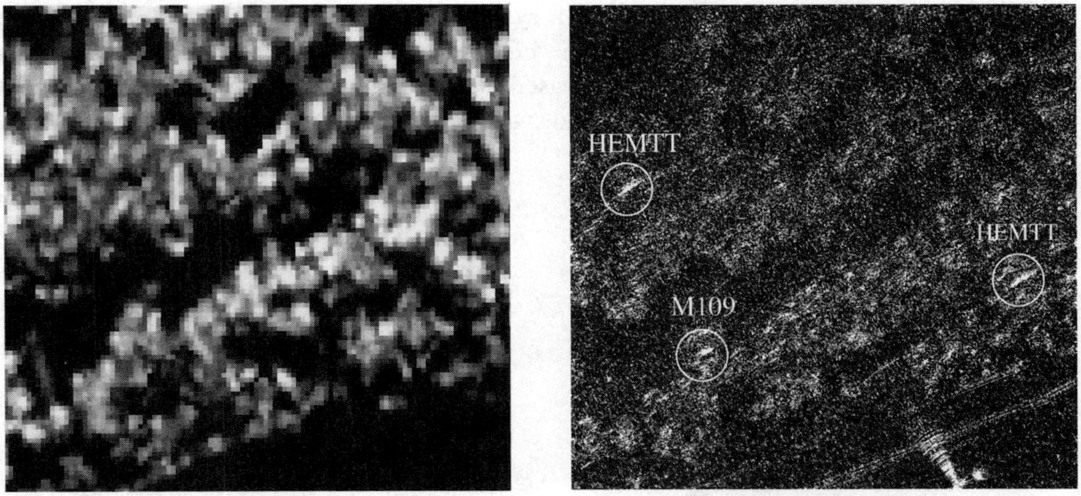

a. X-band 2.5m x 2.5m, HH, IFSARE   b. UHF, 0.33m x 0.66m, HH, UWB P-3

FIGURE 2–7

Comparison of X-band and UHF SAR images—forested hide position [5]

The first image, Figure 2–7a, is an X-band SAR made by the ERIM IFSARE system to provide high area coverage rate DEM with fine elevation accuracy [8]. This image was collected on the same day that several military vehicles were been placed under the foliage area known as "Sherwood Forest." It is apparent that only the tops of the trees are visible in the X-band image.

The second image, shown in Figure 2–7b, was collected with the UWB P-3 FOPEN SAR, also built by ERIM [9]. The three tactical targets under the foliage were revealed only in the UHF image and at horizontal polarization. However, it is clear that false alarm rates would be very high if only the horizontal polarization image were to be used. It should be noted that the strong return in the foreground was from one of the instrumentation trihedrals deployed to calibrate the multiple polarization sensitivity.

This comparison of X-band and UHF SAR provides sufficient evidence to many operational users of the importance of UWB SAR at VHF or UHF for detecting man-made objects under foliage. However, it was as important to quantify the performance with available technology prior to development of an operational system. The next five sections provide the quantitative performance of the experimental FOPEN SAR systems employed from 1990 to 1998 to obtain support for these important system developments.

2.2.1 SRI's FOLPEN RADARs

One of the first FOPEN SAR systems to be built was the FOLPEN series that was developed by SRI. The FOLPEN II and FOLPEN III systems, shown in Figure 2–8, were both based on a very high peak voltage impulse transmitter. This was an effective source of UWB RADAR signal, whose spectral

FOLPEN II **FOLPEN III**

FIGURE 2–8

SRI FOLPEN RADAR platforms employing impulse transmitters ©1989 IEEE [2] ©1991 IEEE [10]

characteristics were determined by the impulse shape and the interface to the antenna assembly. Because the pulses were only a few nanoseconds in length, the average power of this RADAR was very limited.

FOLPEN II was used in early trials for foliage characterization and for demonstrating land mine remediation [2]. The short-range operation was not an issue. As the need for wide area coverage and target discrimination evolved, SRI developed a two-channel polarimetric system (FOLPEN III) that would alternate horizontal and vertical polarization transmissions [10].

The earlier FOLPEN II system was limited to 200 MHz bandwidth, or nominally 1 meter resolution, due to the limited match between the impulse transmitter and the multiple dipole antennas under the wing of the aircraft. The later FOLPEN III system was improved to 0.5 meter resolution, with the closer coupling of the transmitter to the multiple polarization ridge waveguide antenna.

SRI pioneered using the BPA for image formation processing. By combining the aircraft navigation measurement with a differential GPS, they were able to form moderately wide swath images with very good image quality. The later operation of FOLPEN III also included a real-time image formation processor followed by a target detection system [11]. Characteristics of the FOLPEN RADARs are summarized in Table 2-2.

The FOLPEN II RADAR participated in the 1993 Maine collection campaign. The collection scenario included several trucks in a narrow forest road to determine both the clutter characteristics and the ability to detect and characterize man-made objects. The SRI RADAR provided high-quality SAR images as indicated in the figure, found on the SRI Web site. These clutter data were analyzed by MIT Lincoln Laboratory and are included in the clutter scattering and loss characteristics shown in Section 3.3.

Table 2–2 Characteristics of FOLPEN RADAR © 2002 IEEE [11]

Altitude	[Km]	0.9	Pulse Width	[μsec]	3.5 (II), 2.0 (III)
Velocity	[m/s]	100	PRF	[KHz]	100
Frequency	[MHz]	200, 400	ADC	[MHz]	25, 8 bits
Bandwidth	[MHz]	200 (II), 400 (III)	Range Resolution	[m]	1.0 (II), 0.5 (III)
Antenna Type		Array Dipole (II) Crossed Dipoles (III)	Cross Range Resolution	[m]	1.0 (II), 0.5 (III)
Polarization		HH (II), HH, VV, HV (III)	Slant Range	[Km]	2.0
Waveform		Impulse	Swath Width	[Km]	1.0
Peak Voltage	[Mvolt]	1.0	Developed		SRI

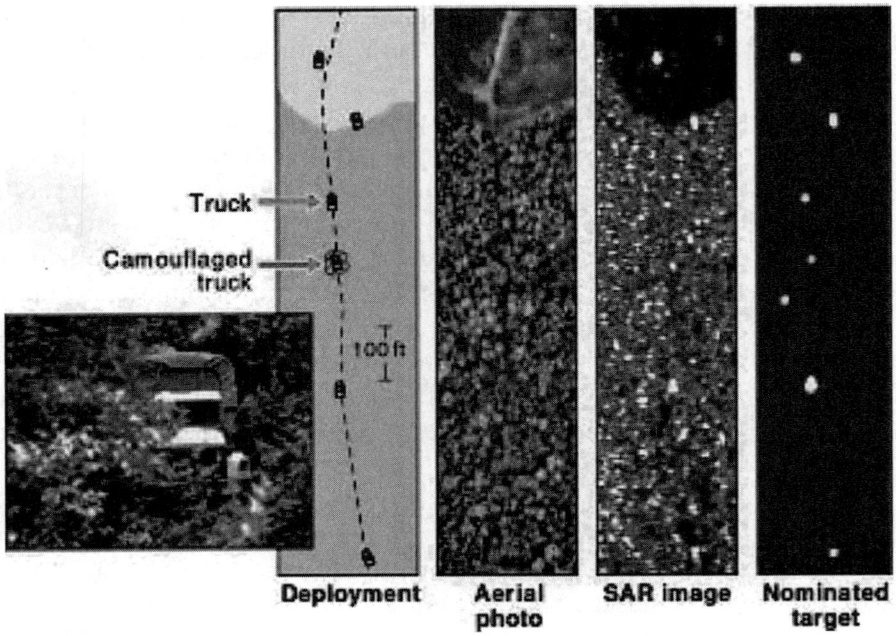

FIGURE 2–9

FOLPEN II detection of targets maine 1993 collection [12]

The major target detection advance at that time was the use of several image processing techniques to discriminate man-made objects from background clutter, as illustrated in Figure 2–9 [12]. The panel titled "nominated targets" illustrates the results of applying several spatial filters to the horizontally polarized data. In fact this is the first know published receiver operating characteristics (ROC) curve on FOPEN target detection. As indicated, the raw constant false alarm rate technique yields false alarm density of over 10 per square kilometer at the 80% detection probability. Many users would consider this level of false alarms excessive. SRI applied two techniques to the data: multipixel phase filtering and a subaperture phase filtering to the data. Both techniques reduced the false alarm density significantly below 1 per square kilometer. The subaperture technique achieved better than 1 false alarm in 10 square kilometers by exploiting the cardinal flash of the large vehicles.These results were encouraging for future development in automatic target detection and characterization performance.

Figure 2–10 illustrates the location and types of targets along the road. This ground truth was used to score the detection probability at several thresholds depending on the false alarm density. This ROC technique is a measure of the effectiveness of image processing technique for detecting targets and

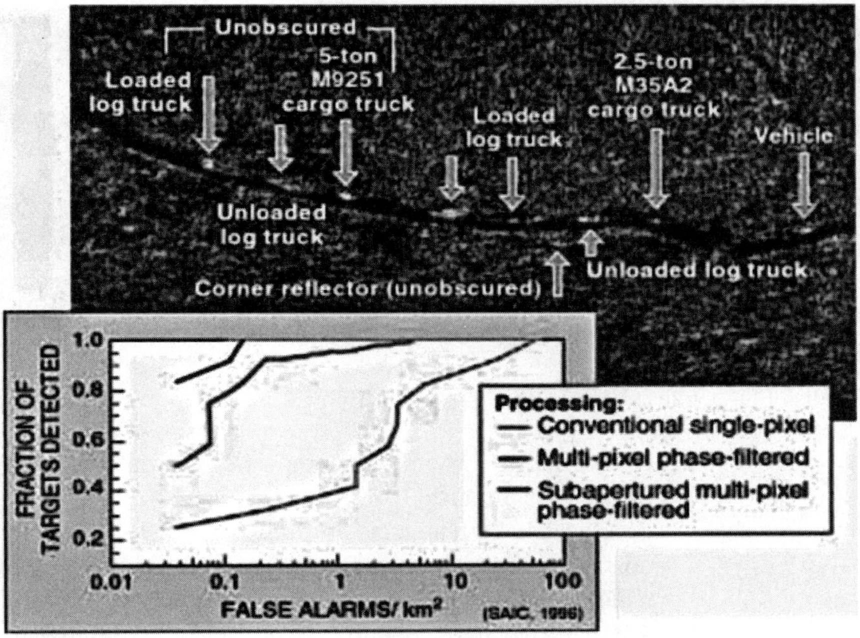

FIGURE 2–10

FOLPEN II receiver operating characteristics from maine collection [12]

discriminating from local clutter. The panel titled "nominated targets" illustrates the results of applying several spatial filters to the horizontally polarized FOPEN SAR data. This is the first-known published ROC curve for FOPEN target detection.

2.2.2 Sweden's CARABAS RADAR

The Swedish National Defence Research Establishment (FOA) developed a unique, low VHF-band FOPEN SAR called coherent all radio band sensing (CARABAS) in the early 1990s. CARABAS used a majority of the short waveband for operation. The motivation for the system development at VHF band is the reduction of speckle, which improves the ability to detect and discriminate man-made targets. When the wavelength is near the Rayleigh limit of the target, the speckle is significantly reduced, and detection is enhanced. Speckle in SAR is reduced by operating over more than an octave bandwidth and with a resolution comparable to the wavelength of the signal [3].

The CARABAS system was flown in two configurations on a Saberliner aircraft, characterized by the two antenna configurations as shown in Figure 2–11. CARABAS I used two inflatable sleeve antennas that trailed the aircraft. This contrasted with the more permanent antenna installation of

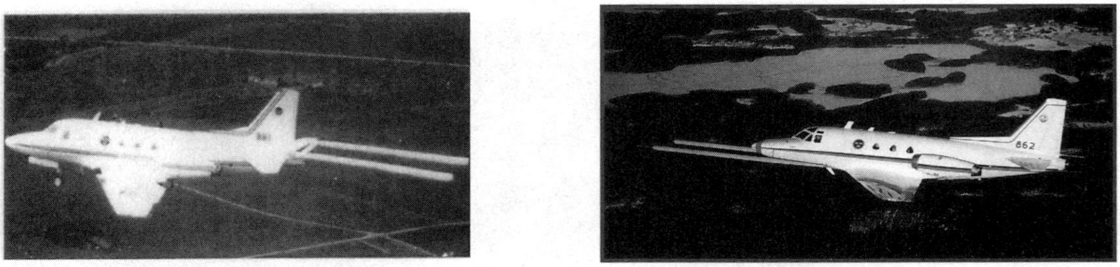

Carabas I (1992) **Carabas II (1996)**

FIGURE 2–11

Swedish FOA carabas VHF FOPEN platforms [3], [13]

CARABAS II, where two composite material antennas were attached to the front of the aircraft [13,14].

Because the antennas are mostly in free space, there would be no natural suppression of the individual pattern backlobes. However, since the two wideband dipoles are placed side by side within a fraction of a wavelength distance, they interact with each other. The resulting backlobe suppression has been measured to be about 10–11 dB in CARABAS-II with true time-delay steering on transmit [15]. The suppression is further increased by digitally combing signals from respective antennas, as shown in Section 7.2.1.2. Measurements have shown that backlobe suppression is improved to 21 dB [16]. The principal system characteristics of the CARABAS SAR system are high power, wide swath width, and efficient detection of targets under foliage, with the principal RADAR characteristics summarized in Table 2–3.

In addition to the unique antenna construction and pattern control, CARABAS-II had several design innovations. By operating in a shared radio band, significant interference sources need to be avoided and excised from the image processing. Therefore, the transmit waveform used a frequency jump

Table 2–3 CARABAS RADAR characteristics [3], [13]

Altitude	[Km]	6.0	Pulse Width	[μsec]	0.1
Velocity	[m/s]	100	PRF	[KHz]	100/2/N
Frequency	[MHz]	20–90	ADC	[MHz]	2.5, 12 bits
Bandwidth	[MHz]	2.5 per pulse	Range Resolution	[m]	3.0
Antenna Length, Type	[m]	5.5, Segmented Dipole	Cross Range Resolution	[m]	3.0
Polarization		HH	Slant Range	[Km]	10–25
Waveform		FJB, N bursts	Swath Width	[Km]	15
Peak Power	[Kwatt]	1.0	Developed		FOA Sweden

burst of up to 37 frequencies, with the first center frequency at 21.25 MHz and a 1.875 MHz step to cover the nominal band 20–90 MHz. The sequence and spacing of the frequency steps were maintained between CARABAS-I and CARABAS-II. However, in the later system, notching in the individual steps was used to avoid the radio frequency interference. A wide dynamic range analog-to-digital converter (ADC) provided very good imagery, even in the presence of RFI [17].

The maximum bandwidth of CARABAS II operation was 70 MHz, yielding a 3 meter range resolution. Typically the collection angle was 60 degrees, providing a corresponding 3 meter azimuth resolution.The high peak power provides moderately long-range SAR maps, and use of BPA image formation processing enables wide swath operation. These characteristics of the VHF system design provided for the high area coverage rate of 1 km^2/sec [17].

The first flight trials with CARABAS-I were conducted in Sweden during 1992. And CARABAS I participated in the 1993 Maine FOPEN data collection,where several large vehicles were assembled in the open and under a tree-lined road to determine the effects of foliage on the detection of trucks. The objective of the 1993 collection was to measure clutter return and attenuation for characterizing the foliage phenomena. The quantitative analysis of these factors will be presented in Chapter 3. However, it is illustrative to look at the same geometry and target array with two frequency bands—FOLPEN II at low UHF band and CARABAS I at low VHF band. Figure 2–12 provides

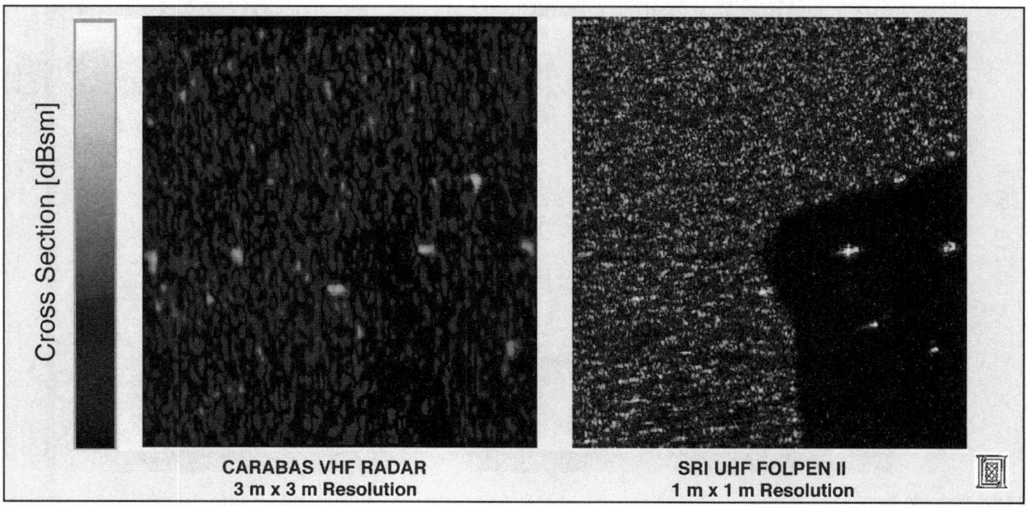

FIGURE 2–12

Comparison of VHF and UHF target detection 1993 maine collection
Source: MIT Lincoln Laboratory [18]

a side-by-side comparison of images from the two collection platforms. MIT Lincoln Laboratory carried out independent image formation processing and calibration on recorded data from both platforms based on the respective SAR system characteristics and recorded navigation data.As such this should be a real comparison of the phenomenology at two collection frequencies [18].

The attenuation on the targets in foliage is significantly less in VHF than UHF, as is the clutter return. However, the targets in the low UHF band image exhibit strong scattering characteristics that were considered as discrimination for target characterization. These two points will be presented in Chapter 3 in more quantitative detail.

CARABAS II also participated in the 1997 Keystone collection in Pennsylvania and demonstrated the improvements in performance of the intervening 4 years. The major difference between the two was the change in the antenna to rigid booms and refinement in signal processing. Other noteworthy differences were (1) average power significantly increased (10 dB) by introducing linear frequency modulation in each pulse; (2) Doppler aliasing above 55 MHz eliminated by increasing effective PRF (one side illumination only and fewer frequency steps); and (3) narrowband notching introduced on transmit.

The images shown in Figure 2–13 provide further evidence of the benefit of VHF on target detection in foliage clutter. They are of different geographic locations and a different array of targets; however, both have essentially the

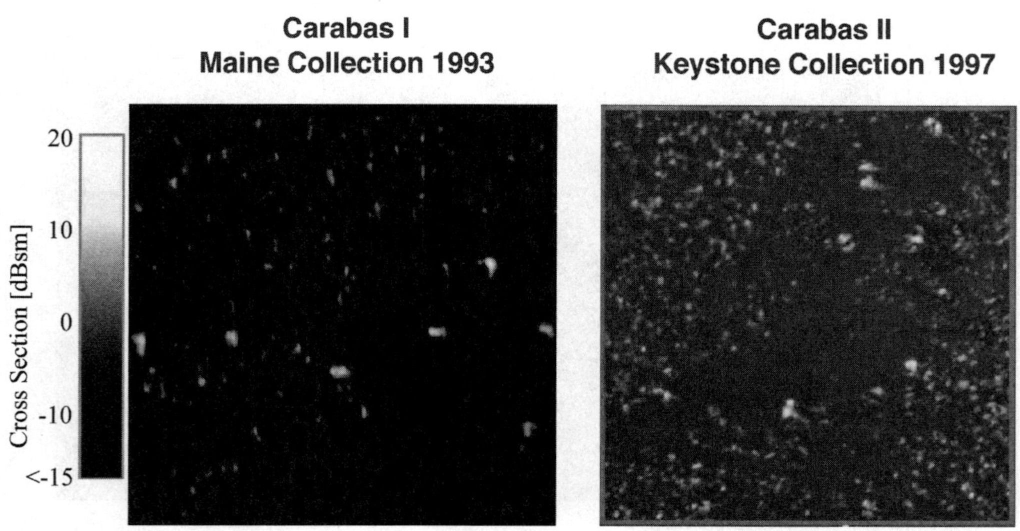

FIGURE 2–13

Comparison of CARABAS I and CARABAS II target detection
Source: MIT Lincoln Laboratory [18]

same image resolution. The image on the left was from CARABAS I and illustrates enhanced target cross section and lower clutter in low VHF band. The trucks were easily discerned from the cultural clutter providing good detection probability. In the image on the right, the focus of the foliage returns appears to be sharper than in the earlier collection. In addition, it is clear that the ability to cancel the RFI has been improved. The area in the middle of the scene is a clearing in the trees, and the noise equivalent σ_0 has been improved by approximately 7 dB. Unfortunately, no quantitative analysis of these observations was carried out.

2.2.3 NADC's P-3 Ultra-Wideband SAR

The Naval Air Development Center (NADC) in Warminster, Pennsylvania, developed a series of multiple bandwidth SAR systems under contract with the Environmental Research Institute of Michigan (ERIM). The last in the series was a UWB UHF SAR system that had full polarization capability [19]. Each of these RADAR test beds was installed on a Navy P-3C aircraft, as shown in Figure 2–14. The size of the platform allowed significant instrumentation and recording for data collection missions. It also had the speed and endurance to collect long data runs at several remote foliage test sites.

Design of the UWB RADAR and development of the ground image formation processor presented a number of challenges caused by the RADAR's large percent bandwidth and wide synthetic aperture integration angles [20]. Several of the unique designs were in critical hardware subsystems. The characteristics of the P-3 UWB SAR are summarized in Table 2–4.

With an available waveform bandwidth of 515 MHz, it was possible to conduct experiments with 0.33 meter range resolution. However, a unique solid-state high power transmitter was needed to support this larger percentage

FIGURE 2–14

P-3 Ultra wideband SAR platform © 2001 IEEE [19]

Table 2–4 ERIM navy P-3C UWB FOPEN characteristics © 1996 IEEE [20]

Altitude	[Km]	7.5	Pulse Width	[μ sec]	26.3
Velocity	[m/s]	135	PRF	[Hz]	500–1200
Frequency	[MHz]	215–730	ADC	[MHz]	30, 6 bits
Bandwidth	[MHz]	515	Range Resolution	[m]	0.33
Antenna Area,	[m^2]	1.0,	Cross Range	[m]	0.66
Type		Flared Notch	Resolution		
Polarization		HH, VV, HV	Slant Range	[Km]	6.24
Waveform		LFM, Notched	Swath Width	[m]	929
Peak Power	[Kwatt]	1.0	Developed		ERIM

bandwidth and high average power. The wide UHF spectrum also presented a new problem, since there are many critical users with sensitive frequencies. To address these issues, chirp waveform modulation techniques were developed to synthesize programmable notches and to avoid interfering with critical users.

The UWB transmitter was a challenge due to the need for matching the output power and impedance to the antenna. The state of the art in solid-state amplifiers was such that the power versus frequency could vary by as much as 8 dB over the band. This affected the UWB SAR waveform in two ways, as illustrated in Figure 2–15: (1) the average power transmitted would be degraded by the low-power components; and (2) the range side lobes of the waveform would be degraded by the amplitude variation over the spectrum.

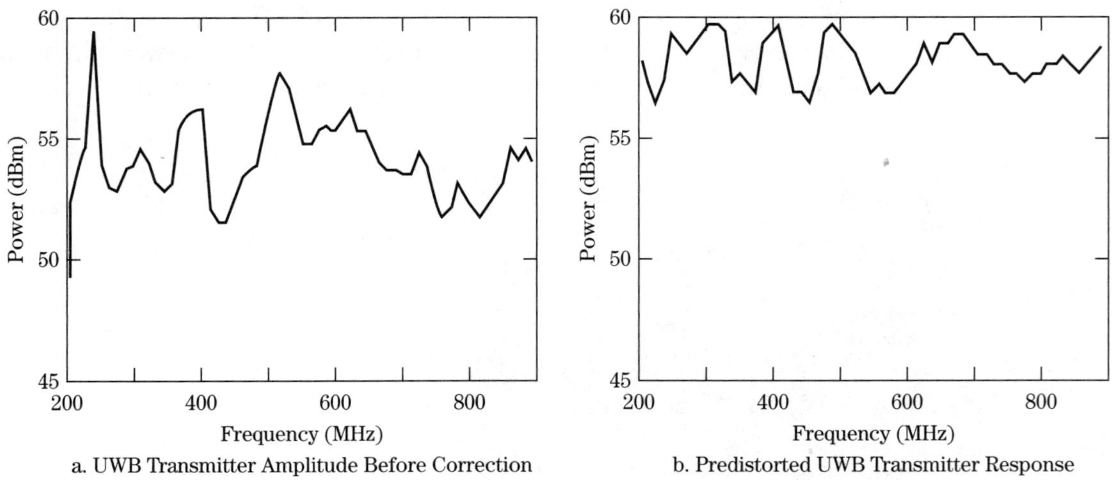

FIGURE 2–15

UWB SAR transmitter amplitude response [21]

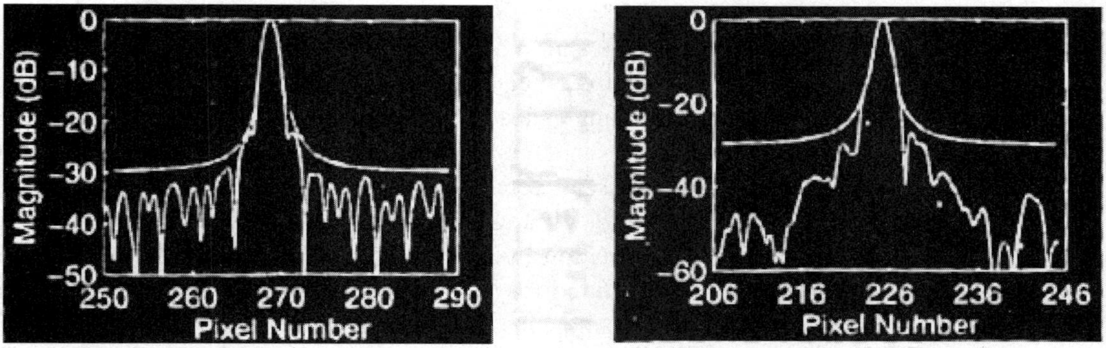

a. Range Impulse Response

b. Cross-Range Impulse Response

FIGURE 2–16

P-3 UWB SAR waveform after transmit power equalization [21]

As a result, a predistortion approach was adopted in an attempt to equalize the power over the band. Figure 2–15 also shows the output power from the transmitter before and after predistortion of the drive power. Because of the improved power spectral density across the 515 MHz bandwidth of the pulse, it was possible to greatly improve the SAR waveform. The measured waveform response after the predistortion technique was applied, in both range and cross-range, is shown in Figure 2–16. The specified peak sidelobe envelope of –30 dB is also indicated in the figure, illustrating the anticipated excellent response of the system [21].

The FOPEN SAR research required full polarization to investigate novel target detection and discrimination techniques. A fully polarimetric antenna with critical size constraints for installation into the P-3 aircraft was developed. The result was a 1 square meter flared notch antenna constructed with temperature and vibration stabilized material to preserve the phase center between each of the polarizations. The constant area antenna provided a beamwidth of 60 degrees at the low end of the band and 18 degrees at the high end of the band. For the 0.66 meter azimuth resolution, this provided 31.7 degrees of integration angle support at the center frequency of 470 MHz [22].

RFI is present from VHF andUHF TV stations, and this inhospitable environment necessitated the development of techniques to remove RFI in the SAR returns while preserving SAR image quality.Since the notched spectrum on transmit is synonymous to a thinned array in the spectral support, the sidelobes in the image response will be degraded. These notches need to be compensated in the range compression filter to obtain adequate image quality. The techniques developed for the P-3 UWB SAR to remove the RFI will be described in Section 5.2.3.

The FOPEN SAR's low frequency of operation, together with its fine azimuth resolution (wide azimuth beamwidth), required the creation of long synthetic apertures lengths or, equivalently, large integration angles. In turn, these large integration angles lead to severe range migration or differential range curvature of the scatterers during image formation, as presented in Section 4.1. Moreover, scatterers at different locations in an imaged scene experience different levels of range migration.This variation makes selection of the proper image formation algorithm critical. Whileit is straightforward to compensate range curvature for a given range bin, it is difficult to compensate range curvature for all range bins simultaneously.

A number of algorithms are available for fine-resolution SAR image formation, as detailed in Section 4.2. Two algorithms that are commonly used to minimize the motion of scatterers across the image are the BPA and the RMA. The former is computationally complex, requiring order of N^3 operations, where N is the number of pixels in the array. The RMA is unique in that it provides an exact solution to the problem of differential range curvature and has a computational complexity on the order of $N^2\log_2 N$. RMA was selected to provide the most efficient image formation that could be integrated into the long-term objective of a real-time onboard processor [22].

A final problem in low-frequency UWB SAR is the presence of dominant interfering radio frequency signals. These signals originate from a number of sources, the most serious being VHF and UHF television transmitters and cellular telephones. To improve the received image, a simple filtering scheme was employed to remove most of the interference energy prior to image formation. An important image quality metric is the multiplicative noise ratio (MNR), or the ratio of the image intensity in a low return area (e.g., water) to the return from bright clutter, such as the foliage. Many factors contribute to MNR, as will be detailed in Section 7.3.1. However, RFI is a major contributor to the background interference affecting target detection. Without RFI rejection, the MNR was –9 dBm^2/m^2, and after RFI rejection it was reduced to –20 dBm^2/m^2.

With approximately 1 km swath, moderate areas could be collected at three polarizations to provide significant data for fully polarimetric SAR characterization. The image shown in Figure 2–17, which was one of the early ERIM performance verification tests, demonstrated the ability of multiple polarizations to provide improvements in target discrimination. The image covers a biological field station in Michigan with an array of trees and other cultural objects. The figure includes some ground truth photographs of the instrumentation and the vehicles used for image characterization. It is interesting to note in the image that the large trihedrals, used for calibrating the polarization channels, have persistent sidelobes in the range dimension (horizontal,

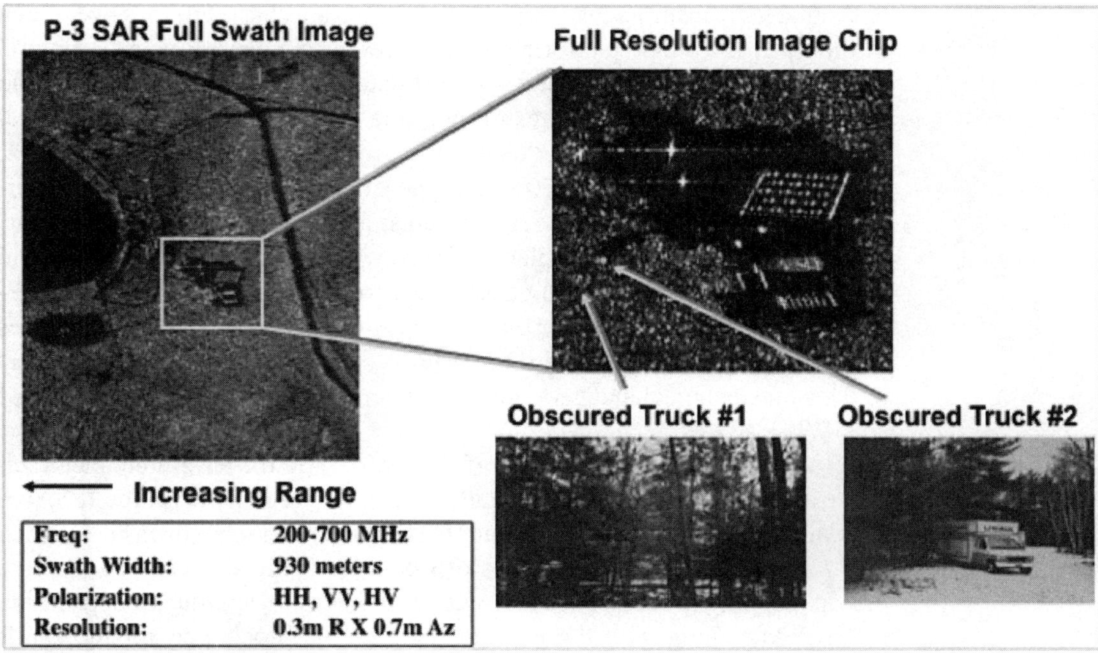

FIGURE 2–17

P-3 ultra wideband SAR image characterization © 1996 IEEE [20]

with far range to the right in this image). These sidelobes are due to notching out of strong RFI sources, and, although quite persistent, are still over 30 dB down. This prompted additional research into methods of RFI suppression to fill notches in the received spectrum.The low noise equivalent clutter return is also evident in the open areas [22].

The P-3 carried a multiple-channel wideband recording system. All of the data were recorded and calibrated after each flight. Significant advances in waveform generation, image formation processing, and automatic target detection and characterization were made with this instrument from 1995 to 2000.

2.2.4 NASA JPL's GeoSAR P-Band Interferometric Mapping SAR

GeoSAR is an interferometric mapping SAR development in the late 1990s to enable digital terrain elevation data (DTED) formation and terrain characterization for a "bald earth." The motivation for the program was a major earthquake in Los Angeles, California, where significant damage was exacerbated by insufficient knowledge of the terrain characteristics over much of the state. If a single platform could collect terrain elevation and structural characteristics at ground level, it was postulated that the loss of property could be avoided by better procedures for building and land-use planning. The development proceeded under a "dual-use" commercial consortium, where JPL provided the technology, DARPA; and the California Department of Conservation provided the end-user requirements; and EarthData provided the aircraft and commercialization plan for worldwide mapping services. It should be noted that the international geosciences community use the notation of P-band instead of UHF band for the frequency of operation. GeoSAR operates in the middle of the frequencies for FOPEN SAR operation and shares many of the same technical and geopolitical challenges for worldwide operation of an UWB RADAR.

NASA JPL had pioneered interferometric SAR for terrain characterization extensively on its AIRSAR platform [1]. At the same time, the IFSARE system had been collecting accurate DTED measurements over large areas, albeit in open areas or on the tops of trees [8]. JPL provided the interferometric mapping processor for IFSARE, and there was significant application to GeoSAR, along with the need to improve the processing for the UWB imaging and interferometric products. The prospect of having two accurate, cross-track mapping RADARs on the same platform was considered to be a major innovation for terrain height and land characterization. The completed system installed on a Gulfstream II is shown in Figure 2–18.

DARPA funded the effort with NASA Jet Propulsion Laboratory to build the dual-frequency, interferometric SAR (InSAR) system: X-band for open

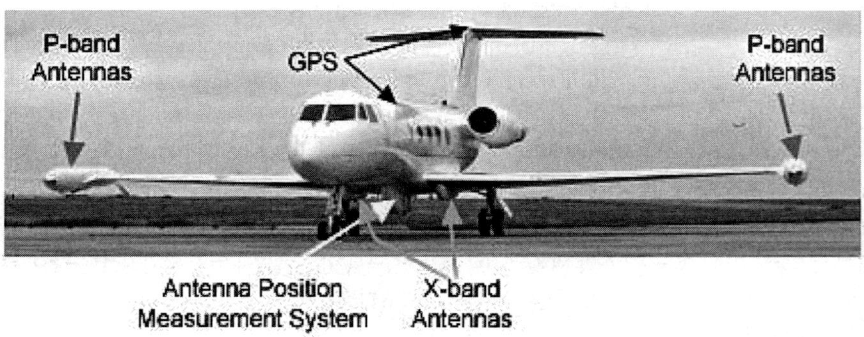

FIGURE 2–18

GeoSAR interferometric mapping platform [24]

Table 2–5 GeoSAR P-band interferometric SAR characteristics © 2001 IEEE [23, 24]

Altitude	[Km]	10–12	Pulse Width	[μsec]	40
Velocity	[m/s]	220	PRF	[Hz]	500, per side per antenna
Center Frequency	[MHz]	350	ADC	[MHz]	360, 10 bits
Bandwidth	[MHz]	80, 160	Range Resolution	[m]	1.0
Antenna Area, Type	[m^2]	0.57, 4-element Array	Cross Range Resolution	[m]	1.0
Polarization		HH, HV; or VV, HV	Slant Range	[Km]	25
Waveform		LFM, Notched	Swath Width	[Km]	10–12
Peak Power	[Kwatt]	4.0	Developed, Operated		NASA/JPL, Fugro Earth Data

terrain DTED, and P-band for DTED below the foliage. The P-band bandwidth of the GeoSAR system, as shown in Table 2–5, would not be as broad as FOPEN SARs used for tactical target detection. However, several innovations in waveform generation, multiple polarization antennas, and signal processing were made. These features provided well-calibrated data for two objectives: DTED under dense forest; and polarimetric characterization of terrain features [23].

The operational objective was to collect four 10 km swaths simultaneously, one at each frequency and one on each side of the aircraft. Efficient image formation, RFI rejection, and DTED formation processing were required for both the civilian and commercial applications for GeoSAR.

The dual-polarization P-band antennas are housed in the wingtip pods as shown in Figure 2–18. Each pod has two antennas, one looking port and one starboard. GeoSAR is a dual-baseline, single-pass system simultaneously collecting both ping-pong and single-antenna transmit interferometric modes. Ping-pong processing is used for improved DTED resolution on relatively flat terrain and in single-antenna transmits processing for rugged terrain. Additionally, the polarization channels on either pod can be used for land use data characterization. Fugro EarthData maintains, modernizes, and enhances the GeoSAR system and continues to provide commercial GeoSAR DTED and land use mapping services worldwide [24].

The use of interferometric SAR had been well developed when GeoSAR was started. The technical challenge was to produce two well-focused images from each antenna, which could be aligned precisely pixel by pixel. Figure 2–19 illustrates the basic geometry used in InSAR processing.

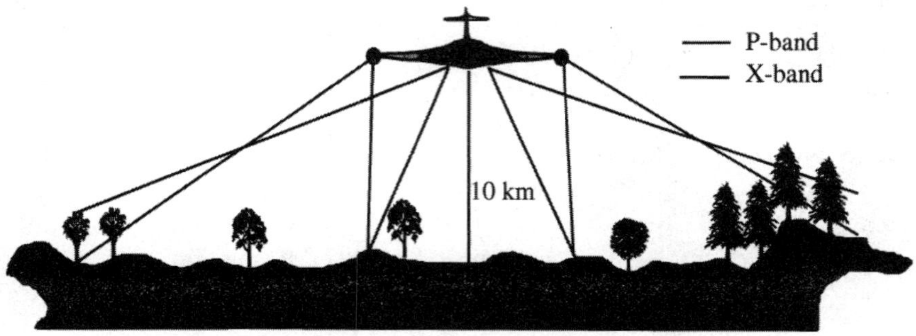

FIGURE 2–19

Two side operation of GeoSAR for P-band and X-band mapping [27]

However, this figure has been simplified by omitting the four additional illumination beams to form an interferometric pair, at the two frequencies and on the port and starboard side of the aircraft.

For each InSAR case, two antennas of area A_1 and A_2 in separate pods illuminate the scene swath. The slant range distance from the phase center of the two antennas to the scene pixel is given by ρ_1 and ρ_2, as shown in Figure 2–20. When the pixel on the ground has a scattering amplitude of A_b and phase φ_b, the signal at the two antennae can be measured as S_1 and S_2 [25]:

$$S_1 = A_b e^{j\varphi_b} e^{j\frac{4\pi}{\lambda}\rho_1} \tag{2.4}$$

$$S_2 = A_b e^{j\varphi_b} e^{j\frac{2\pi}{\lambda}(\rho_1+\rho_2)} \tag{2.5}$$

Based on the accurate knowledge of platform orientation and distance from the image plane, the angle difference between the two signal vectors is used to determine height of the local terrain. Taking A_1 as the reference, the interferometric phase to each pixel is given by

$$\varphi_1 = \frac{2\pi}{\lambda}(\rho_2 - \rho_1) \tag{2.6}$$

The digital elevation map is obtained by measuring the phase to each pixel in the two images formed in a single pass. Since the phase is modulo 2π, this phase must be unwrapped to get accurate height information. More importantly, the absolute baseline **D** between the two phase centers needs to be known within a small fraction of a wavelength. These challenging objectives were obtained by using an accurate laser baseline measurement system with several targets on each antenna pod to obtain the range and orientation of the apertures. To obtain the two-frequency InSAR map for determining the

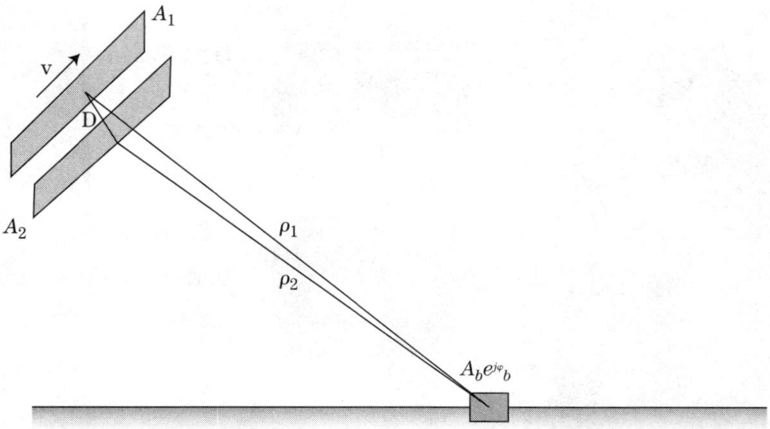

FIGURE 2–20

Interferometric SAR processing geometry [19]

scattering from the ground (as opposed to the tops of the trees), the pixels at P-band and X-band need to be similarly coaligned.

GeoSAR developed several innovations in image processing to make this possible. First, the focusing of the images over wide angles and swath widths are important [26]. If autofocus were used to take out variations in the phase errors due to terrain variation or propagation uncertainties, the absolute accuracy would be degraded. Second, the measurements in each band need to be radiometrically calibrated for estimating the signal correlation and scattering center [27]. Finally, the effects of RFI and transmit notching need to be accounted for in the waveform reconstruction [28]. These developments have been accomplished and are being used to provide commercial imaging services with GeoSAR.

A FOPEN SAR image of the Amazon River is shown in Figure 2–21 [29]. Both the X-band and P-band imagery are combined in false color to illustrate land use. The RGB image is made of a combination of X, P, and P–X returns from the system. For open areas the X-band provides significantly better texture of the return with its shorter wavelength.

However, as expected the X-band images only the tops of the trees, whereas the P-band penetrates the foliage.The plots on the right show three transects through the image. The top traces show the X-band DEM of the tops of the trees. The lower traces provide the derived DEM below the trees (combined X-band and P-band IFSAR processing).

These data show a difference of between 5 meter and 25 meter in the forested area when the X-band and the P-band traces are compared. However, in the open areas, the difference between the two DEMs is small. Thus, there is a definite benefit from the P-band interferometric image in determining the elevation below the treetops.

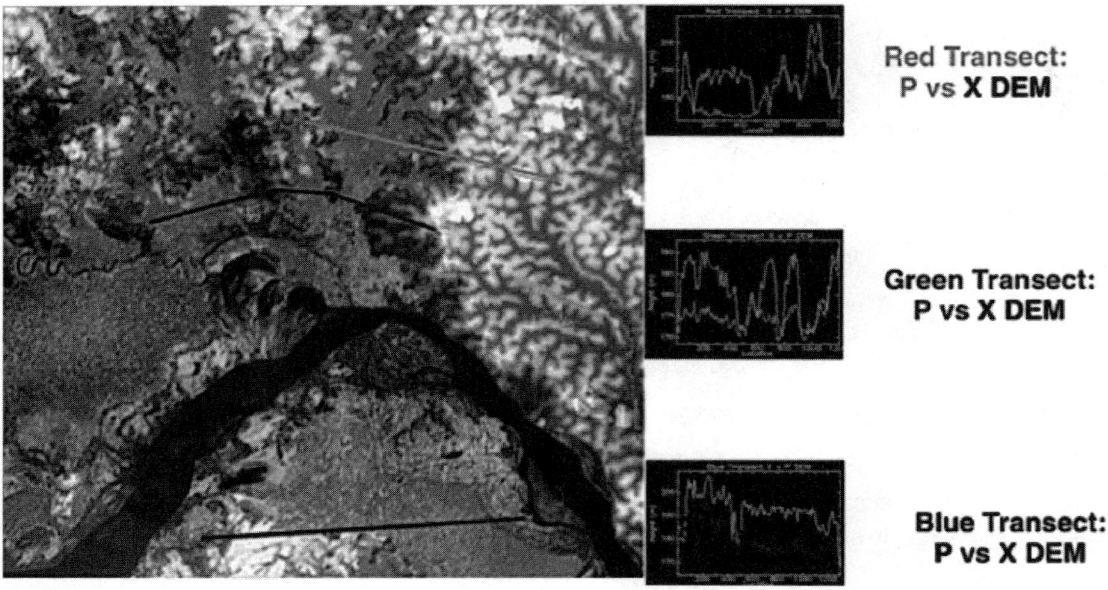

FIGURE 2–21

GeoSAR collection of terrain height characteristics—Amazon River [29]

2.2.5 ARL's BoomSAR

The BoomSAR was an experimental instrument developed by Army Research Laboratory (ARL) in Adlephi Maryland. The developmental effort to investigate critical technologies for penetrating foliage and the ground to detect and characterize hidden objects was started in 1988. The test bed UWB RADAR system was designed to provide controlled imaging over a 1 GHz bandwidth from HF to L-band and fully polarimetric illumination and data recording with parameters summarized in Table 2–6 BoomSAR instrumentation and

Table 2–6 Army research laboratory BoomSAR characteristics © 1995 IEEE [30]

Altitude	[Km]	0.05	Pulse Width	[n sec]	1.0
Velocity	[Km/hr]	1.0	PRF	[Hz]	750
Frequency	[MHz]	40–1200	ADC	[MHz]	60, 8 bits
Bandwidth	[MHz]	50–1100	Range Resolution	[m]	0.15
Antenna Area,	[m^2]	1.0,	Cross Range	[m]	0.3
Type		TEM Horn	Resolution		
Polarization		HH, HV, VH, VV	Range Bins		4092
Waveform		Impulse	Noise Equ.σ_0	[dBm^2/m^2]	−50
Peak Voltage	[Mvolt]	2.0	Developed		Army Research Laboratory

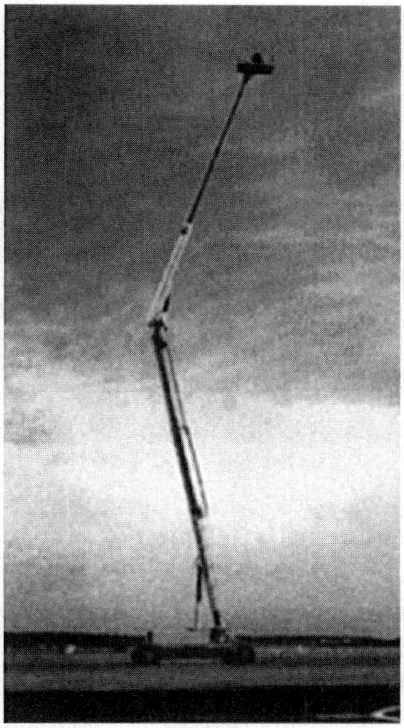

FIGURE 2–22

Army research laboratory BoomSAR [30]

algorithm research focused both on foliage and ground penetration phenomenology, target detection and discrimination and on understanding the interaction of dense foliage on the scattering characteristics of obscured objects [30].

Its 50 meter high boom, shown in Figure 2–22, was controlled over wide geometries to insure accurate measurement of grazing angle effects on foliage loss, clutter characteristics, and complex target scattering. Moreover, the boom and RADAR subsystems were installed on a 50 meter high boom lift platform so the SAR collection would emulate an airborne collection. However, at a 1 km/hour velocity, the images were certainly not collected in what would be considered *real time*. The BoomSAR system operated at several test ranges such as Yuma, Arizona, and Aberdeen, Maryland, where military targets and unexploded ordinance could be characterized in a scientific and operationally significant environment.

The antenna consisted of a set of four TEM horns,which were impedance matched to the impulse transmitter to provide calibrated spectrum and polarization characteristics. Each of the transmit antennas operates from 40 MHz to 1200 MHz with a beamwidth of 90 degrees. This provides the illumination support in both angle and spectrum to satisfy the system range and cross-range resolution objectives. Two of the horns transmit and two receive, with orthogonal polarization, to provide the full polarization matrix.

The system was used extensively to collect high-resolution, fully polarimetric data using the RADAR's UWB waveform and to develop two-dimensional (down-range versus cross-range) images of a controlled swath area. Within the controlled swath area were targets inthe clear, targets under foliage, subsurface targets, and natural and man-made clutter. The system provided the image swaths of up to 300 meters down-range by 1 Km cross-range, with spatial resolution in each dimension of less than 0.3 meter. The high-range resolution was obtained by using waveform bandwidth greater than 1 GHz and comparable high-speed sampling and recording techniques. A 2 megawatt impulse transmitter produced the UWB signal, having a spectral response extending from 60 MHz to over 1 GHz.

Through careful matching between the transmitter and the antenna and attention to receiver dynamic range and match to the ADC assemblies, the instrumented noise equivalent sigma-naught σ^0_{ne} was a very low –50 dBm^2/m^2. To illustrate this impact on image processing, Figure 2–23 shows greater than 60 dB dynamic range image from one of the foliage penetration runs at Aberdeen, collected over the frequency range of 130–1,100 MHz. A number of 42 cm trihedrals are visible in an open region between two areas of trees, and a 50 cm sphere is located at the edge of the woods.The resolution of the RADAR is demonstrated by the pair of lines running between the poles along the lower edge of the image. The first of these lines is the return from the wire

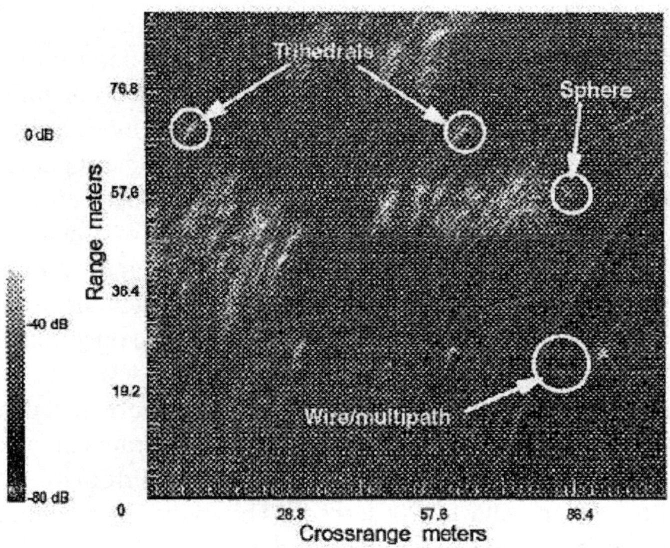

FIGURE 2–23

BoomSAR image from aberdeen MD © 1996 IEEE [31]

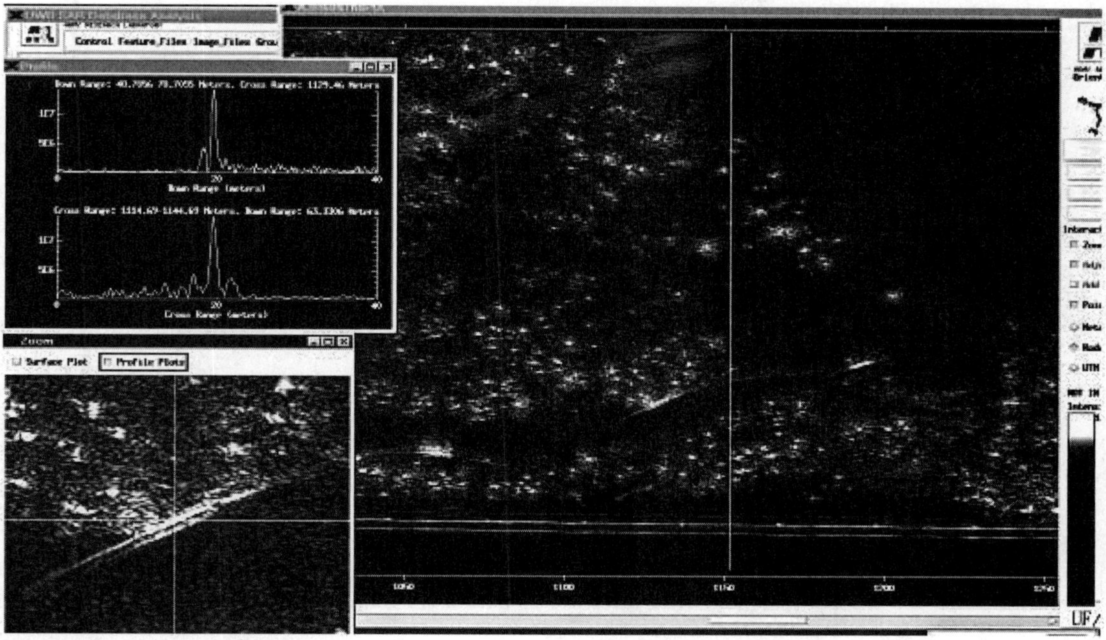

FIGURE 2–24

Army research laboratory image processing evaluation tool © 1999 IEEE [32]

strung between the poles, whereas the second is the multipath return from the ground reflection of the signal [31].

The Army Research Laboratory had a strong in-house team developing algorithms for image formation and target recognition. The wide dynamic range image recordings were processed in a high-performance computer to test and verify performance predictions. Figure 2–24 shows the screen capture of data from the test range at the Army Research Laboratory facility. Significant metrology was built into the analysis tools to quantify the signal processing and target recognition figures of merit [32].

Several critical FOPEN phenomena are shown in Figure 2–24, from the ARL image analysis tool. First, there is an excellent example of a long wire above the ground, indicated by the parallel lines in the near range of the figure. The closest return is the direct path from the RADAR to the wire. The next two parallel lines are the single and double bounce of the return from the ground, respectively. This clear return is a benefit of collecting SAR image over a wide beamwidth, with very fine-range resolution.

The second critical example is based on the return from two similar corner reflectors, one in the open and one 40 meters in the woods. Figure 2–25 shows

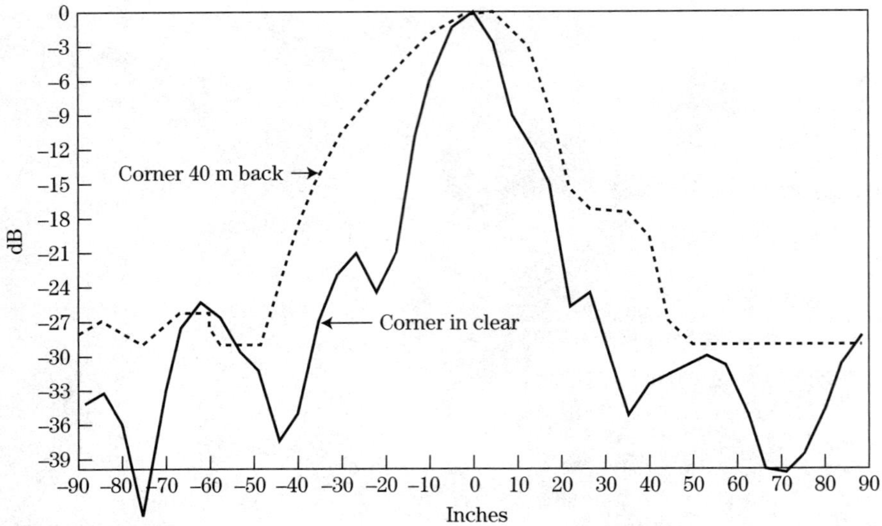

FIGURE 2–25

Impact of foliage loss variation on corner reflector [33]

the cross-range resolution of the RADAR measure from the corner reflector. The narrow resolution is characteristic of the wide-angle SAR collection. However, the return from the corner reflector in the foliage has degraded cross-range resolution due to the variation of loss and blockage of the forest, as a function of collection angle [33].

2.3 References

[1] Details of AIRSAR on NASA Jet Propulsion Laboratory, Pasaden CA, Website, http://airsar.jpl.nasa.gov/

[2] Vickers, R.S., Lowry, R.T., and Schmidt, A.D., "A VHF RADAR to Make Terrain Elevation Models through Tropical Jungles,"*Proc 1988 IEEE RADAR Conference.*

[3] Hellsten, H., "CARABAS-an UWB Low Frequency SAR,"*IEEE MTT-S International Symposium* Vol.3, June 1–5, 1992, pp. 1495–1498.

[4] Robinson, C.A., "Robots Counters Camouflage," *Signal Magazine,* June 2007, p. 40.

[5] Davis, M.E., "Technical Challenges In Ultra-Wideband RADAR Development for Terrain Mapping," *Proc*", Presented at *IGARSS,* Seattle, WA, April 1998.

[6] Carrara, W.G., Goodman, R.S., and Majewski, R.M., *Spotlight Synthetic Aperture RADAR Signal Processing Algorithms*, Artech House, Boston, MA, 1995, Chapter. 2.

[7] IEEE Standard RADAR Definitions, IEEE STD 686-2008, IEEE, New York, May 2008, p. 38.

[8] Adams, G., et al., "The ERIM Interferometric SAR: IFSARE,"*Proc 1996 National RADAR Conference,* Ann Arbor, MI, May 13–16, 1996, pp. 249–254.

[9] Vandenberg, N., et al., "P-3 Ultra Wide Band SAR: System Applications to Foliage Penetration,"SPIE Vol. 2757, Orlando, FL, April 1996, p. 130.

[10] Vickers, R. S., "Ultra Wideband RADAR—Potential and Limitations,"*Proc 1991 IEEE MTTS Conference*, June 1991, pp. 371–74.

[11] Vickers, R. S., "Design and Application of Airborne VHF/UHF RADAR,"*AES Systems Magazine,* Vol. 17, No. 6, June 2002, pp. 26–29.

[12] Stanford Research Institute, Palo Alto CA, "Foliage Penetration Radar", Web site http://www.sri.com/esd/penetratingradar/folpen/folpen.html

[13] Hellsten, H., Ulander, L.M.H., Gustavsson, A., and Larsson, B.,"Development of VHF CARABAS II SAR," *Proc. RADAR SensorTechnology, SPIE* Vol. 2747, Orlando, FL, April 8–9, 1996.

[14] Ulander, L.M.H., Frölind, P.-O., Gustavsson, A., Hellsten, H., Jonsson, T., Larsson, B., et al., "Performance of the CARABAS-II VHF-Band Synthetic Aperture Radar," *Proceedings of IGARSS 2001,* Sydney, Australia, July 9–13, 2001, pp. 129–131.

[15] Murdin, D., Flood, B., Frölind, P.-O., Haapahlati, G., and Ulander, L., "Upgrade of Real-time CARABAS-II Image Formation and Change Detection," Technical Report, FOI-R-2371-SE, Division of Sensor Technology, Swedish Defence Research Agency, 2007.

[16] Hellsten, H. and Ulander, L.M.H., "VHF/UHF Synthetic Aperture RADAR—Principles and Motivation," Proc 1999 IEEE *RADAR* Conference, Boston, MA, April 1999, p. 47.

[17] Ulander. L. and Frolind, P.-O., "Precision Processing of CARABAS HF/VHF-Band SAR Data," *Proceedings* IGARSS, Hamburg, Germany, June, 1999, pp. 47–49.

[18] Courtesy of M. Toups, MIT Lincoln Laboratory, Lexington MA, June 1997.

[19] Lee, R.R., Verdi, J.S., and Soumekh, M., "Enhancements of NP-3 UHF Image Quality Using Digital Spotlighting Technique," Proc 2001 IEEE *RADAR* Conference, Atlanta, GA, May 2001.

[20] Sheen, D.R., Vandenberg, N.L., et al.,"P-3 Ultra-Wideband SAR:Description and Examples," *IEEE AES Systems Magazine*, November 1996, pp. 25–30.

[21] Carrara, W., Goodman, R., Rawson, R., et al., "A Foliage and Ground Penetrating SAR", Proc. 41st Tri Services Radar Symposium, Unclassified, Laurel, MD, June 1995, pp. 187–202.

[22] Goodman, R., Tummala, S., and Carrara, W., "Issues in Ultra-Wideband, Widebeam SAR Image Formation," *Proc. 1995 IEEE International RADAR Conference,* Washington, DC, May 1995, pp. 479–485.

[23] Hensley, S., Chapin, E., Freedman, A., et al., "First P Band Results Using the GeoSAR Mapping System," Proc 2001 IEEE *RADAR* Conference, Atlanta, GA, May 2001, p. 126.

[24] Reis, J.J, Fugro EarthData,Frederick, MD, www.earthdata.com, private communication.

[25] Rosen, P., et al., "Synthetic Aperture RADAR Interferometry," *Proceedings of the IEEE* Vol.88, No.3, March 2000, pp. 333–382.

[26] Madsen, S., "Motion Compensation for Ultra Wide Band SAR," Proc IGARSS '01Vol.3, July 2001, pp. 1436–1438.

[27] Chapin, E., Hensley, S., and Michel, T.R., "Calibration of an Across Track Interferometric P-Band SAR,"Proc IGARSS *2001*Vol. 1, July 2001, pp. 502–504.

[28] Le, C. and Hensley, S., "Removal of RFI in Wideband RADARs," Proc 1998 IGARS Symposium, Seattle, WA, July 1998, p. 2032.

[29] Carson, T. M., "Topographical Mapping in the Equatorial Belt Using Dual Frequency Airborne IFSAR (GeoSAR),"Photogrammeric Engineering & Remote Sensing, August 2008, p. 939.

[30] Ressler, M., et al., "The Army Research LaboratoryUltra-Wide Band Testbed RADARs,"*Proc. IEEE International RADAR Conference,* Arlington, VA, May 1995, pp. 686–691.

[31] Ressler, M., "The Army Research Laboratory Ultra Wideband BoomSAR," *Proc IGARSS* Vol. 3, May 1996, pp. 1886–1888.

[32] Kapoor, R., Banerjee, A., Tsihrintzis, G. A., and Nandhakumar, N., "UWB RADAR Detection of Targets in Foliage Using Alpha Stable Clutter Models," IEEE Trans. AESS Vol. 35, No. 3, July 1999, p. 819.

[33] McCorkle, J., "What's So Special about UWB Propagation," Presentation at 3rd IEEE Ultrawideband Communications Workshop, Atlanta, GA, May 19–22, 2002.

CHAPTER 3

Foliage Penetration Phenomena

The development of foliage penetration (FOPEN) RADAR had a major resurgence in the early 1990s for both military and geoscience applications. A series of data collections was carried out starting in 1990 as a risk reduction for the development of a more reliable FOPEN synthetic aperture RADAR (SAR) system. Several government laboratories including NASA Jet Propulsion Laboratory (JPL), Sweden's Defence Research Establishment (FOA), US Army Research Laboratory (ARL), US Air Force Research Laboratory (AFRL), and the US Naval Air Development Center (NADC) initiated these collections for both scientific and system technology objectives. A combination of instrumentation collection platforms and airborne brassboard RADARs were used to obtain data on surface clutter, foliage scattering losses, and the ability to detect objects and terrain features in forested regions. However, it was deemed to be very important that accurate instrumentation and calibration targets be included in the test to characterize the target and clutter phenomena toward the development of operational and commercial RADAR systems. To this end, the US Defense Advanced Research Projects Agency (DARPA) sponsored MIT Lincoln Laboratory to set up and maintain the scientific and analytic standards. These capabilities would provide the community with a reliable understanding of these measurements' influence on new systems applications [1].

The FOPEN RADAR collections were conducted at wavelengths that spanned from low very high frequency (VHF) band (30 MHz) to C-band

(6 GHz) and foliage types from northern latitude to tropical forests. Instrumentation systems including standard targets, forest characteristics measurements and receiver calibration techniques were developed to characterize the one-way losses at a variety of grazing angles. SAR systems were used primarily to obtain moderate to fine-resolution characterization of clutter type backscatter and propagation loss. More importantly, these test systems were developed with multiple-channel characteristics to evaluate the emerging techniques used by Earth-resource measurements from polarimetric and interferometric scattering. Finally, the impact of background radio frequency interference (RFI) from radio and television transmissions was assessed. These interferences were evident in the early tests as a major limitation to system sensitivity and eventually to operational utility.

The lessons summarized in Chapters 1 and 2 from early FOPEN SAR and ground moving target indicators (GMTI) RADAR experiments, and their limitations, were important. They established sound system design principals and reduced the performance risks for future subsystem and signal processing algorithms. The following sections quantify these lessons and cover the impact on system design including:

- Amplitude and phase scattering by the forest biomass that affects the coherent processing of signals from airborne- and ground-based RADAR systems
- The design of standard targets that provide radiometric calibration of the RADAR returns
- Foliage attenuation as a function of frequency, grazing angle, and polarization
- Clutter backscatter characteristics affecting signal to clutter analyses of targets near and under the forest canopy
- Internal motion of the clutter, especially at tree lines, that affect the coherency and Doppler characteristics for both GMTI RADAR and SAR processing
- RFI environments and its effects on waveform and signal processing design
- Scattering characteristics of man-made targets as a function of size, orientation, and frequency

It was also important to obtain geographically diverse data because of the experiences in early FOPEN tests. Several test campaigns were conducted between 1990 and 1997 (the forest types as summarized in Table 3–1). Tropical data were collected in Panama, Puerto Rico, and Australia. Forest types

Table 3–1 Summary of FOPEN SAR data collections, 1990–1997 ©1995 IEEE [1], ©2001 IEEE [2]

Year	Location	Forest Type	Sensors
1990	Maine	Mixed Northern	JPL AIRSAR
1992	Maine	Mixed Northern	FOLPEN II, P-3
1992	Puerto Rico	Rain Forest	FOLPEN II
1992	Sweden	Deciduous	CARABAS I
1993	Panama	Rain Forest	FOLPEN II, CARABAS I
1993	Maine	Mixed Northern	FOLPEN II, CARABAS I
1993	Australia	Rain Forest	JPL AIRSAR
1994	California	Redwood Forest	FOLPEN II
1995	North Carolina	Coniferous	P-3 UWB
1995	Maine	Mixed Northern	P-3 UWB
1995	California	Varied	P-3 UWB
1995	Michigan	Mixed Northern	P-3 UWB
1996	California	Sierra Mountain	FOLPEN III
1996	California	Sierra Mountain	P-3 UWB
1997	Pennsylvania	Mountain Mixed Northern	P-3 UWB, CARABAS II

were evaluated with data from California, Maine, Michigan, North Carolina, Pennsylvania, and Sweden. In each of these campaigns, MIT Lincoln Laboratory provided the standard corner reflectors along with dihedral and top-hat reflectors to measure the image quality and two-way losses as a function of frequency, grazing angle, and clutter type.

There was a great collection of data and RADAR operation experience in the 1990s that characterized both foliage clutter and propagation in a dense foliage environment. Many of the technical and operational issues observed in the early FOPEN systems were verified. At the same time, analytic and system tools were being created for development of both military and civilian systems applications. The results were very positive but pointed out issues of operation in a dense radio frequency signal environment. First, it was verified that SAR systems were practical at both VHF and ultra high frequency (UHF). More importantly, polarization was shown to be instrumental in characterizing and eventually mitigating the clutter that competed with detection of man-made objects. However, as more systems were built and flown in semipopulated areas, the political issue of spectrum management was raised. System designs needed to factor in not only the removal of strong radiofrequency emissions but also the avoidance of many frequencies in the desired transmit bandwidth.

3.1 Foliage Phase Effects on RADAR Propagation

It was well known that SAR resolution and impulse response (IPR) were important in target characterization. The first FOPEN RADAR technical question posed was whether the foliage amplitude and phase scattering would destroy the ability to form reliable SAR images. To visualize the problem, consider the SAR collection geometry shown in Figure 3–1. The airborne platform ideally flies a straight line, and the RADAR illuminates the area on the ground with an interpulse spacing of approximating one half wavelength in the along-track direction. This geometry was shown in Section 2.1 in developing the cross-range resolution at a point P, with a synthetic aperture length L, and where the aperture subtends an angle θ with the point on the ground.

The foliage scattering is illustrated as the shaded volume above point P. At each incidence angle during the collection there will be an amplitude and phase perturbation of the RADAR propagation between the transmitter location and the point P. If this perturbation were constant, the image formation process would not need amplitude or phase error compensation for the propagation medium. However, if the illumination and receiving angles through the foliage change significantly, the variation of the nonuniform scattering must be understood in terms of creating a high quality IPR function at P.

Several data collections were conducted to measure the phase scattering as a function of frequency, polarization, and grazing angle through foliage. ERIM conducted a campaign using its RailSAR instrumentation [3]. A second collection designed to be more representative of an airborne SAR was performed in Maine. The test setup shown in Figure 3–2 used the MIT Lincoln Laboratory tone generation experimental equipment and employed the NASA

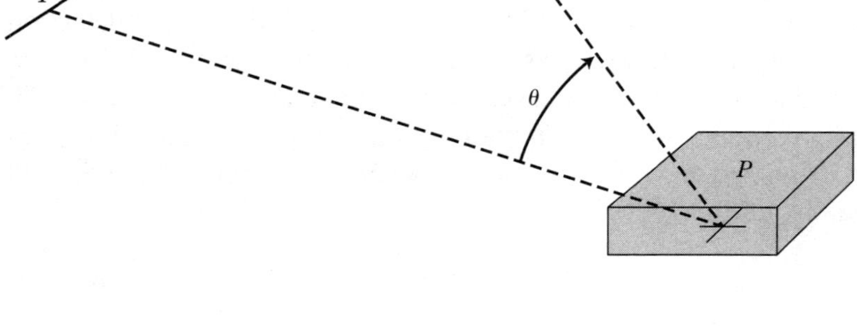

FIGURE 3–1

Collection geometry for FOPEN scattering measurements

FIGURE 3–2

Foliage phase perturbation test experimental setup
Source: MIT Lincoln Laboratory ©1996 IEEE [4]

Jet Propulsion Laboratory AIRSAR platform as a receiver [4]. The tone generator and antennas provided both horizontally and vertically polarized signals from the ground. And the one-way received signal characteristics were collected in the air, at three frequencies (UHF, L-band, and C-band) using the AIRSAR platform. Several passes were made to provide variation of grazing angle through the trees as well as statistical variation of the collection conditions [5].

The graph in Figure 3–3 shows the standard deviation in phase error measured during a series of collection paths and compared with a representative linear path of a SAR image. The phase errors at three frequencies are indicated in the groupings of data points. The circles are for horizontal polarization and the triangles for vertical. The incidence angles measured from the receiver to the signal source were collected at 30, 45, and 60 degrees and were grouped with the corresponding frequency. As the wavelength increased, the phase perturbation was reduced accordingly. So UHF had significantly less phase error during a pass than C-band. Both UHF and L-band phase scattering would support modest waveform resolution and integrated sidelobe levels. Grazing angle also had a marked impact on the RADAR clutter scattering. As long as the grazing angle was greater than 30 degrees, the standard deviation of the phase error was below a quarter of a wavelength.

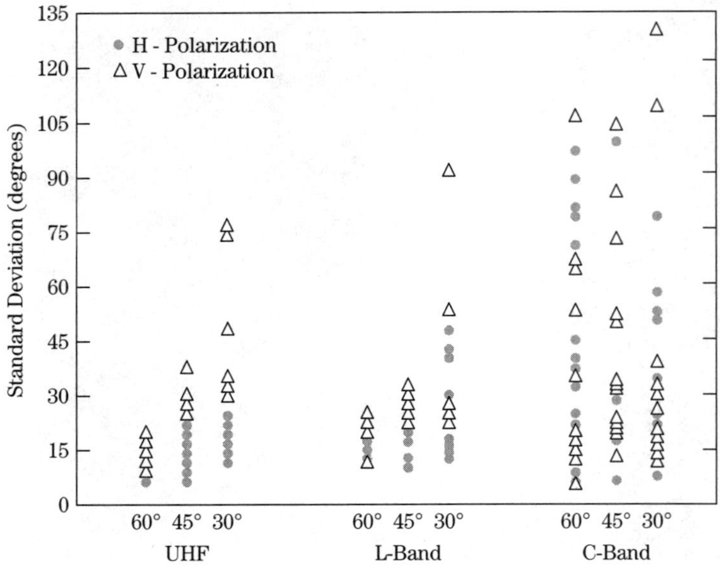

FIGURE 3–3

Phase perturbation versus frequency and incident angles
Source: MIT Lincoln Laboratory © 1996 IEEE [4]

The electromagnetic wave scattering was analyzed in terms of length and diameter of dielectric cylinders, with the trunks of the tree being the dominant scatterer. In all of the phase perturbation measurements, horizontal polarization phase scattering was generally lower than for vertical. This was qualitatively expected due to the vertically versus horizontally polarized electromagnetic wave scattering from the vertical tree trunks. But the separation in polarization phase error is more pronounced at UHF and longer wavelengths, where the large tree trunks dominate in scattering amplitude. At C-band, the tops of the trees provide more contribution to the scattering, where the branches are comparable to the shorter signal wavelength. It should be noted that since these are one-way transmissions from the ground to the air, the effects of ground–trunk interaction on the scattering was not experienced. This feature will be more prevalent in the SAR measurements.

However, it should be noted that these are only one-way continuous wave signals propagating through the foliage, and the signal levels were sufficiently large that phase measurements could be accurately measured. In a real SAR system, the signal will be pulsed, and the airborne platform would have timing errors on the two-way propagation. Both of these effects would contribute to the SAR image formation errors and detract from the utility. Nevertheless, these one-way measurements were important scientific observations that encouraged further SAR system development [5].

The impact of phase errors on image point function can be understood by considering linear array theory for the SAR data collection. The IPR can be described generally by a function $f(u)$, where each sample is collected in the synthetic aperture array, at N incremental locations of nd/λ given by [6]

$$f(u) = \frac{1}{\sqrt{N}} \sum_{n=0}^{N-1} a_n \exp[j(2\pi/\lambda)ndu + \varphi_n] \tag{3.1}$$

where $u = \sin\theta$ is the plane wave incidence angle to the array of N points. The two parameters a_n and φ_n include the amplitude and phase error contributed both by the propagation through the foliage and by other system errors. For uniform amplitude distribution of samples ($a_n = 1$) and no phase error ($\varphi_n = 0$), this is simply the array sampling function:

$$f(u) = \frac{\sin[\pi N(d/\lambda)u]}{\sin[\pi(d/\lambda)u]} \tag{3.2}$$

At broadside to the synthetic aperture $u = 1$, and the half-power beamwidth $\theta_{3\,dB}$ of the array is

$$\theta_{3\,dB} = \frac{0.886\lambda}{Nd} = \frac{0.886\lambda}{L} \tag{3.3}$$

When there are errors in amplitude and phase in the array function (3.1), the IPR will have higher average sidelobes. For small errors and random distribution over many elements, the central limit theorem predicts those contributions will be Gaussian. Ruze has shown that the array functions will become [7]

$$g(u) = \left\{\left[g_0(u) + \sum x_n\right]^2 + \left[\sum y_n\right]^2\right\}^{1/2} \tag{3.4}$$

where $g_0(u)$ is the unerrored amplitude distribution along the array, x_n is the amplitude error distributed around g_0, and y_n is the phase error. For small errors, the function in (3.1) can be expanded in a Taylor series:

$$f(u) = f_0(u)\left[1 + \sum a_n \exp\left(j\frac{2\pi nd}{\lambda}\phi_n\right)\right] \tag{3.5}$$

To quantify the effects on IPR sidelobes, the total mean square error σ_T^2, equal to the sum of the amplitude and phase errors, is examined:

$$\varepsilon_T^2 = \varepsilon_a^2 + \varepsilon_\varphi^2 \tag{3.6}$$

Under the assumption of statistical independence of the error sources, the mean squared sidelobe level (MSSL) that bounds the SAR resolution sidelobe levels is related to the size of the array and total mean square array error [8]:

$$MSSL = \frac{\varepsilon_T^2}{N(1 - \varepsilon_T^2)} \tag{3.7}$$

where the parameters used to characterize the aperture MSSL are

- N Number of elements in the synthetic aperture
- ε_T Total error in the distribution across the N pulse samples
- ε_a Mean error in amplitude distribution a_n across the aperture
- ε_φ Mean error in the phase distributed across the N pulse samples

The errored sidelobe levels are analogous to the far-field response of a linear array of N isotropic radiators. As such, the MSSL is the additional energy in the sidelobes compared to the expected sidelobe distribution of a uniform array. The σ_T^2 term is the energy scattered due to errors in the radiation function that contribute to an increase in the sidelobe level. The denominator of (3.7) is the coherently combined radiation of N elements, in the peak gain direction, and reduced by the total error σ_T^2.

Consequently in the early development of the FOPEN SAR phenomenology, it was important to determine the amount of error introduced by the foliage scattering that limited the SAR image quality. If the attenuation and phase scattering varied significantly, the sidelobes of the image resolution function would be high. Hence, the image quality would be adversely affected by any uncompensated propagation errors.

Figure 3–4 illustrates the effect of mean squared phase error on an array's MSSL from equation (3-5). Reasonable SAR performance (e.g., MSSL

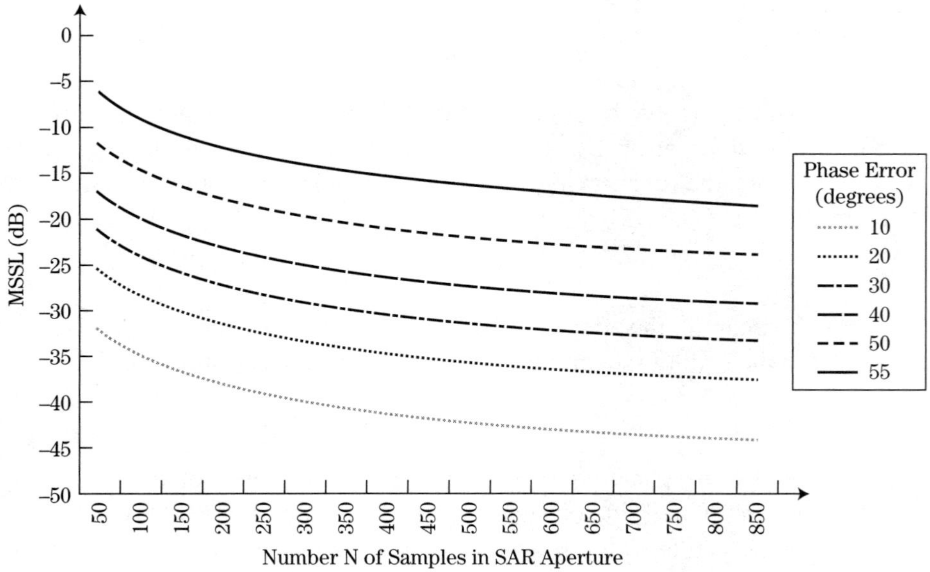

FIGURE 3–4

Mean squared sidelobe level of SAR image with phase errors

<-20 dB) can be obtained if the standard deviation in phase scattering through the foliage is maintained below 45 degrees. This criterion is met when N is large, as is the case for fine cross-range resolution in SAR. It can be seen from Figure 3–4 that this condition is nominally met for VHF and UHF SAR, when N is greater than 200 pulses. However, the error due to phase is not the only challenge for FOPEN SAR images at frequencies above VHF. At these higher frequencies, the amplitude reduction (or loss) through different paths through the foliage is also a major factor in the SAR image quality. Thus both the amplitude and phase variation must be evaluated if well-focused SAR images are to be useful.

3.2 Standard Calibration for FOPEN Measurements

It is important in the characterization of the FOPEN SAR propagation loss (as well as in development of FOPEN SAR systems) that standard RADAR cross-section (RCS) targets be placed in the scene. These targets provide for characterization of the RADAR returns for radiometric calibration, measurement of clutter cross section and the foliage propagation losses, and the polarization vector management [9]. The calibration approach will next be examined in terms of the RADAR range equation.

To illustrate the process, comparison images between VHF and UHF from the 1993 Maine collection are shown in Figure 3–5. The VHF coherent all

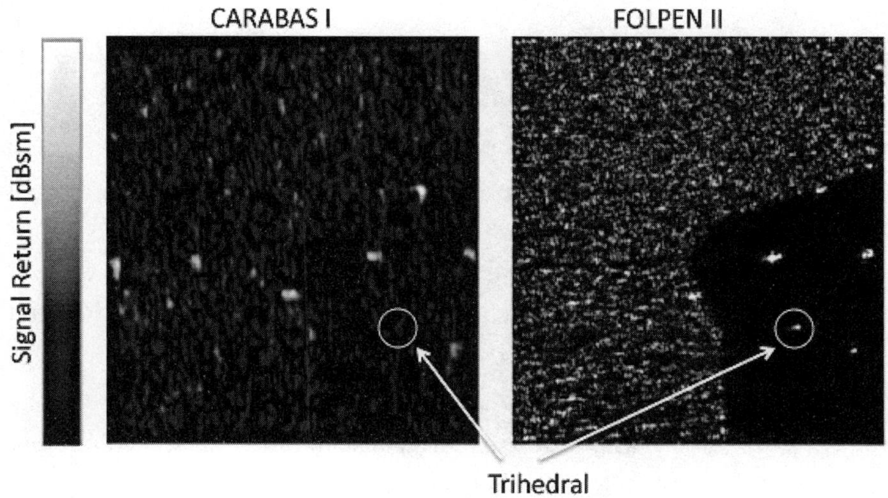

FIGURE 3–5

FOPEN SAR images with calibration targets (Maine 1993)
Source: MIT Lincoln Laboratory © 1995 IEEE [1]

FIGURE 3–6

Top-hat reflector from FOPEN Maine 1993 collection
Source: MIT Lincoln Laboratory [5]

radio band sensing (CARABAS) I RADAR image is shown on the left, along with the FOLPEN II low UHF image on the right. Also indicated in the figure is the location of a top-hat calibration reflector, as shown in Figure 3–6. The return from the top-hat is much smaller in the VHF image, as will be shown in Section 3.3.2.

The RADAR measures a signal return based on system design factors and wave propagation. For a single pulse, the signal-to-noise ratio (SNR) received by the RADAR from a scatterer at range R is given by [8]

$$\frac{S}{N} = \frac{P_T G_T A_R \sigma}{(4\pi)^2 R^4 k T_0 B F_n L_T} \tag{3.8}$$

where the terms have the usual definitions

- P_T Peak power of the transmitter
- G_T Gain of the transmitter in the direction of the scatterer, commonly expressed as $G_T = 4\pi A_T/\lambda^2$
- λ Signal wavelength
- A_R Area of the receiver, normally equal to transmit area A_T
- σ RADAR cross section of the scatterer
- R The range to the scatterer
- kT_0 Boltzmann's constant times the system reference temperature
- B Signal bandwidth
- F_n Noise figure of the receiver
- L_T Total loss from the transmitter to the scattering cell and back to the receiver. For foliage penetration systems, this will include the losses of the forest between the RADAR and the imaged target.

When the RADAR transmits a long, coded pulse and uses pulse compression to achieve improved average power, there is an SNR increase approximated by the ratio of the expanded pulse length τ_I and the compressed pulse length τ_O. Furthermore, in an SAR there is an additional SNR improvement from the coherent integration of the number of received pulses. The frequency of pulse transmission (PRF) and the velocity of the platform v_P approximate the number of pulses during the SAR collection. Therefore, an SNR improvement factor can be approximated by [8]

$$IFactor = \left[\frac{\tau 1}{\tau 0}\right] \frac{PRF\,L}{v_p} \tag{3.9}$$

where the RADAR design factors are

τ_I	Uncompressed pulse width
τ_0	Compressed pulse width
PRF	Pulse repetition frequency of the RADAR
L	Synthetic aperture length
v_P	Velocity of the SAR platform

However, it is important to note that for ultra wideband (UWB) SAR the velocity of the aircraft and *PRF* can vary during the formation of a synthetic aperture length Lsufficient to achieve a cross-range resolution δ_{AZ}. For an SAR system at microwave frequencies, or small integration angles, the synthetic aperture length can be approximated by [8]

$$L = \frac{R\lambda}{2\delta_{AZ}} \tag{3.10}$$

Chapter 4 will introduce the need for a longer effective SAR length which accommodates the wide-angle data collection with its variation in slant range.

And for a pulsed transmitter, the average power radiated can be expressed by

$$P_{av} = P_T \tau_I PRF \tag{3.11}$$

Finally, the time-bandwidth product of the RADAR is normalized with no sidelobe weighting to satisfy

$$B\tau_0 \approx 1 \tag{3.12}$$

When sidelobe weighting is applied to the pulse compression, a correction factor of k_R is applied in the range resolution equation.

When the scatterer is a resolution cell on the earth, the RCS is expressed in terms of the surface reflectivity σ^0, range resolution δ_R, cross-range resolution

δ_{CR}, and the incidence angle ψ go the scattering cell:

$$\sigma = \sigma^0 \delta_R \delta_{CR} \sin\psi \tag{3.13}$$

Substituting these system and coherent integration improvement relationships into the clutter-to-noise ratio (CNR) (3.7) yields the expected CNR for a resolution cell on the ground:

$$\left[\frac{C}{N}\right]_{area} = \frac{P_{av}}{8\pi} \frac{A_T^2 \sigma^0 \delta_R}{kT_0 F_n L_T R^3 \lambda} \frac{\sin\psi}{v_P} \tag{3.14}$$

In characterizing the foliage propagation, many of these terms are unknown and needed to be determined by data collection, and then calibration with known scatterers in the scene. When a discrete scatterer with RADAR cross section σ_T is measured, the RADAR range equation has the form

$$\left[\frac{S}{N}\right]_{point} = \frac{P_{av}}{8\pi} \frac{A_T^2}{kT_0 F_n L_T R \lambda} \frac{\sigma_T \delta_R}{v_a \delta_{CR}} \tag{3.15}$$

When the RADAR views a calibration target having an RCS of σ_T, the return will be a combination of both the target and any clutter that is in the same resolution cell. This is especially true for coarse resolution SAR where the size of the target is a fraction of the illuminated ground area.

When the resolution cell contains distributed clutter, the signal-to-clutter ratio (SCR) of the point scatterer σ_T to the back ground clutter σ^0, and in a resolution cell δ_{CR} by δ_R is obtained from the ratio of (3.15) to (3.14) [9]:

$$\left[\frac{S}{C}\right] = \frac{\sigma_T}{\sigma^0 \delta_{CR} \delta_R \sin\psi L_{fol}} \tag{3.16}$$

One should note the introduction of L_{fol} into (3.16), since the foliage loss can be between the RADAR and the point scatterer. The clutter return is from the volumetric clutter above the scatterer, which have be a different loss factor.

In a practical data collection, the resolution cell can contain both the target and several classes of clutter. Misalignments of the measurements will present errors in the calibration of the sensor and need to be avoided. The practical method to avoid this issue is to sample several points around the area to be measured and to integrate cells. From the measured image response of the known reference reflector one can derive the backscatter coefficient for uniform "clutter" areas in the image. There is an advantage of deriving the backscatter coefficient from an integral over the energy in the image of a reference reflector impulse response. The results are compared to those obtained from the estimated peak value [9].

The total return will be the combination of return from clutter and the target [10]:

$$S_m = S_T + S_C = S_T\left(1 + \left|\frac{S_C}{S_T}\right| e^{j\varphi}\right) \tag{3.17}$$

where $S_m$is the complex RADAR return, and S_T and S_C are the target and clutter contributions, respectively. The measured RCS σ_m is a mixture proportional to the total return. Assuming the RADAR is radiometrically calibrated,

$$\sigma_m = |S_T|^2\left(1 + \frac{|S_C|^2}{|S_T|^2} + 2\frac{|S_C|}{|S_T|}\cos\varphi\right) \tag{3.18}$$

The calibration error will then fall within the bounds

$$\frac{\sigma_m - \sigma_T}{\sigma_T} = \frac{\sigma_C}{\sigma_T} \pm 2\sqrt{\frac{\sigma_C}{\sigma_T}} \tag{3.19}$$

On a decibel scale, the error limits will be given by

$$\sigma_e(dB) = 10\log(1 + \varepsilon_{TC}^2 \pm 2\varepsilon_{TC}) \tag{3.20}$$

where $\varepsilon_{TC} = \sigma_C/\sigma_T$ is the relative error between the measure clutter and the calibration target. If the calibration calls for a 20 dB accuracy in measurement, the difference between the measured return and the target should satisfy

$$\frac{\sigma_C^0 A}{\sigma_T L_{fol}} < 0.01 \tag{3.21}$$

In other words, the measured return from the standard target needs to exceed the clutter return by 20 dBm^2. It should also be noted for the case of measuring clutter return and losses below the foliage canopy, the attenuation in signal from the target through the foliage (L_{fol}) must be included in (3.21).

3.3 Standard RCS Target Characteristics

The UWB characteristics of the RADAR signal, combined with the wide SAR integration angles, make the selection of standard targets important. This section will provide characteristics of the most common instrumentation targets and their design limitations for the collection parameters used for characterization of the clutter and target phenomena. The common targets used for RADAR calibration, as shown in Figure 3–7, are as follows:

- *Calibration sphere*: used for absolute calibration of RADAR sensitivity, independent of wavelength and aspect angle.

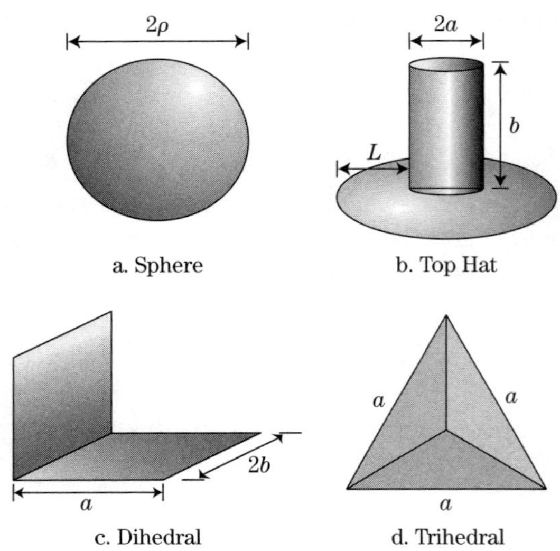

FIGURE 3–7

Common targets for calibrating FOPEN SAR RADAR

- *Top-hat reflector*: used for obtaining a return independent of azimuth and aspect angle
- *Dihedral corner reflector*: used for a single polarization and associated cross-polarization RADAR channel calibration
- *Trihedral corner reflector*: used to calibrate multiple polarization components at the same range and angle resolution cell

These targets are normally deployed in an open area near the forested area to be characterized, as well as within the foliage to directly measure the impact of foliage loss and scattering on the propagation.

The targets rely on specular reflection to establish the peak return for calibration of the RADAR. The peak of the specular return is commonly represented by the projected or effective area A_{eff} toward the RADAR to characterize scattering cross section. With this approximation, the target RCS σ_T is given by

$$\sigma_T = \frac{4\pi A_{eff}^2}{\lambda^2} \tag{3.22}$$

This scattering area depends on the percentage of the projected area perpendicular to the line of sight that reflects at each wavelength [11]. Moreover, the effective scattering area depends on the size and orientation of the targets toward the RADAR illumination, as well as any interactions with surrounding scatterers, such as ground plane and trees between the RADAR and the standard target. As a consequence of these effects, standard targets are placed both in the open and under the foliage to assess any geometric or proximity scattering effects.

Standard targets are also employed to measure the scattering in polarimetric SAR data collections. The radiation returned from a polarimetric FOPEN RADAR will consist of multiple polarization vectors depending on the design of the antenna and the desired scattering effects to be measured. If the system transmits horizontal (H) and vertical (V) polarizations, it can use two separate receivers to measure both H and V polarizations backscatter from the target or SAR scene. An example of a polarimetric backscatter matrix for [H] and [V] transmit polarizations would be expressed as

$$S = \begin{bmatrix} S_{HH} & S_{HV} \\ S_{VH} & S_{VV} \end{bmatrix} \tag{3.23}$$

The convention used in this text is 1st-letter transmit and 2nd-letter receive. Other senses of polarization are used for geoscience RADAR imaging, such as circular and combinations of linear orientations. The details of polarimetric scattering in terms of basis vectors are treated in Section 6.2. The backscatter consideration from (3.23) is presented as in introduction to the standard-target polarimetric characteristics in the following sections.

3.3.1 Metal Sphere

The metal sphere is a commonly used calibration reflector that theoretically provides a constant signature independent of frequency. As a result the reflection of energy is independent of both the orientation and frequency of the return. The metal sphere RCS σ_{MS} is commonly given by [12]

$$\sigma_{MS} = \pi \rho^2 \tag{3.24}$$

However, the size of the target with respect to a wavelength is a determining factor in characterizing the RCS of the target. The electromagnetic scattering will consist of a specular return from the target area, combined with secondary effects of surface waves on the target. Depending on the relative size of the target with respect to the wavelength of the radiation, there can be significant differences in the measured return. These differences are one of the major factors in reduced accuracy in UWB SAR calibration.

An example of the differences in scattering can be obtained analytically for a perfectly conducting sphere and is presented here to illustrate the importance of measuring the standard targets in the local environment. For a perfect sphere, the RADAR return is divided into three regions, as shown in Figure 3–8:

- *Optical*: the specular return is entirely due to the effective area of the target. This occurs when the curvature of the object is large compared with a wavelength (i.e., $\rho \gg \lambda$).

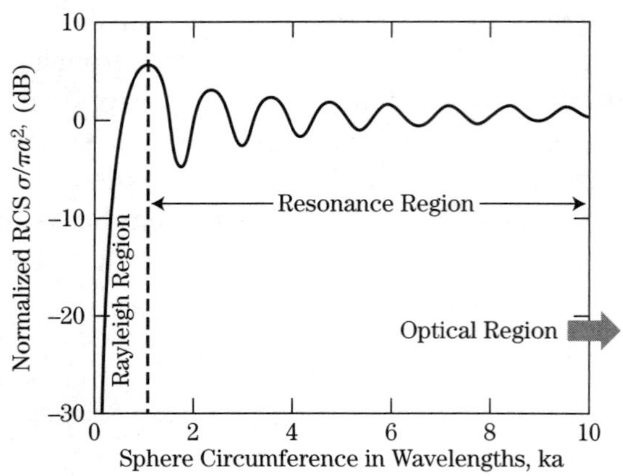

FIGURE 3–8

RADAR cross section of metal sphere [12]

- *Resonance:* the radius of curvature and wavelength are of the same order of magnitude (i.e., $\rho \sim \lambda$). In this regime, the scattering currents vary little along or around the body. The RCS will vary cyclically as a function of wavelength depending on the resonance currents, becoming a difficult effect to characterize.
- *Rayleigh*: the size of the body is small compared with a wavelength, $\rho << \lambda$. In this region, the RCS is proportional to λ^{-4}.

It is important to pick the standard targets in the optical region if at all possible. For operation in VHF to UHF regions of the spectrum, where the wavelength is between 1 and 10 meters, the size of a sphere as a standard target is impractical. In addition, it is important to consider the ground effects on the scattering, and a sphere on the order of 10 meters in diameter would be impossible to deploy.

3.3.2 Top-Hat Reflector

The top-hat is formed by placing a right cylinder on a metallic ground plane. The ground plane provides a perfect reflection of the incident radiation and an efficient double bounce return for the incident radiation. The return has a broad pattern in any plane containing the cylinder axis and a narrow return that depends on the dimensions of the dihedral components, b and L as shown in Figure 3–7b.

The top-hat specular RCS σ_{TH} is given by [12]

$$\sigma_{TH} = \frac{8\pi ab^2L}{\lambda\sqrt{L^2 + b^2}} \tag{3.25}$$

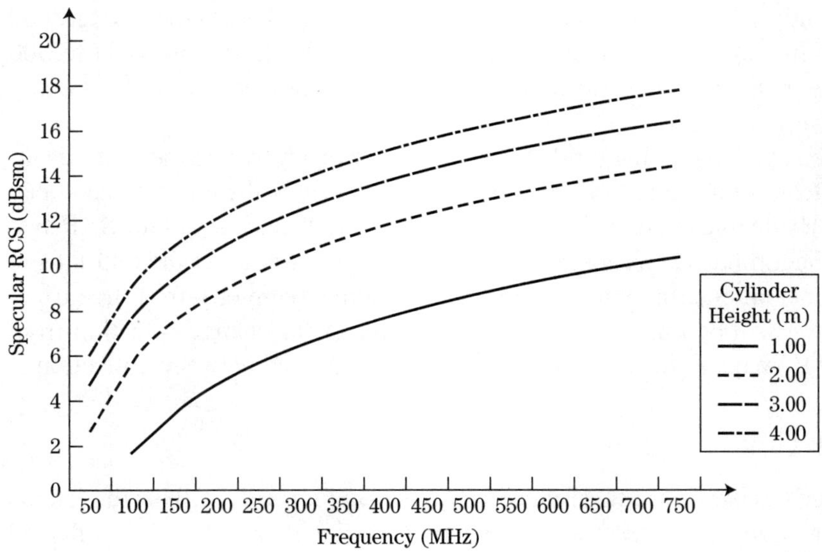

FIGURE 3–9

Specular RCS of top-hat target

For incidence angles in the plane of the cylinder axis, the monostatic RCS can be analytically found as a function of elevation angle ψ from

$$\sigma_{TH} = \begin{cases} \dfrac{8\pi a b^2 \cos\psi}{\lambda}, & for\ \tan\psi \geq \dfrac{b}{L} \\ \dfrac{8\pi a L^2 \tan^2\psi \cos\psi}{\lambda}, & for\ \tan\psi \leq \dfrac{b}{L} \end{cases} \tag{3.26}$$

where the peak response is at the elevation angle ψ_0 given by

$$\psi_0 = \tan^{-1}(b/L) \tag{3.27}$$

In these two cases, the elevation of the RCS return depends on whether the projected aperture is limited by the top-hat height or the width of the ground plane.

The top-hat specular return is given in Figure 3–9 for a 0.25 meter radius cylinder over a 1-meter ground plane, as a function of frequency and for several cylinder heights. The specular returns at lower frequencies have been suppressed, since the height dimension b is less than a wavelength. Even for the case of a 4 meter high tophat at 50 MHz frequency, the specular return can not be adequately estimated from physical optics when the wavelength is 6 meters.

It should also be noted that the values of RCS obtained for a top-hat are modest and would not provide sufficient return under foliage loss. This class of standard target is normally used in the open; where the specular return is sufficiently large that the return can be discriminated from the background clutter. In addition, the return from a VHF band RADAR (30 MHz to 300 MHz) is not sufficiently strong to provide geolocation accuracy, unless very tall targets are used.

The advantage of the top-hat is its axial symmetry, where the return is independent of the azimuth illumination angle. As such, the top-hat in an open field is a valuable target to determine the geolocation accuracy of the SAR system. The top-hat scattering phenomenology is also similar to the return from tree trunks interacting with the ground. The axial symmetry from tree trunk scattering will be important in registering images for change detection from one SAR image to the next, of targets that have moved between collections.

3.3.3 Dihedral Corner Reflector

The dihedral is a right angle reflector as shown in Figure 3–7c. This reflector is another example of a reentrant target, where the incident radiation is reflected by one of the plates and retroreflected back to the RADAR. The scattering effects are most precisely analyzed if the angle between the two arms of the reflector is 90 degrees. Under these conditions, the maximum copolarized RCS σ_D given by [12]

$$\sigma_D = \frac{16\pi a^2 b^2}{\lambda^2} \tag{3.28}$$

The principal advantage of a dihedral corner reflector is its return from multiply polarized SAR illumination. For this class of SAR, both horizontal and vertical transmit polarizations can be transmitted, and the two senses of polarization are received. If the dihedral is deployed with the axis along either the H or V axis, true-sense of polarization will be returned. When the orientation of the dihedral crease with respect to the HV cross-polarization plane is γ, the polarization scattering matrix will provide a measure of both H and V components given by [13]

$$S_D = \begin{bmatrix} \cos(2\gamma) & \sin(2\gamma) \\ \sin(2\gamma) & -\cos(2\gamma) \end{bmatrix} \tag{3.29}$$

For the dihedral corner reflector, the beamwidth in the open dimension (i.e. along the crease) is approximately 30 degrees. In the dimension characterized by the dihedral double bounce reflection, the beamwidth is given by

$$\phi_D = \frac{\lambda}{2b} \tag{3.30}$$

For the dihedral specular reflection, the RCS is shown as a function of dihedral height b and frequency by Figure 3–10. This target is very efficient and

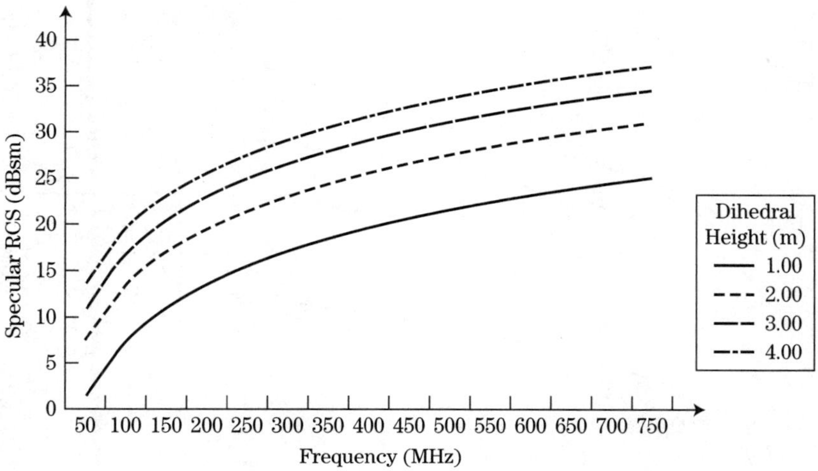

FIGURE 3–10

Specular RADAR cross section of dihedral of 1 m width versus frequency

provides excellent returns for frequencies down to VHF. However, it can be used to measure only one sense of polarization or cross-polarization at a time.

3.3.4 Trihedral Corner Reflector

The trihedral corner reflector shown in Figure 3–7d has two planes of reflection and exhibits the ability to simultaneous scatter both horizontal and vertical polarization energy with analytically understood characteristics. This makes the target very useful in dealing with calibration of fully polarimetric FOPEN SAR operation. Both the senses of linear polarization are received, along with cross-polarization components. It the case of FOPEN SAR propagating through the foliage, measurements of returns from trihedrals in the open and under foliage, and will provide a direct measure the anisotropic scattering of the foliage.

The open area of the trihedral is indicated by the equilateral triangle with sides of dimension a. However, because of the need for the energy to reflect off the internal right angle triangles, the reflector's effective area is provided by one-third of the dimension on each side. As a result, the trihedral specular RCS σ_{TR} will be given by [12]

$$\sigma_{TR} = \frac{4\pi a^4}{3\lambda^2} \tag{3.31}$$

The polarization scattering matrix of the trihedral reflector, with the axes aligned with the horizontal and vertical basis functions, is given by

$$S_{TR} = \begin{bmatrix} S_{HH} & S_{VH} \\ S_{HV} & S_{VV} \end{bmatrix} = \begin{bmatrix} 1 & 0 \\ 0 & 1 \end{bmatrix} \tag{3.32}$$

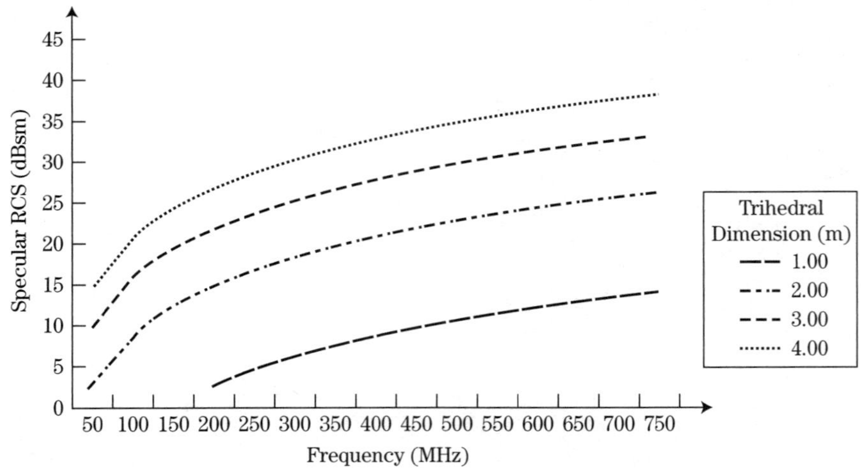

FIGURE 3–11

Trihedral corner reflector specular RCS versus frequency

In this case the copolarization components of the target scattering are returned with equal power. If there are cross polarization components in the radiation, the off diagonal terms of the matrix will provide a measure of this phenomenon.

The specular RCS of the trihedral corner reflector is given by Figure 3–11, for various sizes and as a function of frequency. The 3 dB return is effective over a 40 degree angle, until edge effects and diffraction corrupt the RADAR reflection and calibration. Because of the lower scattering effective area, the 1 meter trihedral is insufficient for VHF. However, a 2 meter trihedral is very effective for UHF-band operation. Sufficient returns are afforded for even high foliage loss of double- and triple-canopy forests. These targets are used to measure the propagation loss of dense forests as well as open-field radiometric calibration of the RADAR sensors.

Because the trihedral corner reflector is so widely used in SAR calibration, extensive analyses have been carried out to characterize the return as a function of the polarization, as well as the surrounding ground plane. For VHF operation, this is very important. The electromagnetic analysis of both frequency and time-domain waveforms has been carried out as a function of incidence angle and polarization. The coordinate system for evaluation of the RCS is given in Figure 3–12.

The approach uses a method of moments calculation of the scattering of incident electromagnetic waveform from a perfectly conducting trihedral suspended above a lossy ground plane. Independent components of the electric and magnetic fields for horizontal and vertical polarized waves can be calculated. Figure 3–13 shows the free-space calculation of the RCS of

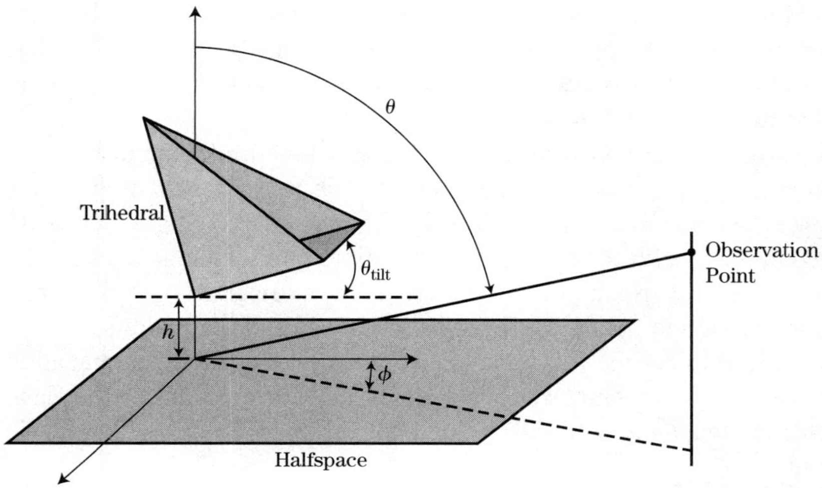

FIGURE 3–12

Geometry of trihedral over lossy half space © 1999 IEEE [14]

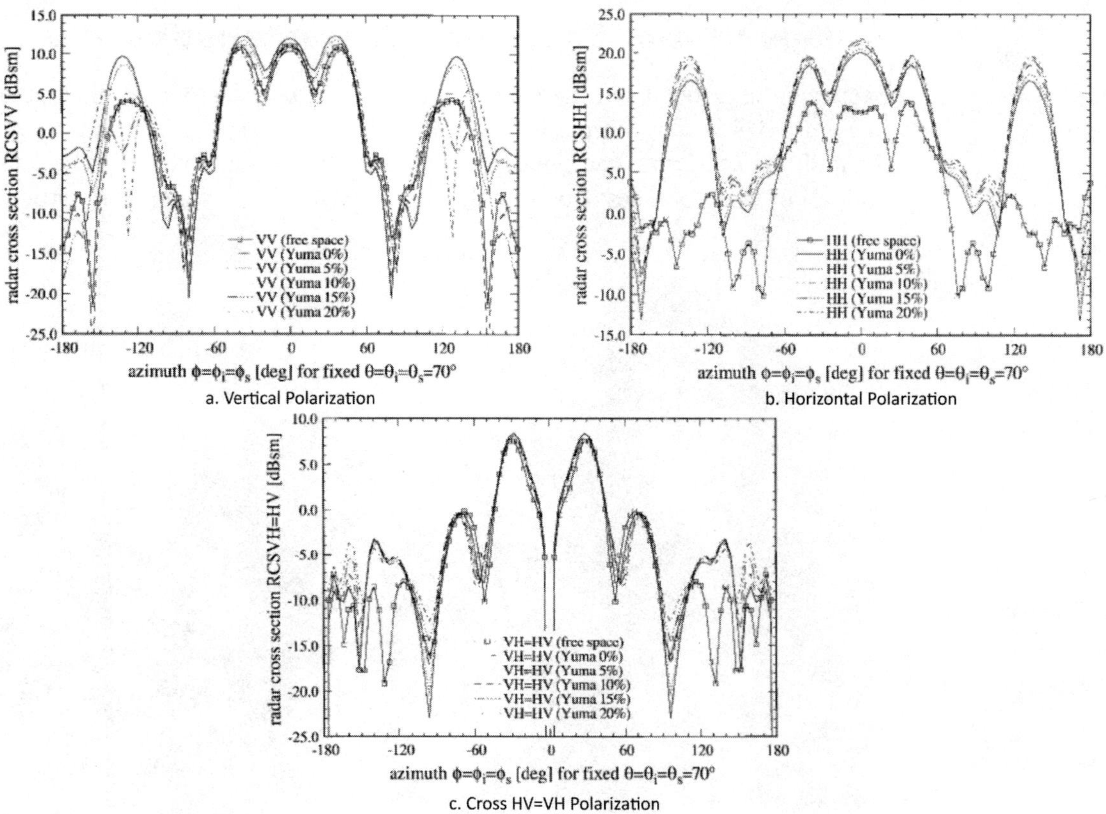

FIGURE 3–13

Method of moments calculation of trihedral RCS over lossy ground plane © 1999 IEEE [14]

$a = 2.44$ meter trihedral, $h = 0.15$ meter above a model of Yuma ground with 0 degrees tilt. The observation is at an elevation angle of $\theta = 70^0$. The soil impact on the RCS is evaluated with 0, 5, 10, 15, and 20% water content as well as for free space [14].

The variation in the RCS of the trihedral as a function of the installation above a ground plane and azimuthally orientation illustrates the importance of detailed modeling for accurate FOPEN calibration. Trihedrals are normally deployed in open fields to provide the calibration return without foliage loss. The advantage of a trihedral is the polarization return from the orthogonal senses. In addition to open field deployment, one or more trihedrals are typically deployed under the forest canopy to measure the foliage propagation loss. However, if the ground plane or the proximity of the trees affects the plane wave propagation, there will be corresponding errors in absolute polarimetric calibration.

3.4 Foliage Clutter Scattering Characteristics

Scattering in a forest setting is a combination of many factors, as illustrated in Figure 3–14. This simple figure illustrates the wave propagation process for multiple scattering and polarized returns from trees and vehicles. In general, all of the processes shown in the figure exist for horizontal and vertical

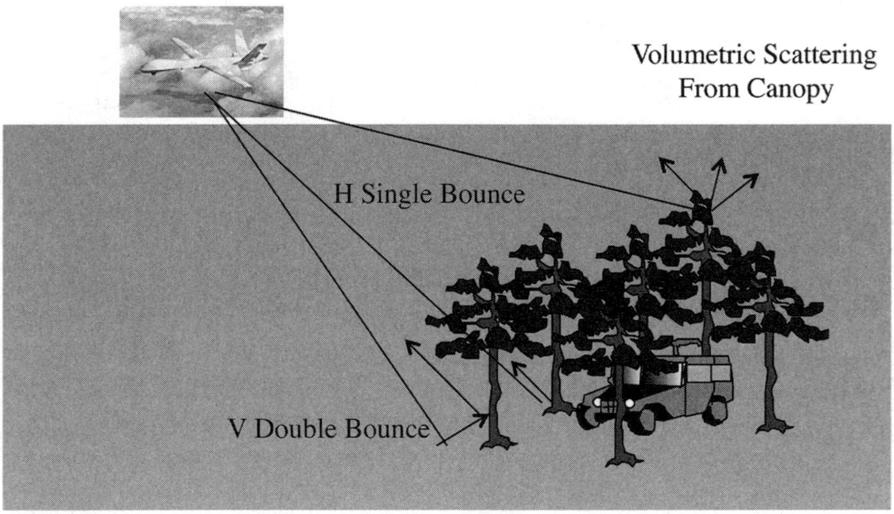

FIGURE 3–14

Scattering of FOPEN RADAR signals from forest and targets

polarization, but with differing strengths of interaction. The basic effects are as follows:

- The vertically polarized return is shown with a double bounce from the ground and the tree trunks. Both horizontal and vertical polarizations will exhibit the double bounce phenomenon. However, the vertically polarized energy will penetrate the ground and interact more weakly with the tree trunk.
- The horizontal polarization is illustrated with a strong single bounce from the ground tree interface. This most accurately resembles the return from a top-hat over a conducting ground plane and has similar characteristics at variable azimuth look angles. A departure from cylandrical-dihedaral scattering will occur when the ground and tree trunk is not at a right angle. As a result, the local terrain slope produces is a strong influence on the magnitude of the return.
- Vehicles will have more complex scattering depending on the number and orientation of panels. These scattering centers can also introduce cross-polarization effects, depending on the orientation to the RADAR. More importantly, the interaction of the surrounding trees and the ground to vehicle dihedral effect is important in the quantitative return from vehicles in dense foliage.
- Finally, the tops of the trees present random scattering to the RADAR depending on the number and orientation of the major branches. A significant effect is to depolarize the incident radiation and to provide a component of cross-polarization. For low UHF-band and VHF, the leaves have little effect on the principal polarization returns but do provide a phase and random fluctuation. These effects tend to be integrated out in SAR but contribute strongly to internal clutter motion in MTI operation.

Early foliage penetration data were collected in 1990 in Maine and Panama using the SRI FOLPEN II and FOA CARABAS II RADAR systems [15, 16]. These two systems were horizontally polarized and hence capable of measuring only the horizontally polarized clutter return along with losses for the VHF and low UHF-band. However, the series of collections were instrumental in answering questions about the backscatter at fine image resolution [17].

The data shown in Figure 3–15 are for variation of clutter backscatter versus frequency. The Panama rain forest data were collected with CARABAS I and FOLPEN II platforms. The Maine collection, which was carried out with the NASA Jet Propulsion Laboratory AIRSAR platform, illustrates horizontal polarization data at UHF, L-band, and C-band. The following plots are exceedance curves giving the percentage of the data that exceed a

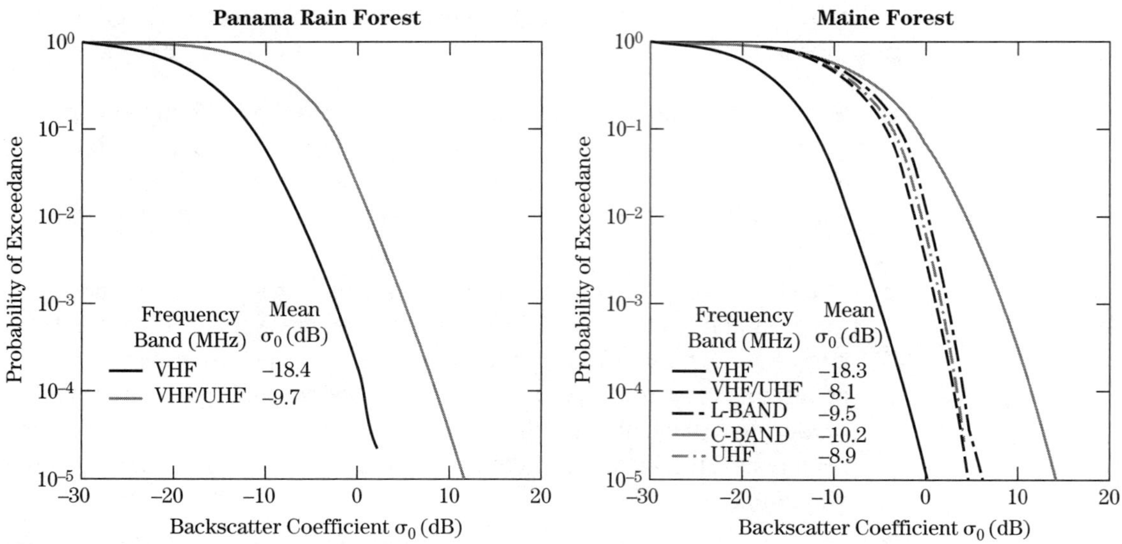

FIGURE 3–15

Clutter scattering characteristics [15]

clutter reflectivity level. These curves do not follow a Gaussian distribution and illustrate scattering at low probabilities of exceedance characterized by large returns. These strong returns are characterized as *spiky clutter*, especially in the bands that can penetrate the foliage canopy and scatter off ground trunk interactions [15].

The analysis of scattering data from Figure 3–15 indicates that the clutter follows a lognormal distribution. The lognormal distributions are generated from the clutter backscatter magnitude statistics—that is, the mean (m) and standard deviation (sd). The lognormal probability density function is defined as [2]

$$f(x) = \frac{1}{\sigma x \sqrt{2\pi}} \exp\left[-\frac{(\ln x - \mu)^2}{2\sigma^2}\right] \tag{3.33}$$

where

$\ln x$ = natural log of the scattering cross section
$\mu = \frac{1}{2}[2\ln(m) - \sigma^2]$
$\sigma^2 = \ln[(sd)^2/m^2 + 1]$

The data in Figure 3–16, which is from the horizontally polarized P-3 UWB RADAR collection in Grayling Michigan, illustrates the importance of examining FOPEN SAR clutter data. The data were taken at three grazing angles—45, 30, and 20 degrees—and are plotted as the lognormal function

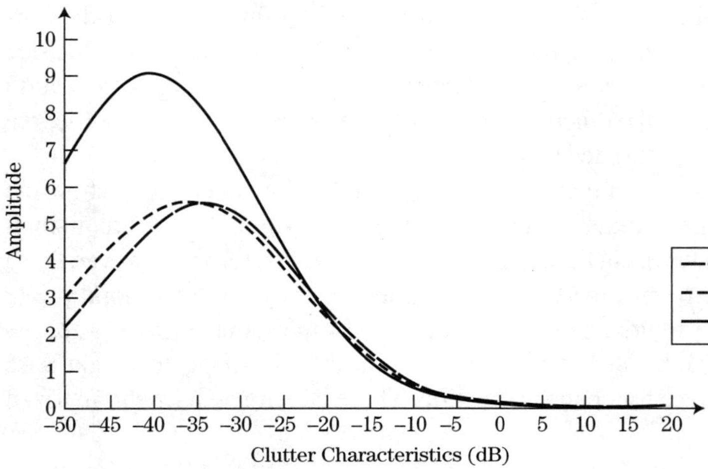

FIGURE 3–16

Log normal distribution of UHF horizontal data from P-3

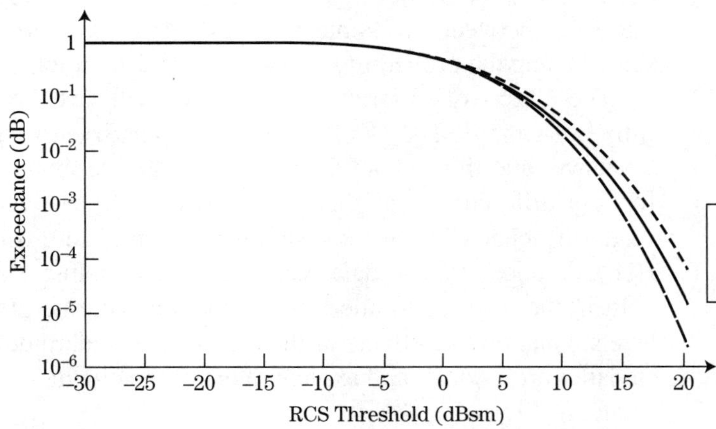

FIGURE 3–17

Probability of exceedance of P3 UWB horizontal polarization
Source: MIT Lincoln Laboratory © 1999 IEEE [19]

$f(x)$ versus the clutter RCS characteristics. It is difficult to compare the three grazing angle returns from just the lognormal function.

However, when the probability of exceedance is examined, as in Figure 3–17, a better insight into the distribution of the clutter is obtained. The probability of exceedance in the data gives a measure of how strong the large scattering centers will contribute to the false alarm density in the target detection. A plot of the data is obtained from the clutter characteristics by computing the quantity (1 minus the cumulative density function). The difference in returns among the three grazing angles is due to a combination of the loss in signal through the foliage canopy and the interaction of the horizontally polarized energy from the ground–trunk interface. At a 20 degree grazing angle, the loss through the foliage is expected to be larger than at 45 degrees. In addition, the dihedral effect will be more pronounced at 45 degree grazing angle than at 20 degrees.

It can be clearly seen that the semi-log plot does not adequately show the effect of the spikey clutter on the scattering distribution. There are a small percentage of scatterers in the scene that exhibit strong returns. However, these effects can directly bias the CFAR detections of targets without appropriate excision of these returns from the statistics. On the more positive size, the strong ground trunk interactions are a signficant advantage in repeat pass alignment of images for change detection. These benefits will be summarized in Section 6.5.3.

The data collections were then needed to carefully examine the interdependencies on scattering due to frequency, polarization, foliage type, and interfaces between different types of clutter. These effects will now be presented using the probability of exceedance data graphs.

The effects of different foliage types will next be examined using the fully polarimetric P-3 UWB SAR. Fully polarimetric operation was critical to answer questions about the backscatter at fine image resolution. The data for four different geographic areas and forest types were collected, and the scattering characteristics are summarized in Figure 3–18 for polarizations HH, HV and VV. The data were calibrated against top-hats and trihedrals both in the open and under trees to obtain accurate metrics on backscatter levels. Quantitative effects of the mean and standard deviation of the clutter statistics are evident, and explain much of the qualitative results illustrated in Figure 3–14.

Grassland areas were used to obtain a comparison of backscatter for open grass versus the trees. A summary of the median return for forest clutter versus grazing angle and polarization is given in Table 3–2. Figure 3–19 shows the exceedance statistics for grasslands versus forests as a function

Table 3–2 Median Clutter Cross Section Return versus Depression Angle and Polarization
Source: MIT Lincoln Laboratory © 1999 IEEE [19]

Channel	Depression Angle		
	20^0	30^0	45^0
VV	0.6	0.5	1.5
HV	0.9	0.9	1.6
HV	0.7	1.3	1.0

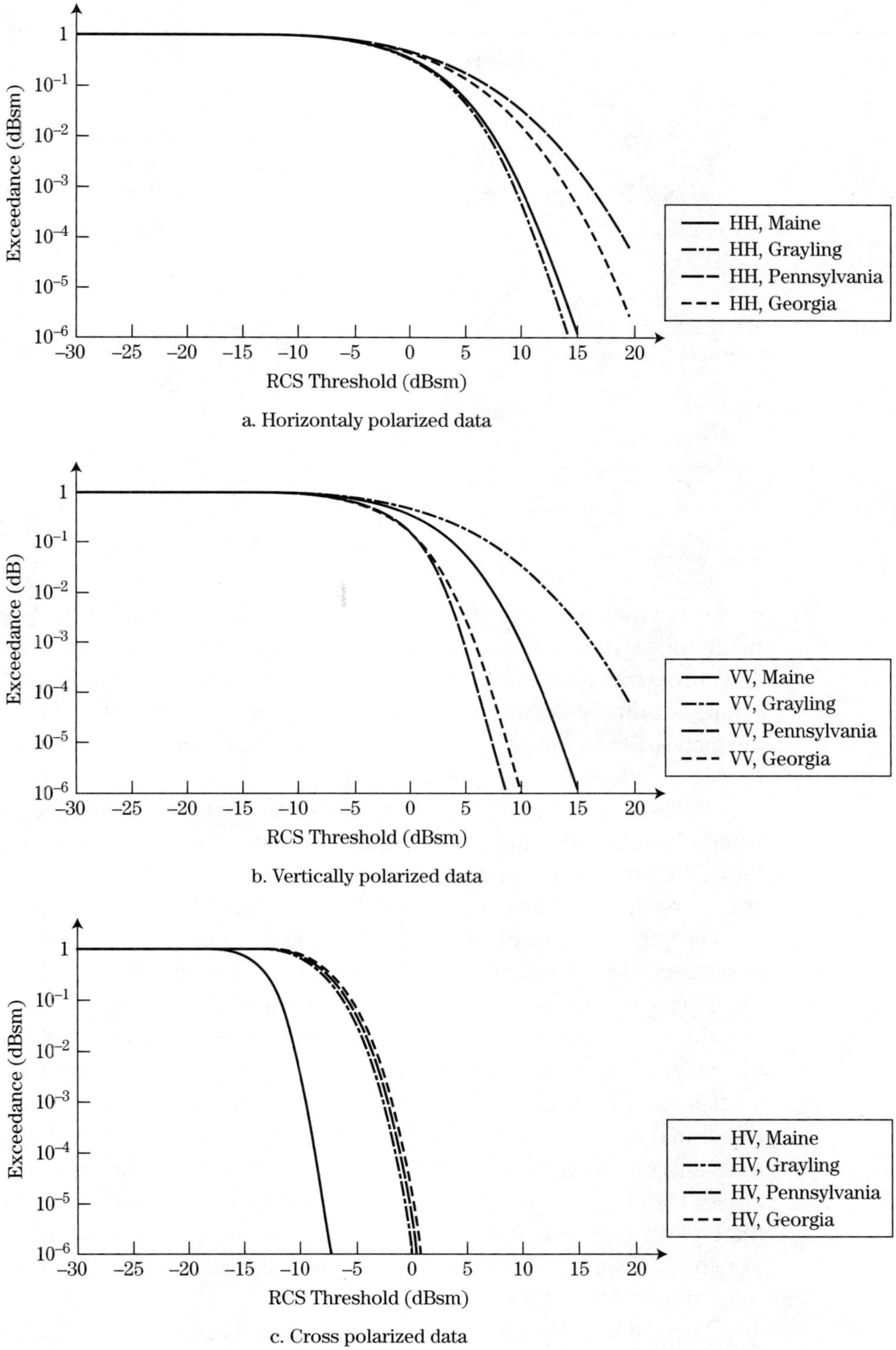

FIGURE 3–18

UHF clutter characteristics at various geographic sites
Source: MIT Lincoln Laboratory © 1999 IEEE[19]

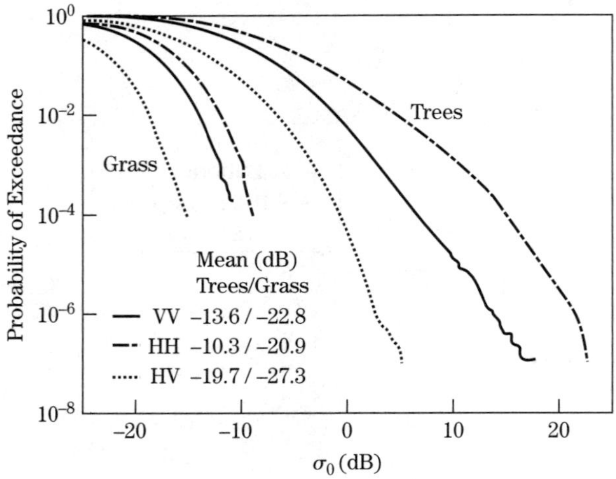

FIGURE 3–19

Tree and grass clutter scattering cross section versus polarization
Source: MIT Lincoln Laboratory [19]

of the polarization. The strong scattering by the forest is clearly shown in this comparison, with approximately 10 dB higher return than grasslands. This difference in clutter returns is important to size the distributed clutter return within the scene. It has been important in determining the land use characteristics in image interpretation as well as aiding in chosing the correct CFAR technique for target detection.

However, it should be noted that the cumulative scattering curves for forests have significant tails in the distribution. For horizontal polarization, the distribution is very pronounced due to the ground–trunk interaction. These strong localized returns will be very target-like at UHF.

The grazing angle difference between the forest versus grasslands is shown in Figure 3–20. It is clear that the scattering in open grass is increased for higher grazing angles, due to reduced forward scattering of the energy. However, there is a close grouping of the scattering in forests. This difference is attributed to the penetration of the radiation, reducing any potential for forward scatter.

The data represented in Figure 3–21 were shown superimposed on the lognormal distribution function for HH and VV polarization and for interior trees and tree lines. Table 3–3 presents the values of σ_T and μ_T for the four cases. The tree lines have significantly stronger returns, due in great part by the unattenuated propagation of the energy to the ground trunk interaction. For the interior trees, the propagation is attenuated, reducing the dihedral scattering of the energy.

One of the important observations in the early FOPEN RADAR systems was the effect of tree lines on the clutter and, more importantly, the variation in both the size and Doppler signature of the clutter. These effects were a

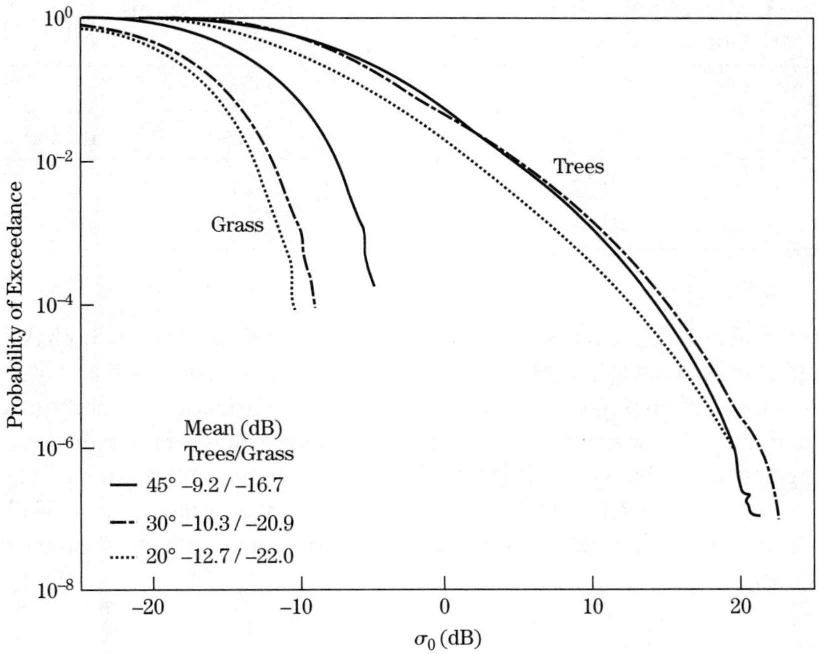

FIGURE 3–20

Tree and grass clutter scattering cross section versus grazing angle
Source: MIT Lincoln Laboratory [19]

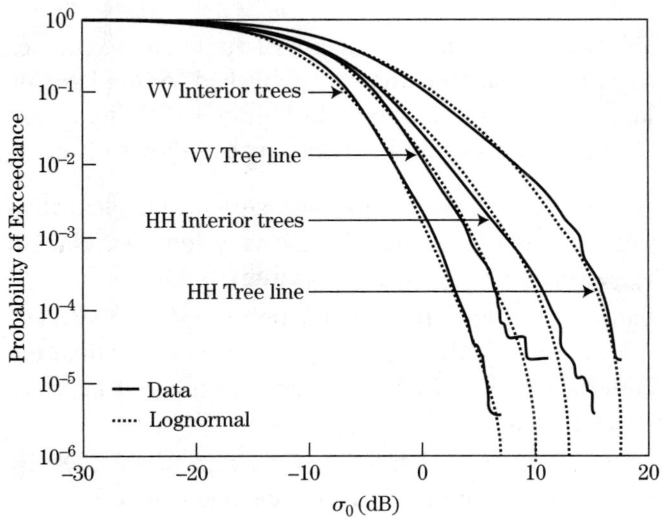

FIGURE 3–21

Comparison of tree line and interior foliage scattering
Source: MIT Lincoln Laboratory
© 1999 IEEE [19]

significant issue with both GMTI RADAR and SAR systems, where stationary returns are important for coherent integration. This set of data, presented in Figure 3–21, focuses on the difference in clutter return from the edges of the forest and the interior clutter.

Table 3–3 Mean Clutter Return for Tree Lines and Interior Regions
Source: MIT Lincoln Laboratory © 1999 IEEE [19]

	HH		VV	
	Mean	Std Dev	Mean	Std Dev
Interior Region	0.22	0.48	0.11	0.13
Tree Lie	0.61	1.78	0.18	0.26

Horizontal polarization is significantly affected by the tree lines in both the mean and standard deviation of the returns. The tree line represents an elongated dihedral, enhancing the return for horizontal polarization. In addition, the ground–trunk interaction is not attenuated, giving much larger discrete scattering centers. What is not shown in the data is the temporal motion of the trees in a line. For SAR this shows up as smearing of the image. For GMTI RADAR, the internal clutter motion is expected to be a significant source of false alarms.

3.5 Foliage Attenuation

After the 1990 FOPEN data collections, a database of attenuation statistics for horizontally polarized foliage attenuation was published for the Panama rain forest and the Maine collections, as shown in Figure 3–22. Significant conclusions provided from the analysis of these data are as follows [20]:

- Low VHF-band (CARABAS I) attenuation was significantly less than at any other frequency. This is explained by the very long wavelength compared with the size of the tree branches and trunks.
- Rain forest attenuation was higher than attenuation in the northern latitude forests. Again the low VHF-band attenuation was lower than the VHF/UHF data from the SRI FOLPEN II system. No higher-frequency data was collected in Panama.
- Microwave frequencies had such high attenuation that future FOPEN system development was concentrated at UHF-band and below.

These data were collected at several grazing angles, from which it was determined that loss for various grazing angles was an important factor for both system design and the operating conditions.

Bessette has characterized the losses from all of the FOPEN data collections from 1990 to 1999 [19]. A statistical analysis of the median two-way foliage loss versus grazing angle and frequency showed regular trends. Based

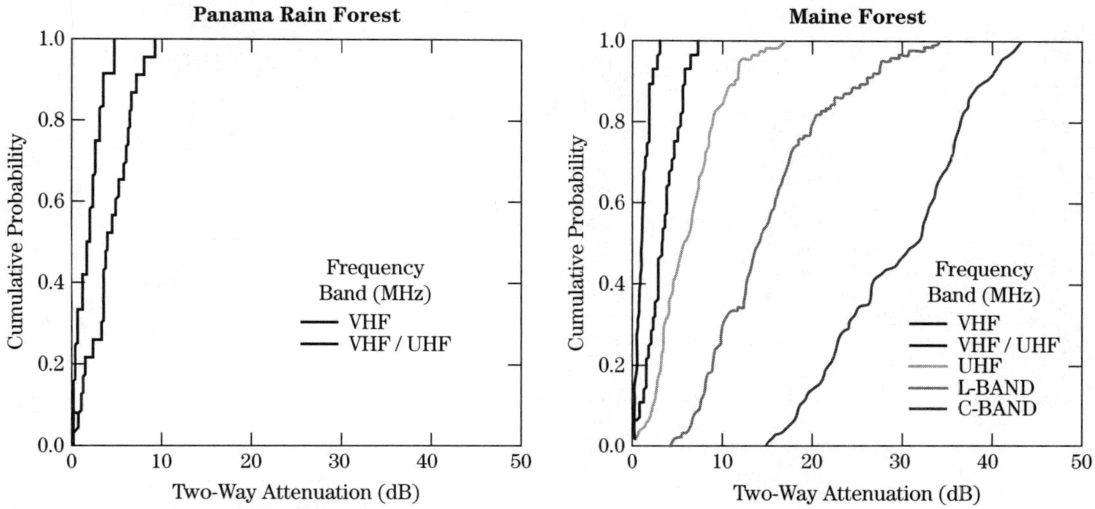

FIGURE 3–22

Attenuation of foliage penetration RADAR signal versus frequency—Panama rain forest and Maine pine forest
Source: MIT Lincoln Laboratory [20]

on these trends, a two-parameter model was constructed for polarization and forest type (single or double canopy) [2].

$$L_{2-way} = \frac{(-\alpha_F)(f_c)^{\beta_F}}{\sin(\gamma_g)} \tag{3.34}$$

where

α_F Foliage attenuation scalar factor
β_F FOPEN RADAR center frequency exponential factor
γ_g Grazing angle to the local clutter patch

Data from several collections were analyzed based on the model in (3.34). The representative values of α_F and β_F were determined and summarized in Table 3–4. The parameters are useful for early system design of a FOPEN

Table 3–4 Foliage attenuation model parameters
Source: MIT Lincoln Laboratory © IEEE 1999 [19]

Median Foliage Attenuation Model for Single and Double Canopy				
Polarization	**Single Canopy**		**Double Canopy**	
	α_F	β_F	α_F	β_F
HH	0.18 ± 0.045	0.53	$.34 \pm 0.085$	0.53
VV	0.30 ± 0.075	0.47	$.71 \pm 0.1775$	0.47

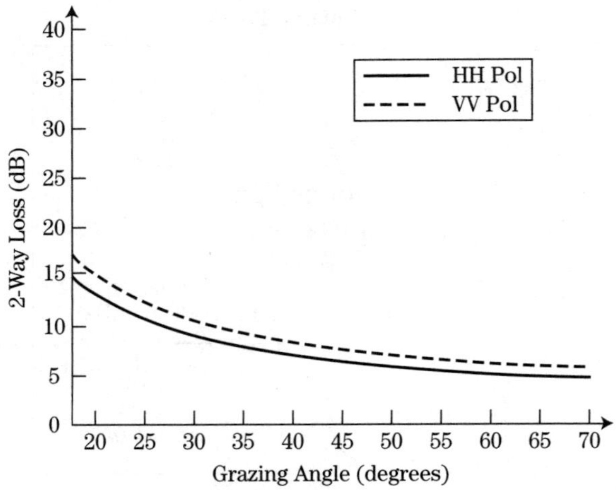

FIGURE 3–23

Single-canopy foliage loss at 450 MHz

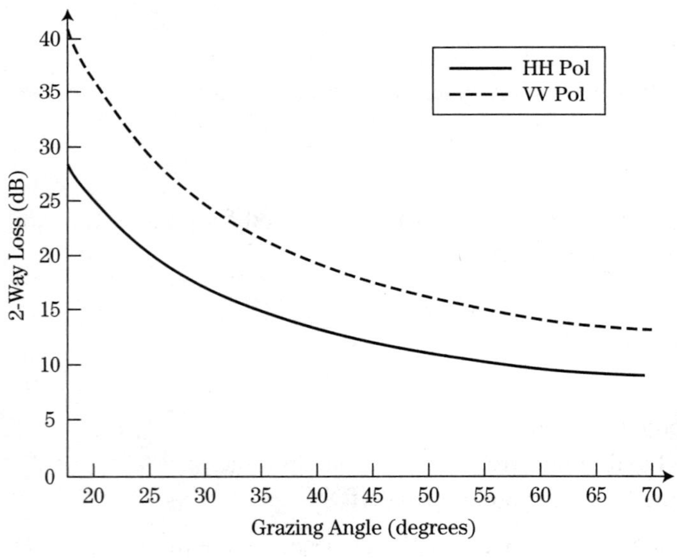

FIGURE 3–24

Double-canopy foliage loss at 450 MHz

SAR system, when representative foliage types and operational grazing angles are needed.

As an illustration of the foliage loss model, Figure 3–23 shows the two-way loss prediction from (3.34) and Table 3–4, for a signal at 450 MHz in both horizontal and vertical polarizations. It can be seen that there is little difference in single-canopy loss between horizontal and vertical polarized propagation. However, there is a measureable increase in the loss at low grazing angles.

In a similar manner, the two-way loss for double-canopy forests, typical of the jungles near the equator, is illustrated in Figure 3–24. The horizontally polarized loss is significantly lower than for vertical polarization. Moreover, the low grazing angle losses are increased to above 20 dB for both propagation polarizations. These significant losses are a major design driver for any FOPEN RADAR that needs to operate in a jungle environment,

Note that in Table 3–4 there are also ranges of parameters. These ranges can be indicative of foliage type (e.g., deciduous or conifer) or of the variations within a particular forest (e.g., heavy old growth or light new growth). In conducting system design, these factors should also be considered in bounding the expected performance for detecting tactical targets or characterizing the terrain beneath the foliage.

3.6 Internal Clutter Motion

The effects of internal clutter motion (ICM) are a problem for both SAR and GMTI FOPEN RADARs. In the case of SAR, any ICM is manifest as a defocusing or smearing of the image and often obscures an object in the region of the tree line. For GMTI RADAR, the ICM is a Doppler effect that broadens the clutter line and affects the MDV. It is important to understand the effects of windblown clutter, especially foliage clutter, on the RADAR system design.

An extensive collection of RADAR low-angle propagation data was obtained in the late 1980s from VHF- to Ku-band. These data are well documented by the Billingsley model and published in a significant treatise on the cause of the internal motion and physics of the scattering. Billingsley showed that a reasonable model for the windblown clutter spectrum probability density function $P(v_c)$ would be a delta-function (i.e. direct current DC) term plus an exponential (i.e. alternate current AC) term [21]:

$$P(v_c) = \frac{r}{1+r}\delta(v_c) + \frac{1}{1+r}\frac{\beta}{2}e^{(-\beta|v_c|)} \tag{3.35}$$

where

v_c	Velocity of clutter Doppler in meters per second
r	Ratio of the DC to AC power
β	Shape parameter of the exponential spectrum
$\delta(v_c)$	Dirac delta function centered a zero-Doppler

Based on extensive data with variation in terrain type and seasonal variations, Billingsley generated empirical models for r and β. The DC/AC ratio model is given as

$$r_{dB} = 63.2 - 12.1\log_{10}(f_{MHz}) - 15.5\log_{10}(w_{mph}) \tag{3.36}$$

Table 3–5 Billingsley exponential shape parameter β versus wind condition
Source: MIT Lincoln Laboratory [21]

Wind Condition (m/sec)	β (s/m)
Light air (0.4–3.1)	12
Breezy (3.1–6.7)	8
Windy (6.7–13.4)	5.7
Gale force (>13.4)	4.3

where

f_{MHz} The RADAR frequency in megahertz
w_{mph} The wind speed in miles per hour

The exponential shape parameter β depends on the wind speed and is given in Table 3–5 for conditions from light air to gale force winds.

The model has found wide applications to RADAR even at higher grazing angles, as long as the parameters of the model can be appropriately modified for the grazing angle. Since the DC term in (3.35) is not a Dirac delta function in nature but exhibits a finite Doppler spread under 0.25 m/sec, it is commonly referred to as a quasi-DC term.

To characterize ICM at FOPEN frequencies, a limited set of data collections were taken at the Redstone Arsenal in Huntsville, Alabama, from a 100 meter high tower. The data were collected with both horizontal and vertical polarization against a tree line and under varying meteorological conditions. As part of the calibration of environmental conditions, the winds were measured above the tree line to determine the effects on the foliage ICM at varying wind conditions [22].

The clutter spread for light wind conditions (i.e., 5.5 mph wind velocity) is shown in Figure 3–25 and is examined in homogenous forest clutter at 19.4 and 22.3 degrees incidence angles. The exponential spreading factor β for the clutter on each side of the quasi-DC term is comparable with the model in (3.36). However, in both of the cases the Billingsley model overestimates the DC/AC factor.

The heuristic explanation of the performance is based on the penetration of the RADAR scattering into the tops of the forest. The large stationary scattering is representative of the tree trunks and large branches, which are somewhat sheltered from the wind. However, the large branches in the crown of the tree will have some modest motion, giving rise to the departure from the quasi-DC function of 0.25 m/sec. The ratio of AC/DC returns is attributed to the amount of biomass that is contributing to the higher Doppler spectrum. For

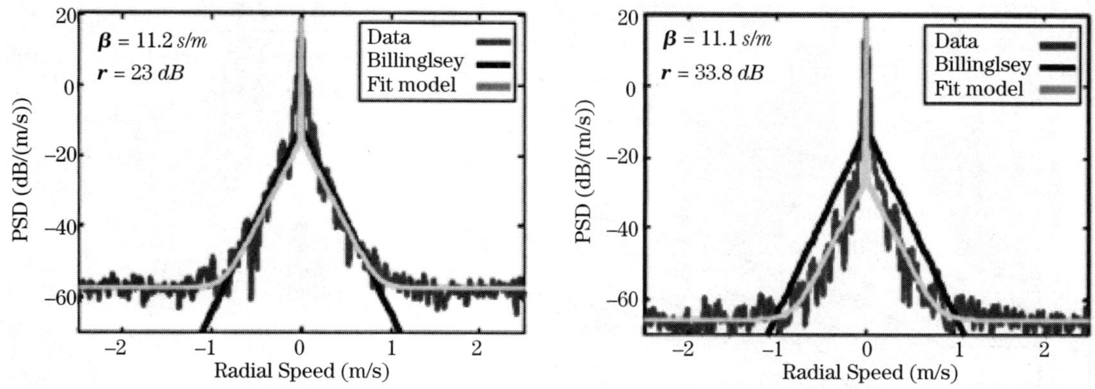

FIGURE 3–25

Forest clutter measured spectral spread for 5 m/sec wind velocity © 2006 IEEE [22]

light wind, the AC component will be low, since the motion is from smaller branches close to the surface of the forest. At higher wind speed, the motion will be transferred to the lower parts of the tree where more attenuation and scattering will take place.

The ratio of DC to AC power in the FOPEN ICM was found to depend on the incidence angle to the tops of the forest. Data across a variation in incidence angles from 12 to 30 degrees were collected. Based on a curve fit of these data, a simple empirical model of the DC/AC ratio dependence on incidence angle is given by [20]

$$r_{dB} = r_{dB}^{Billingsley} + \max(22.7 - 5.8 \csc \psi_I, 0) \qquad (3.37)$$

where ψ_I is the incidence angle from the RADAR to the foliage. The ratio can then be examined as a function of incidence angle, with wind speed as shown in Figure 3–26.

The preceding discussion of ICM should be examined in terms of the limitations of the data collected and the model of (3.33). First, the forests were from a single geographic location. Sections 3.4 and 3.5 illustrated the wide variation in loss and clutter return between the single- and double-canopy forests. The model also assumed that the loss in the foliage followed the csc γ_g model from (3.34). Again, this was derived from penetration in large stands of homogeneous trees. At tree lines or with spatial variations in forest density, an enhanced scattering from lower branches and trunks can be anticipated. More importantly, the wind velocity will affect the motion of tree branches and limbs farther from the tops. These mechanisms will affect the absolute spectrum and the ratio of DC/AC as a function of wind.

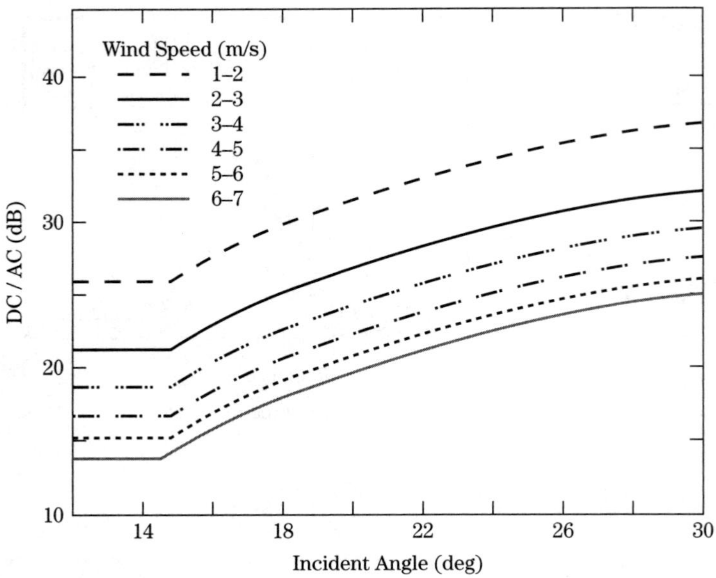

FIGURE 3–26

Empirical model of UHF internal clutter spread DC/AC ratio [22]

The advantage of an empirical curve given by (3.33) is when it is employed in early system analysis. For GMTI RADAR systems it is know that internal clutter motion will affect the Doppler processing and the MDV of any slowly moving targets in the clutter resolution cell.

3.7 Target Characteristics

Several vehicles of differing size were imaged during the Maine 1993 collections with SRI FOLPEN II and FOA CARABAS I. Figure 3–27 illustrates the cumulative target RCS returns from vehicles in the open for two two different FOPEN systems, both at horizontal polarization. The RCS statistics between the classes of vehicles depend primarily on the size and orientation of the vehicle. However, at VHF the target sizes are close to the resonance wavelength, and thus a larger RCS is observed when compared to the same vehicle at UHF. This VHF enhancement is only modest, that is, nominally 2 dB for the median-size vehicles. But the interaction is complicated by the proximity of the ground plane and surrounding foliage, where multipath scattering can either increase or decrease the return. So a quantitative statement about the RCS increase at VHF cannot be made without considering the surrounding bodies.

It is important to note that UHF target returns have a wider variation in RCS. This enhancement, which is attributed to returns at cardinal angles,

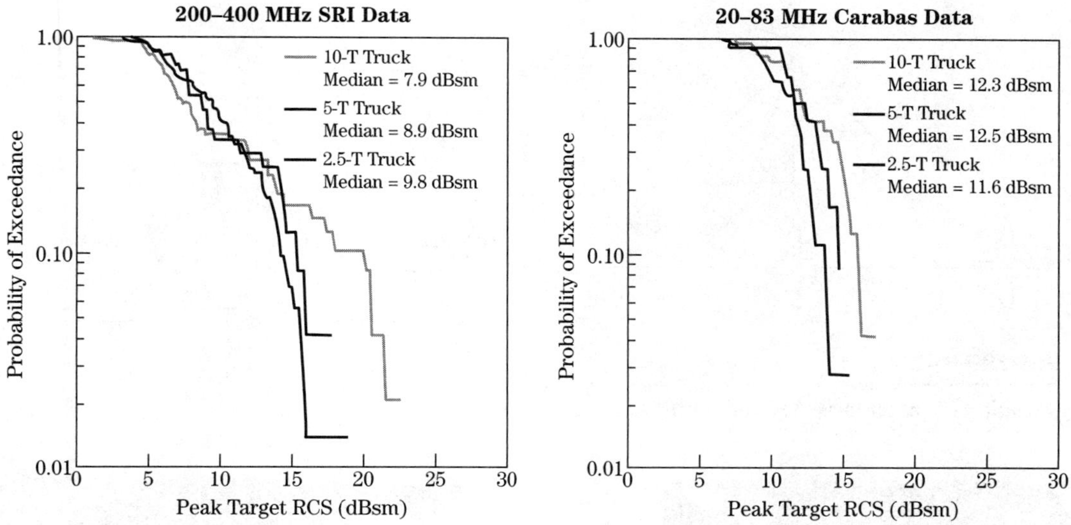

FIGURE 3–27

Variation of vehicle cross section under foliage
Source: MIT Lincoln Laboratory [12]

could be at least 7 dB or more above the corresponding VHF returns. Several researchers have modeled the targets and explain the results in terms of regular conducting shapes over a lossy ground plane.

It is expected that large vehicles have an increased RCS, especially along the illumination directions perpendicular to the large, planar panels (i.e., at *cardinal angles*). This increase in scattering strength is important for improved detection of vehicles and other man-made structures under dense foliage [23]. However, for large dimension surfaces, the increase in RCS occurs over a smaller angle. The explanation is that a large area of return would resemble an antenna of width D and height H. The target would thus have a "beamwidth" of λ/D, as shown in Figure 3–28, and a peak RCS value σ_T of

$$\sigma_T = \frac{4\pi D^2 H^2}{\lambda^2} \tag{3.38}$$

If the image were collected from this angle within the collection geometry, an enhancement would be obtained at the angle around the cardinal return. However, for angles off of the cardinal angles, the return will be reduced. This change in scattering contributes to the scintillation of target returns. One general observation is that for VHF and UHF this scintillation variation with angle is significantly lower than at microwave frequencies. As a result, the correlation for change detection is less sensitive to viewing angle. Various

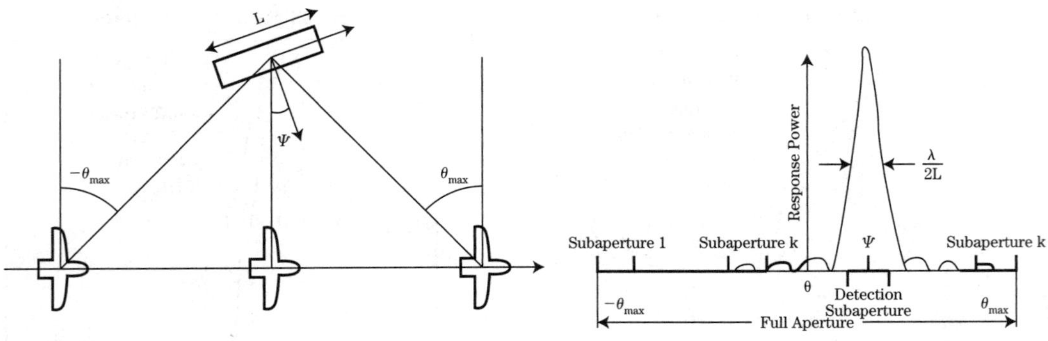

FIGURE 3–28

Variation of target cross section with squint angle [24]

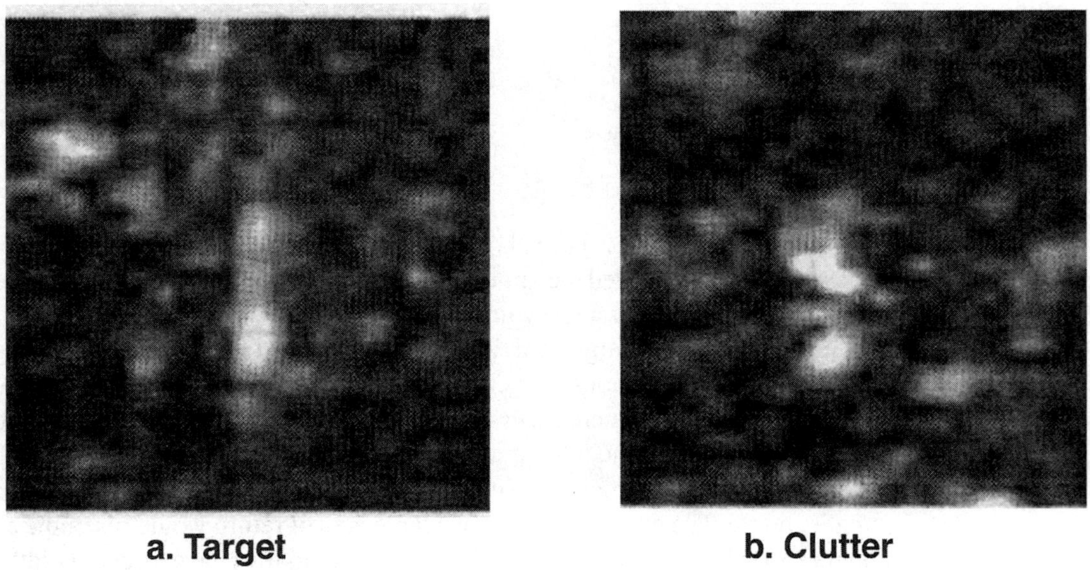

FIGURE 3–29

Comparison of target with clutter return
Source: MIT Lincoln Laboratory [15]

techniques were proposed to provide subaperture processing to fully exploit the return.

The nature of the return would have a regular shape, which would be quite different from cultural clutter, as illustrated in Figure 3–29. This shape could be exploited with morphological filtering and discriminated from large rocks or clumps of trees. However, factors of blockage by trees were known to break up the return from vehicles. So it was recognized that there could be a risk of characterizing natural objects as man-made targets; hence, there was the need for further technical developments in target detection and characterization [23].

A final realization from these efforts was that the wide-angle collection geometry from FOPEN SAR could be an asset. If the imaging angle could be combined to obtain 90-degree total viewing on any one vehicle, there would be an increased probability of obtaining this cardinal angle enhancement in RCS [24].

Several vehicles have been imaged to determine the favored aspect angles for detection. These images were collected at Grayling, Michigan, show the impact of vehicle size and the flash of the returns. Figure 3–30 shows the median target RCS value from a medium truck and a tank along with their variation as a function of grazing angle, aspect angle and polarization. The four cardinal aspect angles were verified for the potential for enhanced detection.

The primary advantage of the P-3 UWB system was its range and cross-range resolution improvement over the CARABAS and FOLPEN systems. In

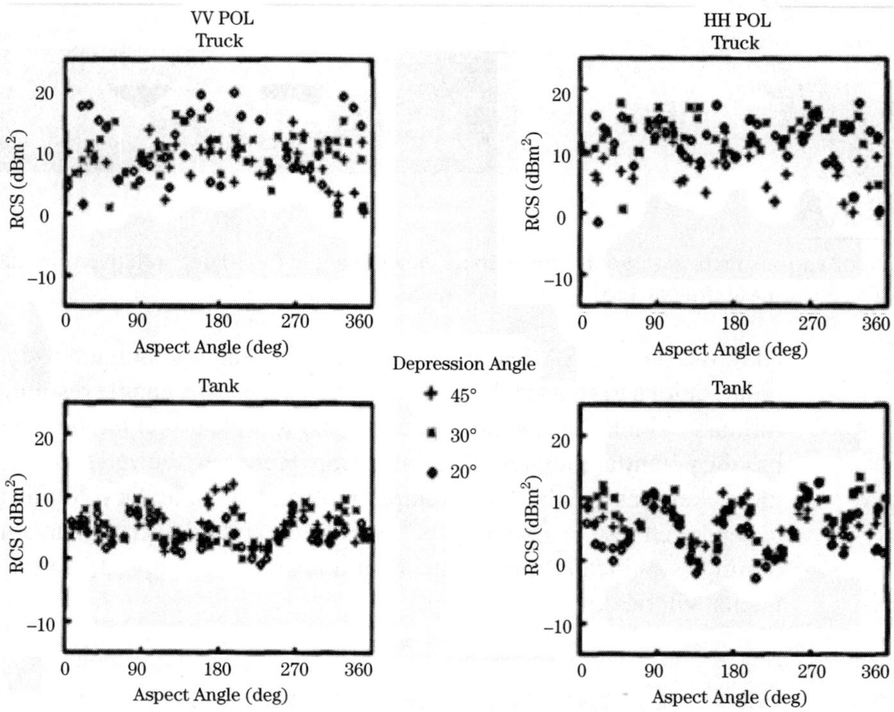

FIGURE 3–30

Detailed RCS of truck and tank versus aspect and grazing angle
Source: MIT Lincoln Laboratory [25]

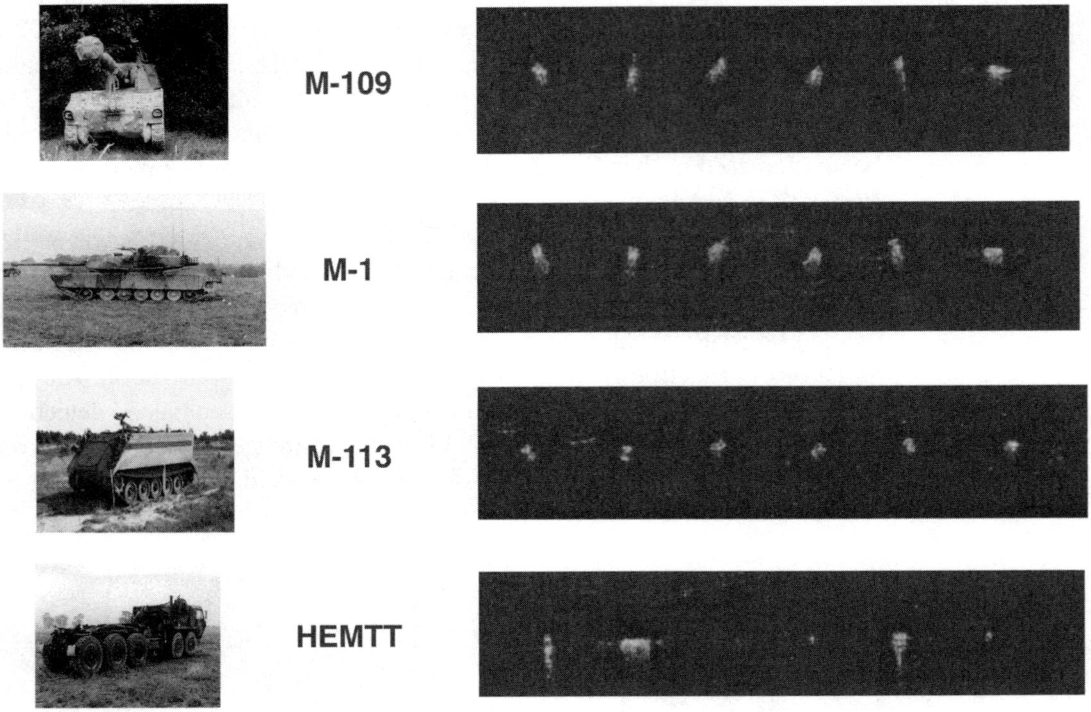

FIGURE 3–31

Comparison of various tactical target returns versus viewing angle (at 30 deg. relative orientations)
Source: MIT Lincoln Laboratory [25]

addition, the P-3 SAR provided images with three simultaneously collected polarizations to be used. Figure 3–31 shows the aspect angle resolution of four military vehicles at several aspect angles. It is clear that the image resolution has the potential for target discrimination. However, the variation in aspect angle makes many of the "target chips" look quite similar. Developing the ability to separate these tactical targets from clutter and to discriminate among them requires significant improvements in image processing [23]. These improvements will be covered in Chapter 6.

3.8 Radio Frequency Interference Spectrum

The VHF and UHF bands are dense with communications, radio, and television transmissions. With the recent rise in popularity of mobile communications, many civilian and military signals fall within the most effective

Table 3–6 Radio frequency allocation bands

Frequency [MHz]	Function
30–50	Mobile Radio
54–88	VHF TV
88–108	FM Radio
108–112	ILS and VOR
116–152	VHF Radio
114–174	Mobile Radio
174–216	VHF TV
225–400	UHF Radio
400–470	Mobile Radio
470–690	UHF TV

frequencies for foliage penetration RADAR. Because of the altitude at which FOPEN platforms operate, these signals can be detected at long distances. Table 3–6 shows the bands that are allocated in the United States for radio, television, and communications in the HF, VHF, and UHF bands. Signals in these bands have licensed use of the spectrum, and any UWB operation must be designed to minimally interfere with these operations. Therefore, when FOPEN SAR systems operate, they need to coexist with these other signals and comply with the established use of the frequency bands.

Figure 3–32 shows one of many collected spectra from the P-3 UWB platform during FOPEN SAR data collections. The figure, which is from data collected near Camp Roberts, California, in 1995, can be considered to be a moderate RFI environment due to the distance from any large metropolitan region. However, the figure does illustrate the number of signals and their relative strengths in that environment. The lower half of Figure 3–32 shows the location of specific frequencies that are allocated for specific functions, especially for emergency communications. For comparison, Figure 3–33 shows the RADAR operation of the P-3 UWB platform superimposed on the RFI spectrum. It can be clearly seen that the two spectra are very likely to cause interference if the system is not designed to accommodate the RFI environment.

Two operational issues need to be addressed in the system design. First, the narrowband emissions from radio and TV stations need to be filtered out before high-quality SAR images can be constructed. Specific algorithms that have been developed to reduce the effect of these signals will be discussed in Chapter 5 on radio frequency interference suppression. Second, many of the communications frequencies are for flight safety or emergency location systems. The National Telecommunications and Information Administration (NTIA) and worldwide treaties dictate that UWB systems must operate at a

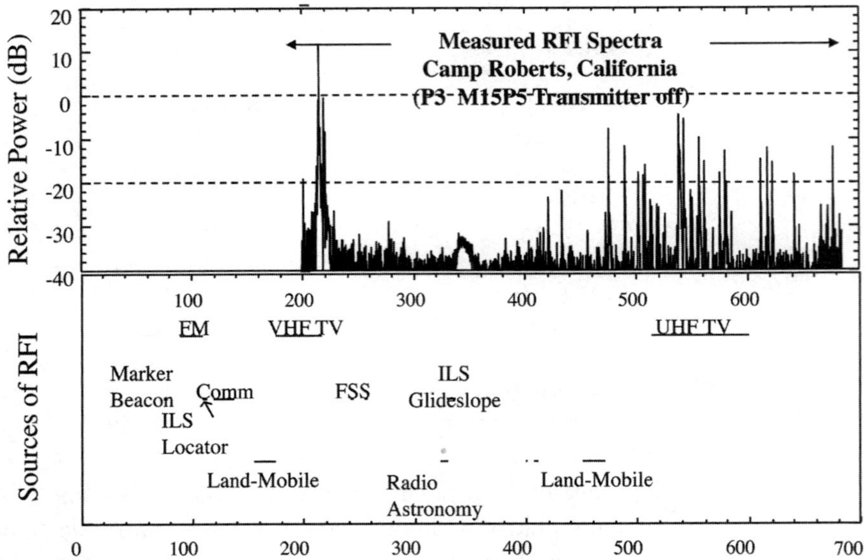

FIGURE 3–32

Radio frequency interference spectrum
Source: MIT Lincoln Laboratory

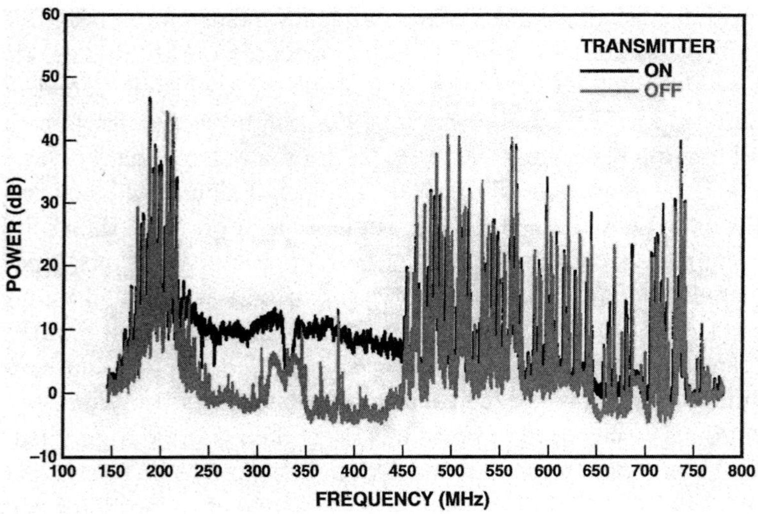

FIGURE 3–33

P3 UWB RADAR signal superimposed with RFI spectrum
Source: MIT Lincoln Laboratory

very low power spectral density or filter out the spectral content of these bands on transmit. The permission to operate FOPEN SARs has been one of the many risk factors in development of systems for use in the United States and Europe. In the early operation of these UWB platforms, the permission to operate could be obtained only with an experimental license, and operation was limited to times of the night when few commercial systems were operating. It was a lengthy and costly process to get an operational license for a platform such as GeoSAR to operate without severe restrictions on the geography and time of day. Chapter 6 will address these electromagnetic compatibility factors in the system design, with particular emphasis on notching the transmitter.

3.9 References

[1] Binder, B. T., Toups, M. F., Ayasli, S., and Adams, E. M., "Foliage Attenuation and Backscatter Analysis," *Proc 1995 IEEE International RADAR Conference,* Washington, DC, May 1995, pp. 158–163.

[2] Bessette, L. A. and Ayasli, S., "Ultra Wideband P-3 and CARABAS II Foliage Attenuation and Backscatter Analysis," Proc 2001 IEEE *RADAR* Conference, Atlanta, GA, May 2001, p. 357–362

[3] Sheen, D. R., Malinas, N. P., Kletzli, D. W., et al., "Foliage Transmission Measurements Using a Ground Based Ultrawide Band (300–1300 MHz) SAR System," IEEE Trans Geoscience and Remote Sensing Vol. 32, No. 1, January 1994, p. 118.

[4] Fleischman, J. G., Ayasli, S., Adams, E. M., and Gosselin, D. R., "The 1990 Foliage Penetration Experiment: Part 1 Foliage Attenuation and Backscatter Analysis of SAR Imagery," *IEEE Trans Aerospace and Electronic Systems* Vol. 32, No. 1, January 1996, pp. 145–155.

[5] Toups, M. F., Ayasli, S., and Fleischman, J. G. "Analysis of Foliage Induced Synthetic Pattern Distortions from the July 1990 Foliage Penetration Study," MIT Lincoln Laboratory Project Report STD-50, July 1993.

[6] Cheston, T. C. and Frank, J., "Phased Array Radar Antenna," in *RADAR Handbook* (2d ed.), Ed. Skolnik, M. I., McGraw Hill, New York, 1990, Chapt. 7.

[7] Ruze, J., "Physical Limitations on Antennas," MIT Research Laboratory of Electronics, Technical Report 248, October 1952, Cambridge MA, p. 20.

[8] Cutrona, L. J., "Synthetic Aperture Radar," in *RADAR Handbook* (2d ed.), Ed. Skolnik, M. I., McGraw Hill, New York, 1990, Chapt. 21.

[9] Gray, A. L., Vachon, P. W., Livingstone, C. E., and Lukowsk, T. I., "Synthetic Aperture RADAR Calibration Using Reference Reflectors," IEEE Trans. G*eoscience* R*emote Sensing* Vol. 28, No. 3, May 1990, pp. 374–383.

[10] Moller, D., Siqueira, P., Hensley, S., and Chapin, E., "GeoSAR Calibration and Characterization Flight Plan," Unpublished GeoSAR Memo, NASA Jet Propulsion Laboratory, Pasadena, CA, August 1998.

[11] Ulaby, F., Moore, R., and Fung, A., *Microwave Remote Sensing; Active and Passive vol 2*, Artech House, Dedham, MA, 1982.

[12] Knott, E. F., Shaeffer, J. F., and Tuley, M. F., *RADAR Cross Section,* Scitech Publishing, Raleigh, NC, 2004, Chapt. 6.

[13] Doerry, A. W., "Reflectors for Calibration of SAR Performance Testing," Sandia Report SAND2008-0396, Sandia National Laboratories, Albuquerque, NM, January 2008.

[14] Geng, N., Ressler, M. A., and Carin, L., "Wide-band VHF Scattering from a Trihedral Reflector Situated above a Lossy Dispersive Half Space," *IEEE Tran. Geoscience and Remote Sensing* Vol. 37, No. 5, September 1999, p. 2609–2617

[15] Fleischman, J., Toups, M. F., and Ayasli, S., "Summary of Results from a Foliage Penetration Experiment with a Three Frequency Polarimetric SAR," SPIE Vol. 1693, April 1992, p. 151.

[16] Downing, R. J., "FOPEN 1993 Portage Maine Data Collections Vol. I Vehicle Ground Truth," MIT Lincoln Laboratory Project Report STD-67, November 1994.

[17] Toups, M. F., Ayasli, S., and Fleishman, J. G., "Part II Analysis of Foliage-Induced Synthetic Pattern Distortions," *IEEE Trans Aerospace and Electronic Systems* Vol, Vol. 32, No. 1, January 1996, pp. 145–155.

[18] Binder, B. T., Toups, M. F., Ayasli, S., and Adams, E. M., "SAR Foliage Penetration Phenomenology of Tropical Rain Forest and Northern US Forest," Proc. 1995 IEEE International *RADAR* Conference, Washington, DC, May 1995, p. 158.

[19] Bessette, L. A., Crooks, S. M., and Ayasli, S., "P-3 Ultra Wideband SAR Grayling Michigan Target and Clutter Phenomenology," Proc. 1999 IEEE *RADAR* Conference, Boston, MA, May 1999, p. 125.

[20] Toups, M. F. and Ayasli, S., "Results from the Maine 1992 Foliage Penetration Experiment," SPIE Vol. 1942, Orlando, FL, April 1993, p. 66.

[21] Billingsley, J. B., *Low-Angle RADAR Land Clutter—Measurements and Empirical Models*, SciTech. Pub., Norwich, NY, 2001.

[22] Franz, J. and Jao, J. K., "UHF Windblown Clutter Measurements and Modeling," *Proc. 2006 IEEE RADAR Conference,* Verona, NY, May 2006, pp. 39–43.

[23] MacDonald, D., Chang, C. F., Roman, J., and Koesel, R., "Automatic Detection and Cueing for Foliage Concealed Targets," SPIE Vol. 2757, Orlando, FL, April 1996, p. 152.

[24] Allen, M. R., "Efficient Approach to Physics Based FOPEN SAR ATD/R," *SPIE* Vol. 2757, Orlando, FL, April 1996, p. 163.

[25] Davis M.E., "Technical Challenges In Ultra- Wideband RADAR Development for Terrain Mapping", Proc IGARSS 1998 Seattle WA.

FOPEN SAR Image Formation

Image formation processing for synthetic aperture RADARs has been covered in many advanced texts [1,2]; hence, derivation of their details will not be reproduced here. The principal objective in this chapter is to develop the primary concepts that affect the design of the ultra wideband (UWB) SAR RADAR image formation processing. UWB waveforms required for operation at very high frequency (VHF) and ultra high frequency (UHF) remain a major challenge for FOPEN SAR. All subsystems that contribute to the transmit, receive, and image formation chain must consider their UWB characteristics for the images to be useful in detection, characterization, and geolocation of the objects under the foliage. This chapter will discuss the algorithms for image formation; the details on the hardware design to support these algorithms will be covered in Chapter 7.

FOPEN SAR has many of the same characteristics as high-resolution microwave SAR, (i.e., wide bandwidth, range curvature, and fine motion compensation for SAR collection geometries). However, to obtain the maximum image resolution, UWB waveforms and large integration angles are required, as previously covered in Section 2.1. These extremes in data collection make increased demands not only on the amount of processing for image formation but also on the motion measurement and compensation to focus over the full image and insure geospatial accuracy.

Because of the wide imaging angles, the collection must include not only the range extent of transmit and receive data for range resolution in the swath width, but also the projected data range extent for the wide angles, to support the along-track image resolution. The next section will describe the impact of integration angle on cross-range resolution and the impact of range curvature on the amount of data collection required for image formation.

The two most common algorithms for FOPEN SAR are outlined, along with their impact on the image formation processing. These two algorithms were developed for the early FOPEN SAR systems covered in Section 2.2 to attain the high-resolution image characteristics. The first is the backprojection algorithm (BPA), which is a straightforward convolution, over the collection window in range and cross-range, of each image pixel with the time-delayed transmit waveform [3]. The second is the range migration algorithm (RMA), which entails data transformation during the synthetic aperture to account for range curvature and enables two-dimensional fast Fourier transform (FFT) processing [1]. Both algorithms require extensive signal preprocessing and motion compensation to maintain focus over moderate swath areas. The amount of processing will be quantified to assist in selection of appropriate real-time processors. Needless to say, in the 15 years since the initial FOPEN SAR platforms were developed, the ability to meet these processing throughput challenges has been greatly improved.

4.1 FOPEN SAR Collection Geometry

The integration angles required as a function of frequency and platform speed are an important system consideration for FOPEN SAR. They have been useful in explaining the concept of operations for VHF and UHF SAR platforms, in comparison to the previous microwave SAR systems. If one needs to see below the tops of the trees, plans for operating the platform for lengthy collections during a single pass are required. Because of the long integration times for collecting fine image resolution, other areas of interest within the horizon of the platform are unavailable for surveillance.

Section 2.1 developed the relationship between the integration angle θ_I and the cross-range resolution δ_{CR}. Because of the assumption on range to the scene being large compared with the SAR integration length, the cross-range resolution is independent of range and L_{SAR}. However, this assumption is not valid for UWB SAR, as seen in Figure 4–1. The more exact expression for cross-range resolution is given by [1]

$$\delta_{CR} = \frac{\lambda_c\, k_{CR} R_C}{2\, L_{eff} \sin \alpha_{DC} \cos(\theta_I/2)} \tag{4.1}$$

where

λ_c	Wavelength at center frequency
k_{CR}	Resolution spreading factor due to aperture weighting
α_{DC}	Doppler cone angle
θ_I	Azimuthal SAR image collection angle
L_{eff}	Effective SAR integration length
R_C	Slant range from image point to center of SAR collection

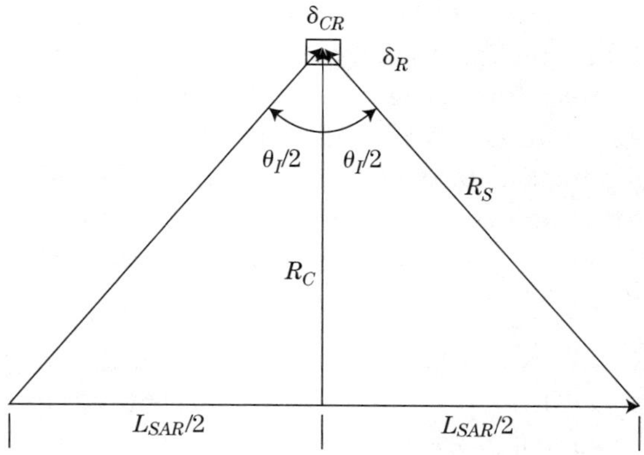

FIGURE 4–1

Geometry for UWB SAR resolution calculation

In Figure 4–1, the imaging is assumed to be broadside for which $\alpha_{DC} = \pi/2$. The $\cos(\theta_I/2)$ transforms the slant range to the center of the SAR image collection. And the L_{SAR} is changed to an effective SAR length L_{eff}, which can be solved from (4.1), using the cross-range resolution from (2.3) and trigonometric manipulations [4]:

$$L_{eff} = \frac{k_{CR}\lambda_c R_c}{2\delta_{CR}\sqrt{1-\left(\dfrac{k_{CR}\,\lambda_c}{4\,\delta_{CR}}\right)^2}} \tag{4.2}$$

Figure 4–2 provides a comparison of microwave versus UWB FOPEN SAR requirements on integration angle versus frequency. Based on (4.1) with

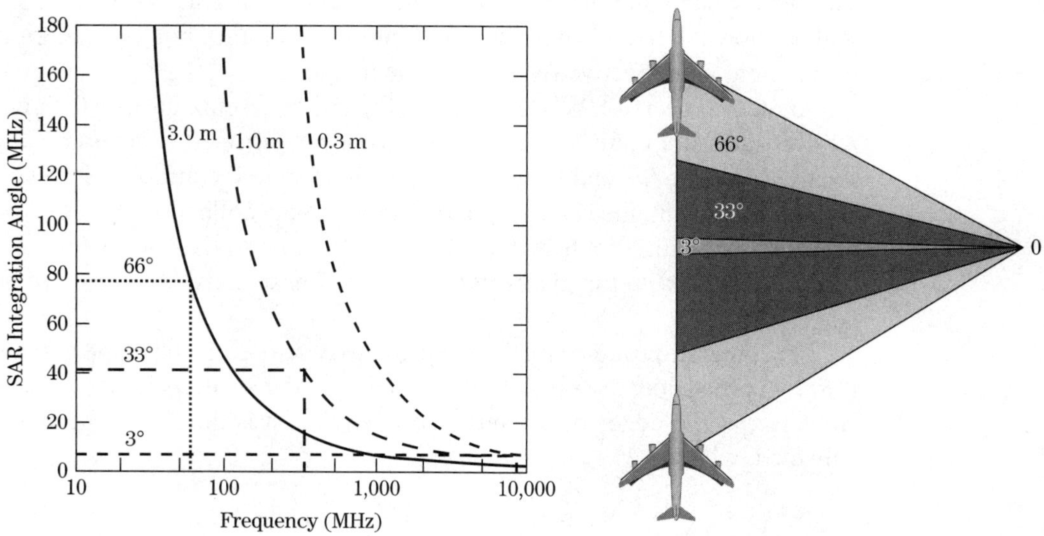

FIGURE 4–2

SAR integration angles versus frequency [5]

$k_{CR} = 1.2$, a microwave (10 GHz) image at 0.3 meters cross-range resolution requires a 3.0 degree integration angle. By contrast, the required integration angle is 33 degrees for 1 meter resolution at UHF (350 MHz) and 66 degrees for 3 meters resolution at VHF (50 MHz) [5].

The along-track data collection time t_I is similarly determined by the effective SAR length and the speed of the platform v_P, by

$$t_I = \frac{L_{eff}}{v_P} \tag{4.3}$$

However, if the SAR collection uses an ultra-wide integration angle, then the value of L_{eff} from (4.2) must be substituted into (4.3) to determine the appropriate coherent processing time.

Figure 4–3 compares the integration times at VHF and UHF for a variety of standoff ranges, when collected from a platform traveling at 150 m/sec. For microwave frequency SAR, a cross-range resolution below 1 meter can be easily obtained in less than 2 seconds at a range of 10 km. This time is contrasted with 35 seconds with UHF FOPEN SAR. At VHF, it is a challenge to obtain a 3 meter cross-range resolution, and the required collection time will be greater than 100 seconds.

The impact on the system is significant. First, because of the large integration angles, an entire pass of the aircraft is consumed in collecting a single swath of data. In addition, the variation of the terrain can be large over this geographic area, which directly affects both the focus and geospatial accuracy of the image. These factors greatly amplify the demands on the motion measurement system and the motion compensation (MOCOMP) processing. As will be shown in Section 4.3, there are limitations on the choice of algorithms for efficient and effective focusing of the image.

Another aspect of UWB SAR is the effect of range curvature on the image. As illustrated in Figure 4–4, data must be collected from the beginning of the swath through to the end to support the resolution in the middle of the swath length. As a result, the slant range extent of the data collection must be larger than the maximum swath width; such that the support data at the end of the swath coincide with the pixels in the middle. These considerations will now be quantified.

The integration angle and the range to the swath define the range curvature ΔR_0. At broadside, the range to the far edge of the swath is indicated as R_0. However, at the edges of the integration angle, an additional ΔR_0 must be collected, where

$$\Delta R_0 = \left[\frac{1}{\cos(\theta_I/2)} - 1\right] R_0 \tag{4.4}$$

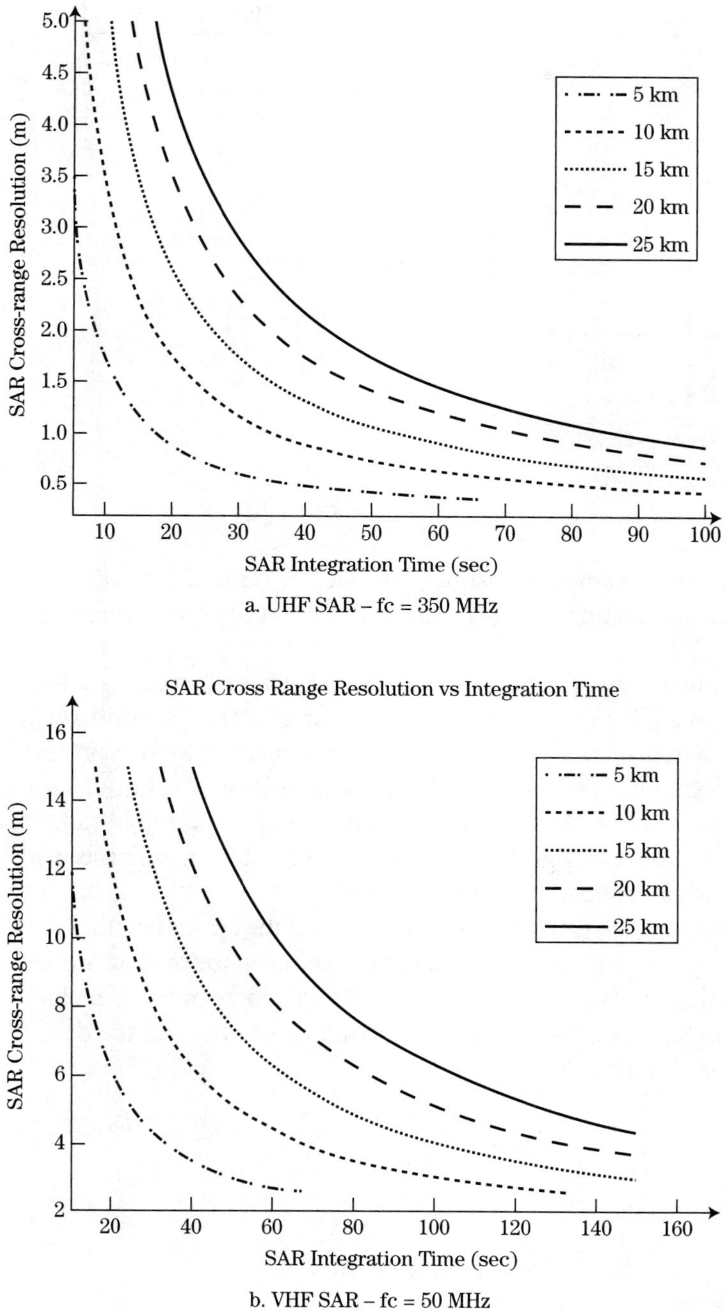

FIGURE 4–3

FOPEN SAR integration time, $V_p = 150$ m/sec

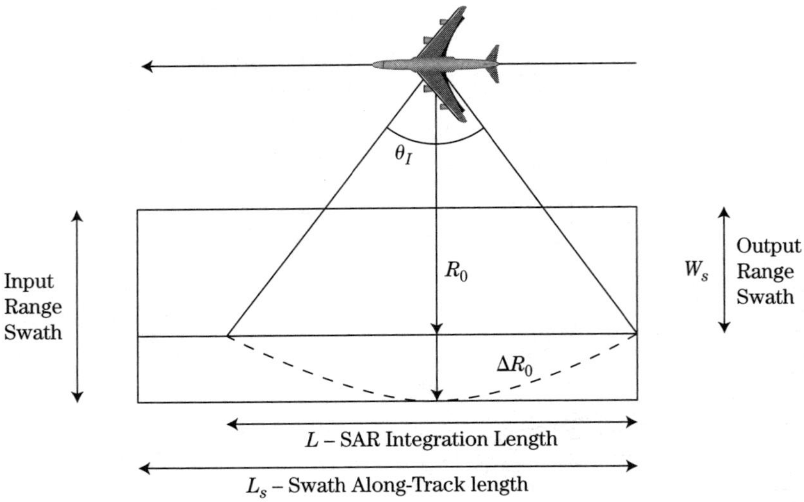

FIGURE 4–4

Effect of integration angle on range swath collection

These additional range samples must be collected for the entire swath length L_S plus the SAR integration length L, to fully support the cross-range resolution.

Figure 4–5 quantifies the impact on additional range samples as a function of range to the edge of the swath. At a nominal 50 km standoff range, an additional 2.4 km of range samples must be collected to support a 33 degree integration angle at UHF. Similarly, an additional 9.8 km of range samples is required for 66 degrees at VHF. These additional range samples must be factored into the design of the SAR waveform, the signal processing requirements, and the onboard memory storage.

The first aspect of UWB SAR to be considered is the swath length, where a sufficiently long integration aperture must be collected to form the desired cross-range resolution. The number of pulses N_P to be integrated, which can be approximated based on the range to the swath center R_C and the desired resolution δ_{CR}, is given by

$$N_P = \frac{k_S \, R_C \, \lambda_c}{2\, \delta_{CR}^2} \tag{4.5}$$

where:

N_P	Number of pulses in the synthetic aperture
k_S	Oversample ratio for data collection
R_C	Range to swath center
λ_c	Wavelength at center of band
δ_{CR}	Cross-range resolution.

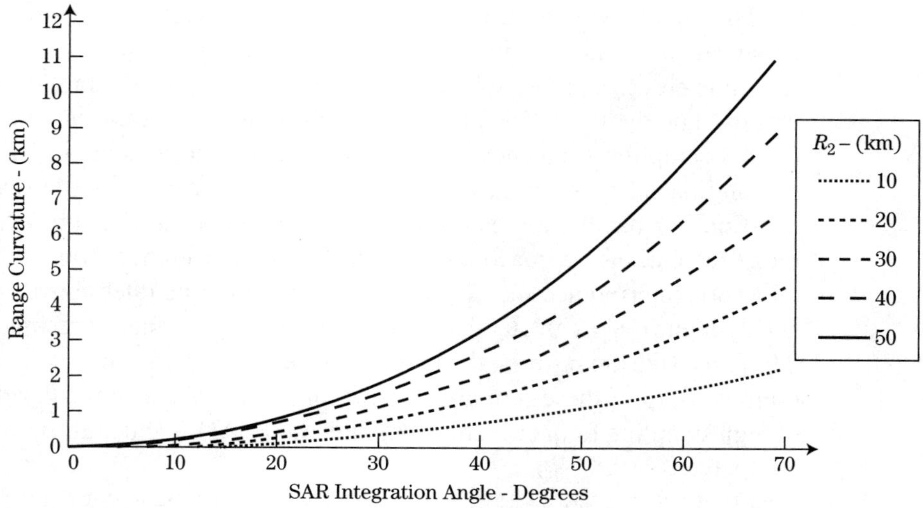

FIGURE 4–5

Range curvature as a function of SAR integration angle

The number of range cells N_R to be processed can be expressed by the width W_S of the input SAR swath and the added range ΔR_0 to accommodate for range curvature:

$$N_R = \frac{(W_S + \Delta R_0)}{\delta_R} \tag{4.6}$$

For UHF and VHF SAR the number of pulses can be very large, due to the long collection times. The along-track data collection is similarly determined by the resolution and the speed of the platform.

It should be noted in Figure 4–4 that the area in the desired swath L_S is not fully covered by the SAR integration length L, if only the slant range $R_0 + \Delta R_0$ is used. Only the pixels at the minimum range to the swath have the full sample support for resolution δ_{CR}, within the integration angle θ_I. As a result, it is common to break the swath up into smaller along-track segments and to form several subapertures. In addition, those subswaths are often further divided when there is excessive departure from a straight line to satisfy the MOCOMP requirements. Section 4.4 covers the impact on platform motion on the subswath design. These subswath approaches are commonly used in strip-map SAR to remove image glint. However, the reasons for multiple subapertures and the impact on the image formation processing can have a significant impact on the UWB FOPEN SAR implementation.

The early experimental FOPEN SAR designs had limits to the swath width, either due to limitations in the waveform processing or to considerations of real-time processing throughput. *Stretch* processing (i.e., deramp-on-receive) was used on the P-3 UWB platform to reduce the analog-to-digital converter (ADC) sampling frequency and dynamic range requirements [6]. However, the range swath was limited to less than 1 km because of the need to substitute fast-time for bandwidth, as discussed in more detail in Section 4.2.4. With range resolutions approaching 0.3 meters and interference from short-range television interference signals, the ADC need sampling rates above 500 MHz and dynamic ranges greater than 60 dB (i.e., 10 bits). Furthermore, images over 2 km on a side are pushing the real-time processing requirements to over 100 GFlops. Both of these limitations are being alleviated with the development of high sampling frequency, high dynamic range ADC, and high-performance computers.

Figure 4–6 shows the history of ADC development over the past two decades. In 1990, ADCs with dynamic range of a modest 48 dB could be obtained only at 50 MHz sampling speed. This explains the use of stretch processing in several FOPEN systems. However, by 2005 a 10-bit ADC at 500 MHz was available. This represents an improvement of 12 dB in dynamic range and 10 times improvement in sampling rate in 10 years. Furthermore, the rapid advance in digital signal processing hardware (commonly referred to

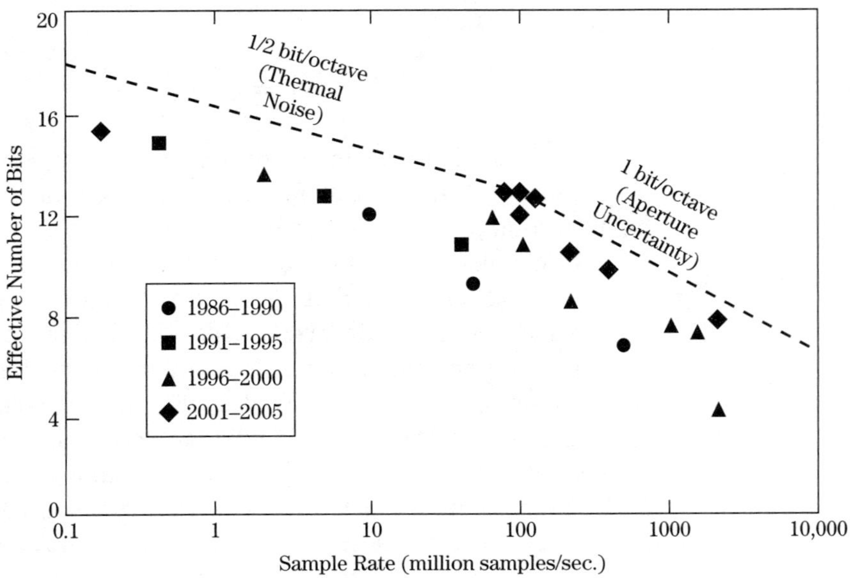

FIGURE 4–6

History ADC speed and effective number of bits (1986 to 2005) [7]

as Moore's law) continues to improve the processor throughput by a factor of two every 18 months or, equivalently, by two orders of magnitude per decade [7]. These subsystem advances will be instrumental in the development of high area coverage rate, real-time FOPEN SAR sensors for the objective of small, unmanned platform operation.

4.2 FOPEN SAR Waveform

To obtain the requisite ground range resolution, the waveform must have a sufficient bandwidth B. The range resolution δ_R in the ground plane is given by [1]

$$\delta_R = \frac{k_R c}{2B\cos(\gamma_g)} \tag{4.7}$$

where:

k_R Range resolution degradation due to weighting for sidelobes
γ_g Grazing angle

To achieve a cross-range resolution δ_{CR} the scene must be illuminated within the SAR collection over an integration angle θ_I.. If the antenna has a physical width D, the illumination angle will ideally follow:

$$\theta_I \geq \arcsin\left(\frac{k_{CR}\lambda}{D}\right) \tag{4.8}$$

If the SAR antenna is fixed in azimuth angle, the minimum cross-range resolution will be determined by the azimuth antenna pattern. However, if the antenna is gimbaled to maintain illumination on the scene or the platform can fly a circle around the scene (e.g., circle SAR), the cross-range resolution will not be limited by the antenna pattern. In the following performance derivation, it is assumed that the antenna is fixed, looking normal to the velocity vector.

Note that the convention on cross-range resolution is to specify integration angle in (4.8) at the center wavelength. Figure 4–7 illustrates the ideal spectral support required for obtaining both range and cross-range resolution, based on (4.7) and (4.8). It is clear that for UWB operation, the wavelength λ varies significantly during the waveform transmit (and receive) time. One can also expect that the beamwidth of the aperture will also vary with wavelength due to practical details of the antenna design over these wide frequency excursions. As illustrated by the shaded region in Figure 4–7, the ideal integration angle is approximately linear in wavelength. So the convention of using the center wavelength λ_c should not introduce significant error in the imaging processes. However, the system design for the transmitter and antenna characteristics

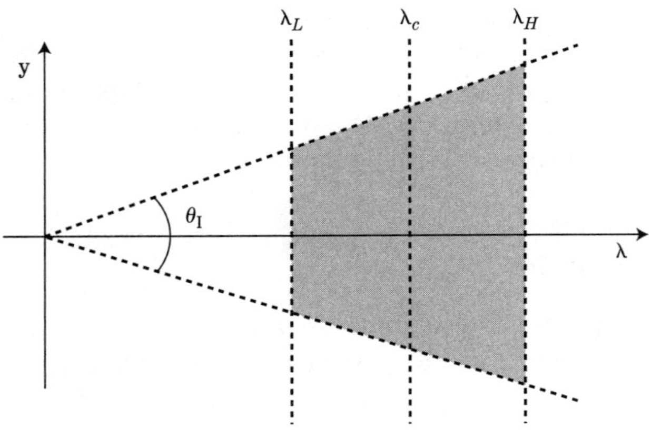

FIGURE 4–7

Waveform support for SAR image quality

over the UWB SAR collection bandwidth does not typically follow these ideal conditions. An efficient RADAR system needs to characterize the frequency variations in the components and to compensate for them in image formation processing. The consideration of these factors is covered in Section 7.2.

The development of the SAR imaging requirements is normally conducted at an arbitrary squint angle α_{DC} with respect to the aircraft velocity vector. However, since FOPEN SAR involves very wide integration angles, the imaging is generally accomplished with a squint angle of 90 degrees with respect to the velocity. In cases where a different squint angle is desired, the integration time $t_{I,sq}$ and resolution $\delta_{CR,sq}$ should be obtained by projection onto the broadside dimension:

$$t_{I,sq} = \frac{t_I}{\sin \alpha_{DC}} \tag{4.9}$$

and

$$\delta_{CR,sq} = \frac{\delta_{CR}}{\sin \alpha_{DC}} \tag{4.10}$$

The imaging problem can be viewed in terms of frequency versus the along-track set of coordinates as shown in Figure 4–7. At each pulse in the SAR collection, the time and cross-range units will vary along a different vector, as shown in the figure. More importantly, the image support will vary in the along-track and cross-track geometry, due to the projection of the waveform onto the surface of the ground. It is important to consider range curvature during both waveform design and motion compensation.

The SAR system transmits a series of pulses as it creates the synthetic aperture, as illustrated in Figure 4–8. The position of the SAR at each pulse is illustrated by a coordinate u_n in the y-axis. In this simplified coordinate system, the illumination is in a two-dimensional coordinate system, and the flight

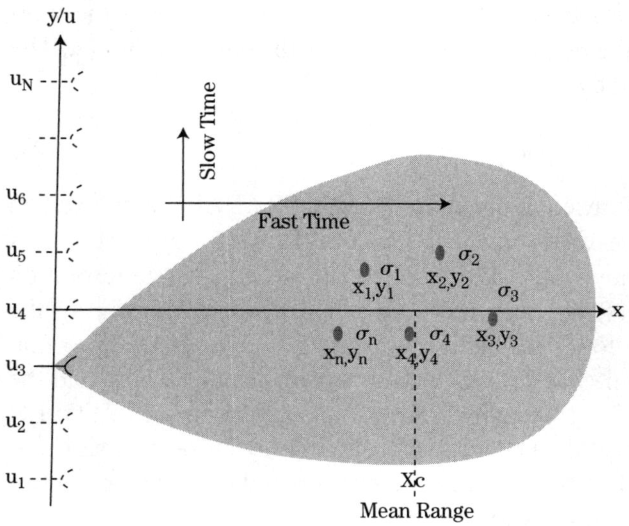

FIGURE 4–8

SAR image formation coordinate system [2]

path is along a straight line on the y-axis. In a real system, the flight path will not be linear; hence, the vertical projection of the beam on a nonplanar surface must be considered. Each scatterer in the scene is illuminated by the RADAR transmit beam, as shown by the scattering variable σ_n, located at positions (x_n, y_n). In the SAR image there is a field of scatterers that are physically separated. The ability to characterize these objects is determined by the range and cross-range resolution of the RADAR and the interaction of the electromagnetic scattering in the UWB SAR imaging process.

In Figure 4–8 there are two "time coordinates": *fast-time and slow-time*. Fast-time t is the time for a single RADAR pulse to travel to the scene and return. Slow-time is the pulse-to-pulse variation of the n-th pulse, within the pulse repetition interval T, expressed by the relation

$$\hat{t} = t - nT \tag{4.11}$$

Each RADAR pulse will illuminate this field with signal intensity set by the peak power and the transmit antenna gain. The received signal will depend on the reflectivity of the imaged surface and of any objects in the field.

Notice in Figure 4–8 that the transmit coordinate system u_n moves with the velocity of the aircraft. The image formation challenge is to coherently add up these signals during the SAR integration time. If the signal transmission and reception geometry conform to a uniform linear array, the processing is simplified. However, due to aircraft accelerations typically experienced during the very long FOPEN SAR collections, uniform aperture spacing cannot be achieved. It is the function of the MOCOMP processing to correct the signal phase history.

The transmit waveform can take many forms. The most basic is the short pulse waveform where the range resolution is set by the pulse width τ_{imp}. The range resolution is given by

$$\delta_{R,pulse} = \frac{c\tau_{\text{imp}}}{2} \tag{4.12}$$

The UWB impulse SAR, used in the early FOLPEN RADARs, employed an unmodulated short pulse waveform [8]. The waveforms used in this series of RADARs provided the bandwidth for moderate- to high-range resolution SAR (i.e., 1 to 5 meters) but required a very high peak power during the transmission due to the low duty factor. As a result, the average power that could be transmitted limit was limited by the very short pulse width and the achievable pulse repetition frequency (PRF). The average power of FOPEN SAR systems has received the benefit from using a long pulse, or a series of pulses, combined with UWB frequency modulation and the development of UWB pulse compression processing.

4.2.1 Linear Frequency Modulation Waveform

The Linear frequency modulation (LFM) technique was successfully demonstrated on several early FOPEN SAR systems, including the P-3 UWB SAR [9] and GeoSAR [10]. Each pulse has a modulation that changes the frequency of the signal as a function of time, as illustrated in Figure 4–9. The most basic LFM waveform has the full transmit bandwidth B applied to the pulse. In contrast, the coherent all radio band sensing (CARABAS) RADAR used a burst of narrowband pulses spread over the desired range resolution bandwidth but at a higher transmitter PRF [11]. The full bandwidth B was the obtained by

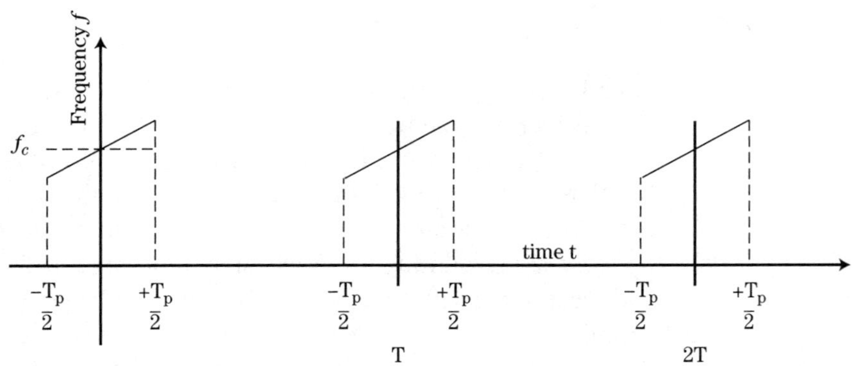

FIGURE 4–9

Progression of linear frequency modulated pulses

preprocessing the range pulses to align the frequencies in slow-time. This approach is a form of frequency jump burst (FJB), which is used when either component or environmental conditions do not allow UWB operation.

The waveform considerations for LFM are developed as follows, and FJB waveforms are developed more thoroughly in Chapter 5. In both cases, the pulse-compressed slant-range resolution, with sidelobe weighting factor k_R, satisfies

$$\delta_{R,PC} = \frac{k_R c}{2B} \tag{4.13}$$

In either case, the SAR processing requires that the signal be coherently processed to a time baseline that is coincident with the transmit pulse and to a point on the surface that corresponds to the image point. In Figure 4–8, this point is shown as $(0, y_c)$ the center of the desired image. For the general consideration of FOPEN SAR image formation, the LFM waveform will be used as a baseline. Consideration of more demanding waveforms, such as FJB and notching on transmit, on image range compression are treated in Section 6.2.

The transmit signal is formed by modulating the instantaneous transmit frequency f over a pulse width of T_p by the linear progression [12]

$$\begin{aligned} \omega = 2\pi f &= 2\pi f_c + \pi\gamma_p(t - nT) \\ &= \omega_c + \pi\gamma_p\hat{t} \end{aligned} \tag{4.14}$$

where

- ω Instantaneous radian frequency
- f_c Center frequency of LFM modulation
- γ_p LFM modulation index $[B/T_P]$ in radians per second

The phase history of SAR pulses is expressed in terms of both the fast-time variable t (e.g., range) and the slow-time index n (e.g., cross-range). Using this convention, the phase history of the transmit waveform is expressed by

$$s_T(n, t) = a_0 rect\frac{\hat{t}}{T_p} \exp[j2\pi f_c t + \pi\gamma\hat{t}^2] \tag{4.15}$$

where

- T_P Modulated pulse width
- $rect()$ Function equal to 1 for $\hat{t} \in [-T_p/2, T_p/2]$ and 0 otherwise.
- a_0 Signal amplitude

One should also note that the transmit amplitude a_0 is assumed to be constant. This constancy is for an ideal system, where the transmit amplitude

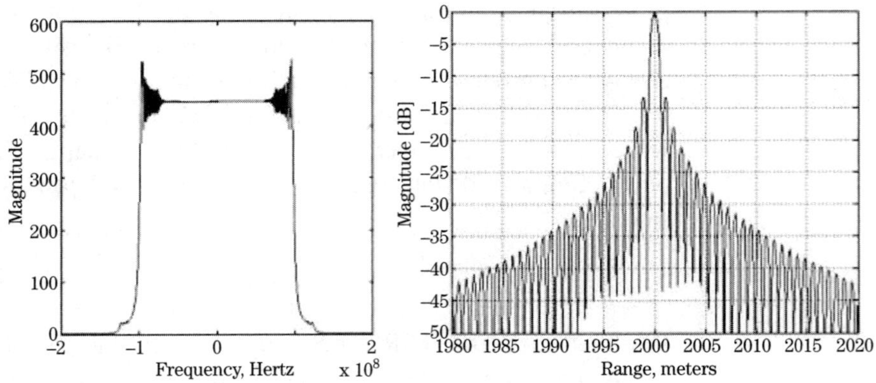

a. LFM Transmit Spectrum

b. Range-Compressed Pulse

FIGURE 4–10

LFM waveform for $f_c = 400$ MHz, $B = 200$ MHz and $T_p = 10\,\mu$sec

does not fluctuate from pulse to pulse or across the frequency band. For more exact image processing, the waveform model must take into account any changes or modulation of the transmit function caused by either intended amplitude changes or unintended changes due to transmit and antenna power-spectral characteristics.

A representative FOPEN SAR LFM pulse having a center frequency of 400 MHz and a bandwidth of 200 MHz, which is modulated over a pulse width of 10 μsec, exhibits the characteristics in frequency and fast-time domains shown in Figure 4–10. The slant-range pulse resolution, with a uniform weighting fast-time aperture, is 0.75 meters.

4.2.2 SAR Phase History

The SAR operates by collecting and combining a series of coherent pulses as shown in Figure 4–8, yielding a phase history to be processed for image formation. This phase history changes continuously as a function of time along the SAR integration path. It is normal to define a *reference point* at a range R_c to calibrate the coherent integration process. The round trip delay to the antenna is $2R_c/c$. For a strip map SAR, R_c will be the constant range to the center of the strip map, assuming a straight flight path. For a spotlight SAR mode, the central point of the image (x_c, y_c) will be constant, and the calibration, or motion compensation range, will vary as the SAR image is collected. In general, the reference phase function for SAR image construction is a constant unity amplitude, given by [1]

$$s_{ref} = \exp[j(\omega_c(t - 2R_c/c) + \pi\gamma_p(\hat{t} - 2R_c/c)^2)] \tag{4.16}$$

The received signal is obtained from the return from each point $R_t(x_t, y_t, z_t)$ in the scene with a signal amplitude of $a_T = \sqrt{\sigma_T}$. Each point has a round-trip time delay from the transmit to receive of t_d. Three factors make up the time delay: (1) motion between successive transmit pulses; (2) motion during transmission of a pulse; and (3) motion between transmission and reception of a pulse. Platform motion between and during pulses does occur and contributes to a space invariant defocusing of the image. This effect can be removed with autofocus up to a limit on the change in spatial frequency during the pulse. In the extreme, the effects need to be compensated using inertial sensors on the RADAR. For the current discussions, the motion during the transmit pulse is ignored in the image formation process.

The two-way time delay to each target point (i.e., *pixel*) in the SAR image is given by

$$t_d = \frac{2R_t}{c} \tag{4.17}$$

The distance R_t is the two-way distance from the transmitter to the target and back to the receiver and varies during each point in the SAR collection path. This distance is shown in Figure 4–8 and is given by

$$R_t = \sqrt{(x_a - x_t)^2 + (y_a - y_t)^2 + (z_a - z_t)^2} \tag{4.18}$$

The coordinates $R_t(x_t, y_t, z_t)$ are the true three-dimensional location of the scattering points on the surface. The points $R_a(x_a, y_a, z_a)$ are the measured locations of the antenna phase center on each received pulse. Note that R_a will have errors due to motion compensation inaccuracies as well as geolocation errors in the exact position with respect to the image surface. Given these coordinate definitions, the received signal phase history is expressed by

$$s_R(n,t) = a_T rect\left[\frac{\hat{t} - 2R_t/c}{T_p}\right] \exp\left[j\omega_c\left(t - \frac{2R_t}{c}\right) + j\pi\gamma_p\left(\hat{t} - \frac{2R_t}{c}\right)^2\right] \tag{4.19}$$

Figure 4–11 depicts several targets used to assess image formation steps and effects of waveform and motion compensation on the FOPEN SAR design. The characteristics of the notional system are summarized in Table 4–1. There are three targets on each primary range and cross-range axes, and additional targets are distributed within the 100 meter by 100 meter image area. In addition, σ_{14} is placed $3\delta_R$ separation from σ_5, and σ_{15} is placed $3\delta_{CR}$ from σ_2. These additional scattering targets will provide an assessment of the resolution characteristics of the waveform and the image processing algorithm and motion compensation capability to maintain image quality.

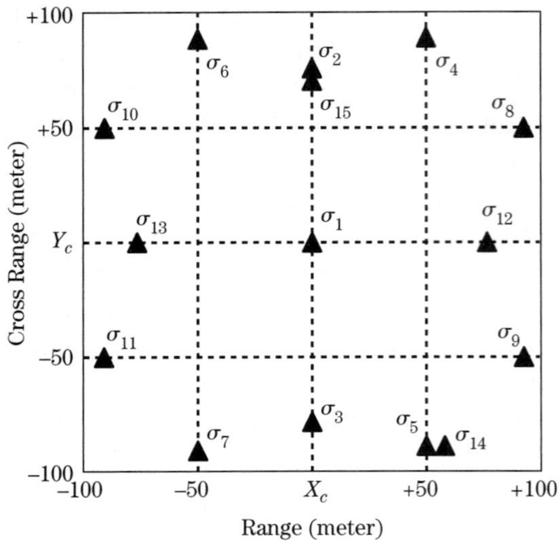

FIGURE 4–11

Location of example targets for image quality assessment

The final step in defining the SAR phase history is to consider the demodulation of the signal. This is accomplished by digitally mixing the reference signal in (4.16) with the received signal in (4.19). The process can be performed either before or after pulse compression, depending on the SAR image formation algorithm and the method of waveform modulation used. If the waveform varies in the fast-time dimension such that compensation to a fixed range will distort the signal spectral characteristics, this factor needs to be either removed prior to SAR demodulation or incorporated into the image formation algorithm.

There are two techniques for obtaining the final signal suitable for digital conversion for image reconstruction. One technique is to mix the incoming signal with a coherent local oscillator at the center frequency f_c. This approach yields a baseband signal with the full bandwidth of the transmit modulation. As a result, the ADC is required to sample the signal at sufficiently high rates to enable recovering the in-phase (I) and quadrature (Q) signals for digital signal processing. This ADC rate also needs to be sufficiently high to eliminate any

Table 4–1 Summary of exemplary FOPEN SAR parameters for analysis

f_c	400 MHz	δ_R	0.75 m
B	200 MHz	δ_{CR}	0.83 m
L	1200 m	θ_I	32 deg
x_c	2000 m	x_0	100 m
y_c	0 m	y_0	100 m

aliasing of the signal. The second technique mixes the incoming signal with a chirp signal at a center frequency f_c but with a modulation rate different from γ_p. This technique, which is called *stretch* processing, enables ADC sampling at a lower bandwith. It will be covered in Section 4.2.4.

If ADCs are available with sufficient sampling rate and dynamic range to process the SAR and RFI signals, this results in a baseband signal that can be processed entirely with digital techniques. The intermediate frequency signal phase return is given by

$$s_{IF}(n,t) = a_t \, rect\left[\frac{\hat{t}-t_d}{T_P}\right] \exp\left[-j\frac{4\pi(f_c+\gamma_p\hat{t})}{c}(R_t-R_0) - j\frac{4\pi\,\gamma_P}{c^2}\left(R_t^2-R_0^2\right)\right] \tag{4.20}$$

This phase history function serves as the basis for image formation processing to be covered in the following sections.

4.2.3 Sampled Fast-Time Formulation

The SAR phase history formulation in the previous sections has considered discrete pulses and continuous time. However, once the ADC sampling has taken place, the fast-time variables will also be discrete. This enables the use of digital signal processing for image reconstruction. The signal phase function is digitally sampled for image formation, along with radiofrequency (RF) interference removal and compensation for waveform match filtering.

The ADC samples the signal at discrete times $t_s(n,m)$ [1]:

$$t_s(n,m) = \frac{m}{f_s} + nT + \frac{2R_0}{c} - \frac{T_P}{2} - \frac{T_s}{2} \tag{4.21}$$

where the time sequence has the following features:

- f_s ADC sampling frequency
- m Sample number of M samples within the sampling interval
- n Pulse number of N_p pulses during the SAR image collection
- T Pulse repetition interval
- T_p Width of the transmit pulse
- T_s Width of the range swath interval ΔR_S for image formation

The total sampling time is the time required to cover the range window plus the length of the transmit pulse:

$$T_S = T_s + T_p = \frac{2\Delta R_S}{c} \tag{4.22}$$

The total collected signal to form a SAR map on an (x_i, y_j) grid is then the summation of N_R samples from N_p pulses:

$$s_t(x_i, y_j) = a_t \sum_{n=0}^{N_P-1} \left[\sum_{m=0}^{N_R-1} \exp\left[j\,\Phi_t(n, m)\right] \right] \tag{4.23}$$

where $\Phi_t(n, m)$ is the phase of the return from a target at R_t:

$$\Phi_t(n, m) = -\frac{4\pi \gamma_P}{c}\left(\frac{f_c}{\gamma_P} + \frac{m}{f_s} - \frac{T_p}{2} - \frac{T_s}{2}\right)(R_t - R_0) + \frac{4\pi \gamma_P}{c^2}(R_t - R_0)^2 \tag{4.24}$$

The SAR baseband return from discrete targets shown in Figure 4–11 is given in Figure 4–12.The process of image reconstruction is to digitally sum the returns from each pulse over the range interval $\Delta R_s = W_S + \Delta R_0$, as indicated in Figure 4–4 using the distance from each image point to the reference as given in (4.18). However, because of the motion of the transmitter from pulse to pulse and the impact of range curvature on the range and Doppler return from each cell on the image plane, it is required that the SAR processor compensate in the digital signal processor for the nonplanar and anisotropic returns.

The finest resolution in the waveform is obtained with uniform waveform weighting, that is, $a_t = 1$ for all samples. However, the compressed pulse sidelobe levels of the waveform would be limited to -13 dB with uniform weighting. In conventional SAR, weighting of the range or cross-range image pulse response is needed to reduce the effects of strong clutter within the region of the image cell. This is also important for FOPEN SAR where the

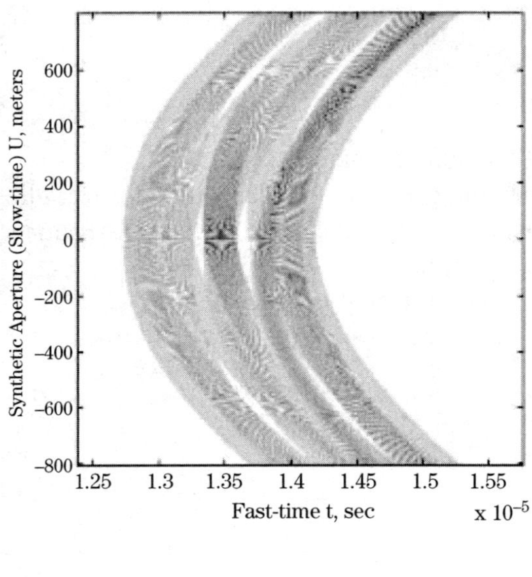

FIGURE 4–12

Target SAR baseband returns prior to range compression

Table 4–2 Weighting functions and impact on resolution and sidelobes [12]

Weighting Function	SNR loss, dB	Resolution Width	Peak Sidelobe dB	Far Sidelobe Falloff
Uniform	0	0.886/B	−13.2	6 dB/Octave
Taylor	1.14	1.25/B	−40	6 dB/Octave
Hamming	1.34	1.33/B	−42.8	6 dB/Octave
Cos^2	1.76	1.46/B	−31.7	18 dB/Octave

interference is due to the spiky clutter of trees as well as the attenuation of the signal returned from the dense biomass. Table 4–2 summarizes the most common weighting factors, along with the resolution degradation factors k_R and k_{CR} used in the definition of range and cross-range resolution.

4.2.4 Stretch Processing

The bandwidths needed for range resolutions of finer than a few meters place significant demand on the ADC sampling frequency. During the development of FOPEN SAR systems, few ADCs could provide the needed dynamic range above more than 100 MHz sample rates. This need for dynamic range is particularly true in areas for which there is with strong RFI from radio and TV stations. However, stretch processing is a well-known SAR waveform technique for processing large time-bandwidth product signal using narrow-band processing techniques. This section will develop the concepts for stretch processing as applied to LFM waveforms.

Stretch processing relieves the signal processor bandwidth problem by trading intermediate frequency bandwidth for reduction in range swath width. By using a swept frequency window deramping technique, targets over the entire waveform pulse repetition interval (PRI) can be processed. For a FOPEN SAR, the image is obtained over a small swath of ground, but with a fine image resolution to distinguish the objects [5].

Stretch processing does not mitigate the bandwidth requirements on the rest of the RADAR. Specifically, the transmitter must be capable of generating and amplifying the wide bandwidth signal, the antenna must be capable of radiating the transmit signal and capturing the return signal, and the receiver must be capable of heterodyning and amplifying the full bandwidth signal.

Figure 4–13 depicts a functional block diagram for stretch signal processing to convert the RF signal $s(t)$ to an intermediate frequency s_{IF}. It consists of a mixer, an LFM sweep generator, timing circuitry, and a narrowband ADC to sample the signal at the reduced bandwidth. The received LFM signal is as given in (4.15).

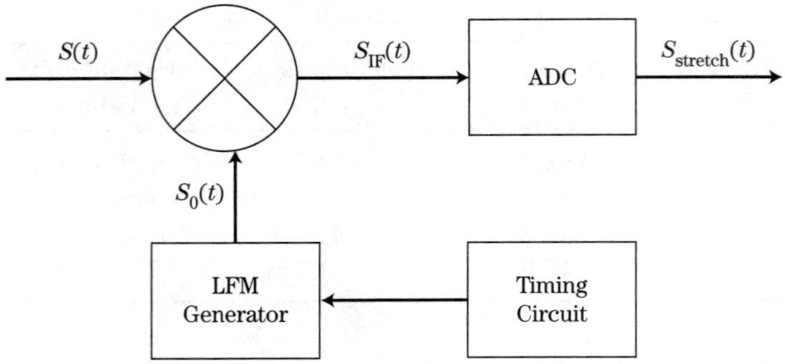

FIGURE 4–13

Signal conversion for stretch processing © 1971 IEEE [6]

The normalized heterodyne signal generated by the LFM generator is

$$s_0(t) = \exp[j\pi\gamma_p(t - t_0)^2]rect\left[\frac{t - t_0}{\tau_h}\right] \tag{4.25}$$

In (4.25), $t_0 = 2R_0/c$, which is the range delay to which the stretch processor is "matched," is usually centered in the region of $\tau_T = 2R_t/c$, that is close to the scatterers that are to be resolved. Furthermore, the duration of the heterodyne signal τ_h and must satisfy $\tau_h > \tau_T$.

Notional sketches of $s(t)$ and $s_0(t)$ are shown in Figure 4–14, where the frequency of each signal is shown over the nonzero time of the signal. Since $s(t)$ and $s_0(t)$ are LFM signals, we note that their frequencies increase linearly over their respective durations, with the same slope of γ_p. The top figure corresponds to the case where the target range delay τ_R is greater than τ_M, and the lower figure corresponds to the case where the range delay is less than τ_M. It is noted that when $\tau_R > \tau_M$ the frequency of $s_0(t)$ is greater than the frequency of $s(t)$; and when $\tau_R < \tau_M$, the frequency of $s_0(t)$ is less than the frequency of $s(t)$. Furthermore, the size of the frequency difference between $s(t)$ and $s_0(t)$ is linearly proportional to the difference between τ_R and τ_M.

Figure 4–14 indicates the approach to set the value of τ_h, the duration of the heterodyne signal. Specifically, τ_h should be selected so that $s(t)$ is completely contained within $s_0(t)$ for all expected values of τ_R, relative to τ_M. It can be seen from Figure 4–14a that the following condition should be met:

$$\tau_{R,\max} + \tau_T/2 \leq \tau_0 + \tau_h/2 \tag{4.26}$$

Similarly, from Figure 4–14b, there is a complementary condition:

$$\tau_{R,\min} - \tau_T/2 \geq \tau_0 - \tau_h/2 \tag{4.27}$$

From these two conditions, τ_h needs to satisfy

$$\tau_h \geq \Delta\tau_R + \tau_T \tag{4.28}$$

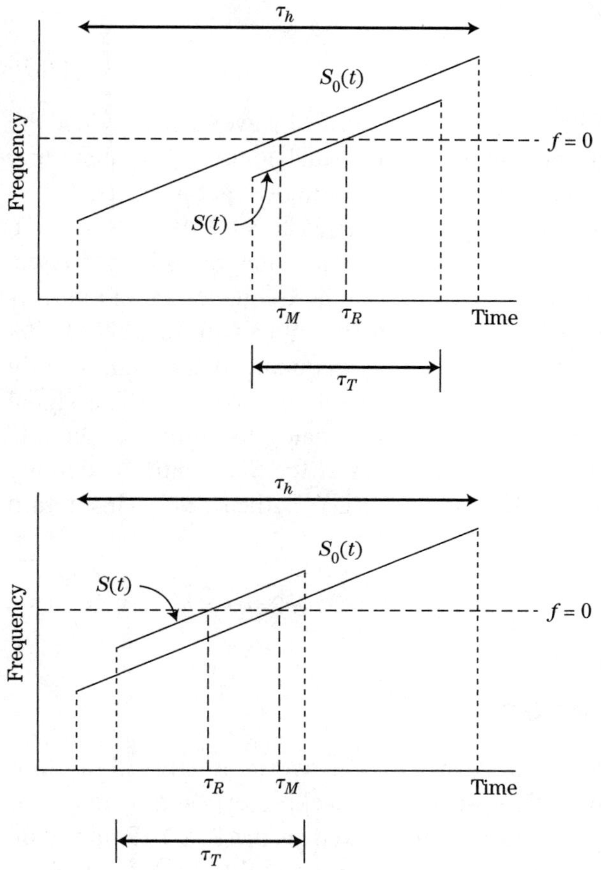

FIGURE 4–14
Stretch processing conversion from bandwidth to time

where $\Delta\tau_R$ is the range delay extent over which we want to use stretch processing. If τ_h satisfies the above constraint and $\tau_M - \tau_{R,\min} \leq \tau_R \leq \tau_{R,\max} + \tau_M$, then $s_0(t)$ will completely overlap $s(t)$, and the stretch processor will provide approximately the same SNR performance as a matched filter.

Given that $s(t)$ and $s_0(t)$ satisfy the previously given requirements, the output of the mixer can be written as

$$s_{IF}(t) = \exp[j\pi\gamma_p(\tau_M^2 - \tau_R^2)]\exp[j2\pi\gamma_p(\tau_R - \tau_M)t]rect\left[\frac{t - \tau_R}{\tau_T}\right] \quad (4.29)$$

The first exponential term of $s_{IF}(t)$ is simply a phase term. However, the second exponential term tells us that the output of the mixer is a constant frequency signal, with a frequency that depends upon the difference between the target range delay $-\tau_R$ and the range delay to which the stretch processor is tuned $-\tau_M$. Thus, the target range is determined by the frequency output from the mixer

$$\tau_R = f_m/\gamma_p + \tau_M \tag{4.30}$$

An amplitude taper can be implemented on the LFM waveform to reduce the range sidelobes via a weighting function on the matched filter. A similar taper to the stretch processor can be applied as an amplitude taper to $s_{IF}(t)$.

An example of stretch processing is the ERIM P-3 UWB SAR system, covered in Section 2.2 [13]. The transmit chirp pulse length τ_R is 26.5 μsec, and the signal bandwidth B is 515 MHz. Thus, the frequency ramp rate γ is 19.4 MHz/μ sec. An intermediate frequency (IF) filter with bandwidth B_{IF} of 120 MHz is centered at f_m of 1250 MHz, a frequency that is significantly higher than the center frequency of the transmitted waveform. The signal is subsequently downconverted to baseband for analog-to-digital conversion. The duration of the chirp tone t_0 from the start of the SAR until the deramp tone frequency equals the center IF. For the P-3 UWB, the stretch range return τ_T lasts for

$$\tau_T = \frac{B_{IF}}{\gamma_p} = \frac{120 \text{ MHz}}{19.4 \text{ MHz}/\mu\text{sec}} = 6.2\ \mu\text{sec} \tag{4.31}$$

4.3 SAR Image Formation

This section will illustrate the requirements on two alternative algorithms for FOPEN SAR image formation processing: backprojection and range migration. Both algorithms have been well covered in the RADAR literature and were implemented during the processing of the FOPEN SAR data from the experimental platforms described in Chapter 3. Although there are several other algorithms for UWB SAR image formation, these two techniques clearly illustrate the solution to the problem of high image quality with wide integration angle and high differential range curvature. It is very important to provide well-focused imagery for target feature detection and characterization, along with wide instantaneous focus for area coverage rate. The BPA provides the wider focus of images during a collection but at the cost of $O(N^3)$ processing operations per image pixel [3]. The RMA requires $O(N^2 \log_2 N)$ processing per pixel, which can be a significant savings in real-time processing for high area coverage [1]. However, RMA suffers from defocusing of the image when platform cross-track motions are significant. Research is continuing to find more efficient algorithms to enable real-time processing as well as expanding the area of focus.

As with any image formation process, there are assumptions on the geometry and conditions for focus. An aircraft cannot always fly a straight path, and the image plane is not always planar. The final section of this chapter will illustrate the effects of flight path wander during SAR integration and will present potential remediation techniques.

4.3.1 Backprojection Algorithm

The BPA was the first successful image formation technique for UWB SAR [3, 8]. BPA is a time-domain convolution of each point on the image surface with the RADAR waveform in both slow- and fast-time. To form a SAR image, a time-domain correlation is performed between the transmit waveform and each individual scatterer location. This correlation step for an image of N_A pixels in the along-track dimension, and N_R pixels in the range dimension using N_P transmitted pulses, requires $N_P N_A N_R$ complex operations. The advantage of this approach is there are no approximations to the range and Doppler frequency contributions to image formation. When the transmitter location is accurately known at each pulse and a geographically stabilized image plane can be maintained on the earth's surface, the image will be well focused. The most accurate georegistration and focus of the image will be obtained when the image pixels are distributed on a digital elevation model (DEM) corresponding to the imaged region. If these image formation criteria can be maintained over the wide geometry, there will theoretically be no degradation of the image focus.

However, the image formation process is computationally expensive and was determined not to be suitable for real-time implementation in early FOPEN SAR systems. As the speed of computers and the combined accuracy of global positioning system and inertial navigation system (GPS-INS) have improved, the BPA has become a serious contender for a real-time airborne collection system. Furthermore, research in *fast* BPA processing has evolved an approximation base on subapertures that reduces the computation to by a factor of $\sqrt{N_p}$ [14,15].

Figure 4–8 provides the coordinate system used for understanding the BPA waveform transmission. The processing flow for the BPA is shown in Figure 4–15, along with an estimate of the processing requirements for each stage. The first block is needed to compensate for nonuniform location of the transmitter during RADAR transmission. Each pulse is transmitted at a time determined by T, the pulse repetition rate of the RADAR. These points, which are nominally established to be on half wavelength centers, are determined from the velocity of the aircraft and the center wavelength λ_c of the RADAR. Based on the inertial measurement system, the position and velocity of the aircraft are measured and compared with the desired transmit position.

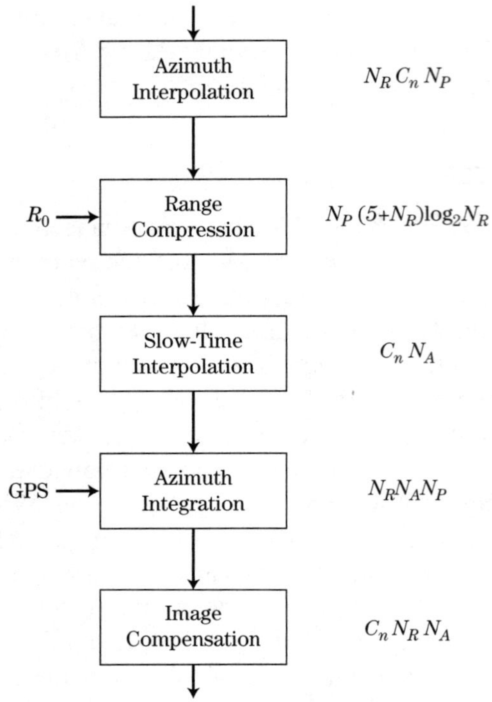

FIGURE 4–15

Algorithm flow for back projection image formation

The SAR image is formed by correlating the transmit waveform $s_T(-t)$ with each pixel on the image surfaces:

$$s_M(t, u) = s(t, u) \otimes s_T(-t) \tag{4.32}$$

where $\otimes$ is the convolution symbol. The time-domain correlation image is given by

$$f(x_i, y_i) = \int_u s_M[t_{ij}(u), u]du \tag{4.33}$$

where the time delay from the SAR platform position u to each pixel in the image is given by

$$t_{ij} = \frac{\sqrt{x_i^2 + (y_i - u)^2}}{c} \tag{4.34}$$

Therefore, the time delay to each point on the surface must be calculated at each position along the synthetic aperture flight path. It can be appreciated that knowledge of the position, velocity, and orientation of the SAR platform is needed at each point in the SAR map. Moreover, having a DEM of the terrain is highly desirable to correctly calculate the time delay to the surface. Otherwise, if the image is formed only in a plane there can be issues in the focus and position for each voxcel of the SAR image.

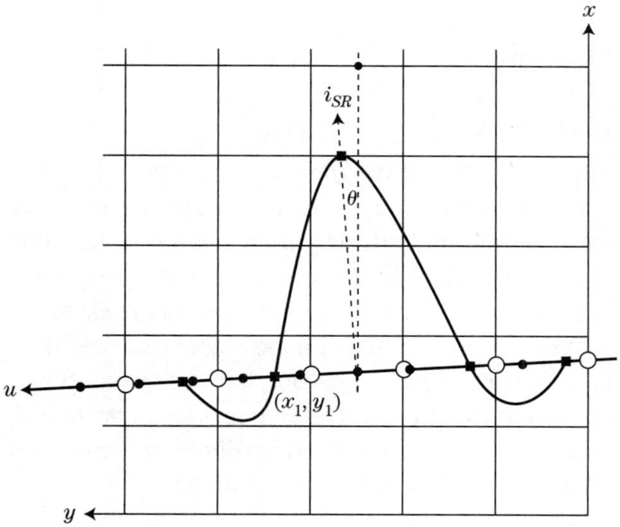

FIGURE 4–16

Interpolation function for along-track samples [1]

Because the distance from the transmitter to each point on the ground varies with slant range and angle, it is important to interpolate the pulse-compressed data on a pulse-by-pulse basis, along the slant range dimension, to each point on the ground. Figure 4–16 shows the basic interpolation used for resampling data in a generic slant range and angle (i_{SR}, θ) coordinate system into the image coordinate system (x_i, y_j). The interpolation is accomplished by upsampling the phase history function from the nouniform sample spacing to a uniform spacing coordinate system:

$$y(m) = \sum_{k=-\infty}^{\infty} \hat{x}(k)h(m-k) \tag{4.35}$$

where

$$\hat{x}(m) = \left(\begin{array}{ll} x\left(\dfrac{m}{M}\right) & m = 0, \pm M, \pm 2M \ldots. \\ 0 & \text{otherwise} \end{array} \right. \tag{4.36}$$

$$h(m) = \text{sinc}\left(\frac{m}{M}\right) \tag{4.37}$$

The solid circles along the u-axis give the along-track samples. For the phase history is given by the unit vector i_{SR} at an angle θ with respect to the image coordinate system (x, y). The open circles are the resampled values of the function $y(m)$ on the image y-axis.

The operations complexity C_P for data interpolation is given by [1]

$$C_P = 0.5N_I[(L_f - 1)DSR + 1] \tag{4.38}$$

where

N_I Number of output points
L_f Length of the resample filter
DSR Downsample ratio of output to input samples

Next, the RADAR waveform in the fast-time dimension needs to be compressed so that the required range resolution is obtained. This requirement for early pulse compression differs from both strip-map and polar algorithm SAR processing [1]. To get the necessary spatial correlation of the signal at each image point, it is important to perform range compression prior to the time-domain correlation of the transmit signal with the signal received from each point on the image surface. The pulse compression will be referenced to a range delay $2R_c/c$ that corresponds to the central reference point in the image.

The pulse compressed signal $s_{PC}(x, t)$ is formed by digitally convolving the signal phase function with the reference chirp function [2]:

$$\begin{aligned} s_{PC}(u, t) &= \sum_{N_s} s(u, t)^* s_0(u, -t) \\ &= \sum_{N_s} s(u, t)^* \exp\left[j\omega_c \left(t - \frac{2R_c}{c} \right) - j\gamma_p \left(t - \frac{2R_c}{c} \right)^2 \right] \end{aligned} \tag{4.39}$$

The process is accomplished through the use of an FFT in the range dimension on each pulse, multiplied by the time delay of the slant range R_C to the center of the scene and then taking an inverse fast Fourier transform (IFFT) to return to the spatial domain. The range-compressed returns from the 15 example targets are shown in Figure 4–17. The amount of range curvature in the SAR signal collection is easily seen in the figure.

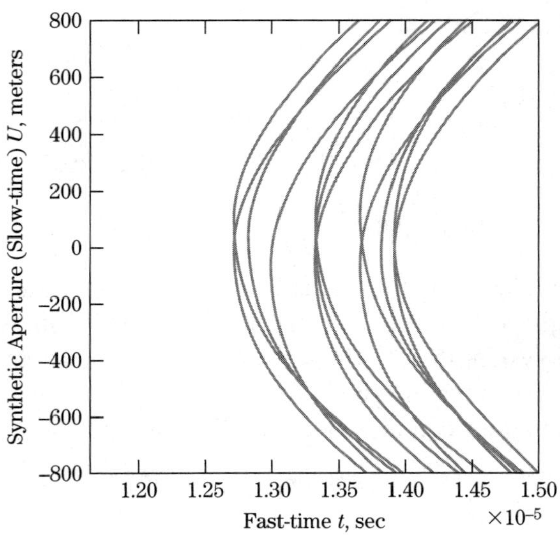

FIGURE 4–17

Range compression for 15-target example

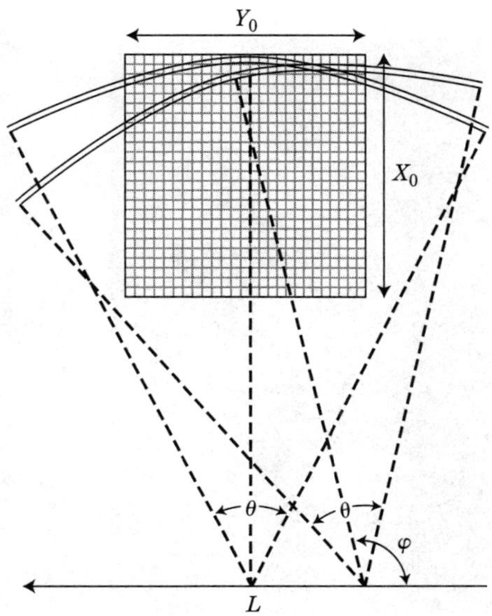

FIGURE 4–18

SAR range resolution superimposed on image x, y plane

The image is coherently added at each image location on a pulse-by-pulse basis, as depicted in Figure 4–18 until the entire synthetic aperture has been collected. However, it is important to resample the input data $s_M(t, u)$ accurately to the individual image locations $[x_i, y_j]$, or else image resolution and focus will be degraded in the processing. The third processing step is to upsample the fast-time waveform to enable this alignment. Figure 4–18 illustrates the range resolution of a the waveform transmitted from two points on the synthetic aperture within the image area of (X_0,Y_0). Depending on where the RADAR phase center lies along the aperture of length L, an accurate map of each pulse and range bin must be added to the appropriate pixels. Typically, the resampling process will use 10 to 20 samples on each side of the cell that is to be corrected. Next, the nearest point to the image resolution grid is chosen, and the value is accumulated. This process is repeated for each pulse and location of the SAR platform.

Figure 4–19 illustrates the BPA image formation for the target locations shown in Figure 4–11. The images of two of the single targets (σ_1 and σ_{11}) have been expanded, providing the image pulse response with no defocus of the image. This is as expected since the motion compensation was exact; hence, there were no spectral notches in the transmit pulse. Two additional expanded images are shown for the target pairs (σ_3 and σ_{14}) and (σ_8 and σ_{15}). For the latter two pairs of targets, the physical separation is three times the range and cross-range resolution, respectively. These images appear to be barely separated, as would be expected. It should be noted that no aperture weighting was applied to these images. With moderate weighting for range or cross-range

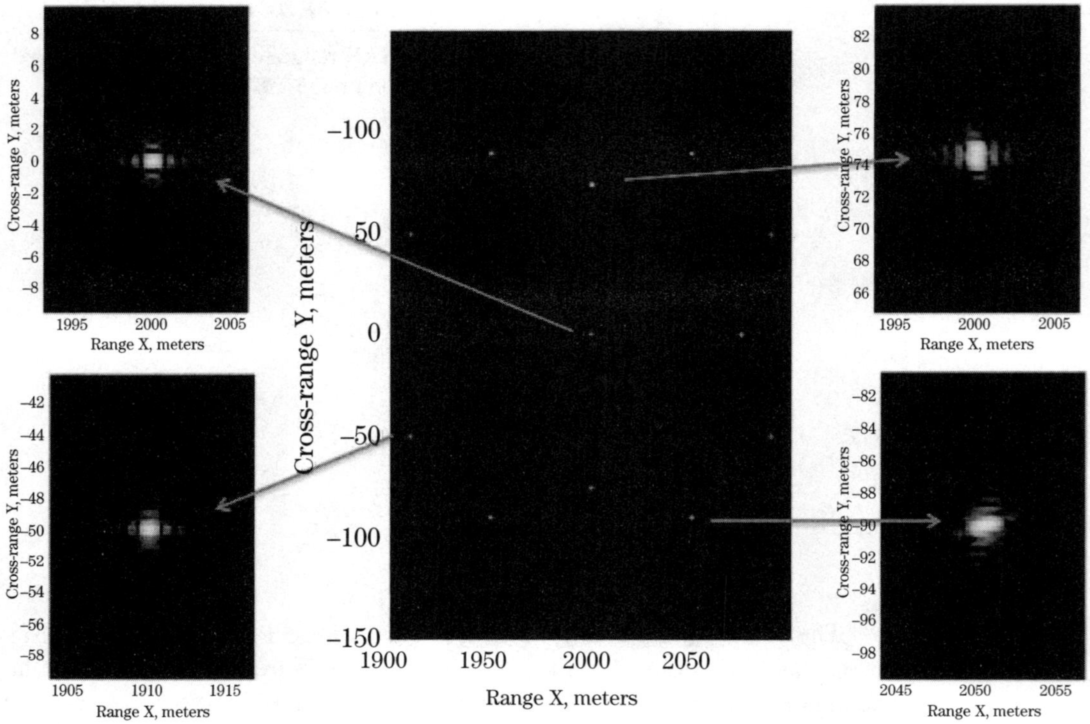

FIGURE 4–19

Reconstructed UWB image with back projection algorithm

sidelobes, the image resolution will be degraded by a factor between 1.2 and 1.6, as indicated in Table 4–2.

The BPA processing requires detailed and accurate knowledge of the system characteristics. The range delay to each point in the antenna beam and for each pulse in the synthetic aperture depends on several variables:

- The amplitude and phase in both the wavenumber (K_X, K_Y) dimensions affect the integration at each pixel in the scene.
- The focus will depend on accurate knowledge of the orientation and velocity of the aircraft and antenna as well as the image plane on the surface of the earth.

Because of the wide antenna angles, Doppler ambiguities need to be characterized and compensated. If this is not done, any Doppler-ambiguous returns at wider angles will fold into the image as artifacts. For VHF, the Doppler frequencies are rather low. As a result, the Doppler ambiguities might be an issue only for a FJB waveform, which has a large number of frequency steps. If these factors can be properly accomplished, the BPA will provide the largest depth of focus because each pixel is focused to the position, velocity, and propagation delay to the RADAR. Early in the development of FOPEN SAR, efficient algorithms were sought to lower this operations count.

Several development efforts have been reported that reduce the computational complexity of the BPA class of algorithms [14,15]. The approaches are generally based on segmenting the apertures into subapertures and compensating for range curvature within the image by applying higher-order corrections. The algorithm reproduces images generated by standard backprojection pixel for pixel to any required tolerance, but it runs roughly $\sqrt{N_P}$ times faster for an $N_A \times N_R$ pixel image. Furthermore, fast backprojection retains the advantages of standard BPA: perfect motion compensation for any flight path, low artifact levels, unlimited scene size, perfect focus for arbitrarily wide bandwidths, and integration angles.

4.3.2 Range Migration Algorithm

The RMA developed for SAR is based on signal processing in oil exploration seismic processing. RMA is designed to accommodate wide angle and dispersive media [16]. It compensates for the range curvature of the transmission through a series of signal phase transformations. More importantly, it uses FFT operations to limit the number of operations for image formation processing. The processing steps for RMA are shown in Figure 4–20, along with the operations count for each stage of process flow [17].

It was well known that the convolution operation can be accomplished with FFT operations. However, the signal has a range curvature in both the time and frequency domains. The principal difference is that in the time domains the amount of curvature for each point scatterer varies with range and the point of closest approach to the radar track varies with the along-track position. After the fast-time FFT, all of these phase histories overlap. Because of the wideband waveform, it is normal to first transform the fast-time domain representation of the signal phase history to the spatial frequency domain, where K is the wavenumber in the spatial frequency domain.

The first step is to transform the phase history via an along-track FFT. This step, which is done prior to range compression and azimuth dechirp functions, removes the along-track dependency of the range curvature. Figure 4–21 illustrates the transformation via RMA to compensate for range curvature, along with the factors that size of the signal processing steps.

After the along-track Fourier transformation, the signal phase history still has a range-varying curvature. The image is first focused along a straight line at the middle of the image. Along this line there is no range curvature in the

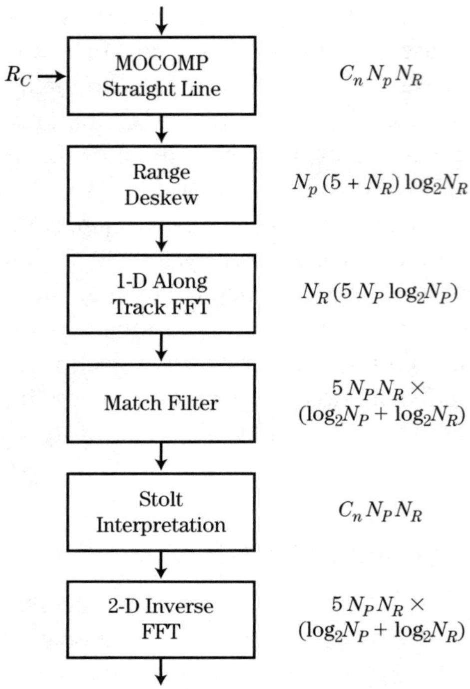

FIGURE 4–20

Image processing flow—Range migration algorithm © 1991 IEEE [16]

direction of the K_X mapping. However, there is a curvature at closer range and longer range that needs to be compensated for. A mapping in phase function that will straighten the range curvature along the center of the image is given by [1]

$$\Phi(K_X, K_R) = -X_t K_X - R_B\sqrt{K_R^2 - K_X^2} + R_s K_R \tag{4.40}$$

where the parameters are:

- K_X Azimuthal spatial frequency
- K_R Range spatial frequency
- X_t Along track position of platform vs slow time
- R_B Minimum range from platform to scatterer
- R_s Broadside slant range from platform to scene center.

The second step is to apply a two-dimensional phase compensation to correct the range curvature of all scatterers at the same range as scene center. Matched filtering of the SAR phase is carried out to remove the last two terms in (4.24). The phase of the matched filter, which removes the linear phase with slant range R_s, is given by

$$\Phi_{mf} = -K_R R_s + R_s\sqrt{K_R^2 - K_X^2} \tag{4.41}$$

The signal phase is now given by

$$\Phi_2(K_X, K_R) = K_X R_t + (R_B - R_s)\sqrt{K_R^2 - K_X^2} \tag{4.42}$$

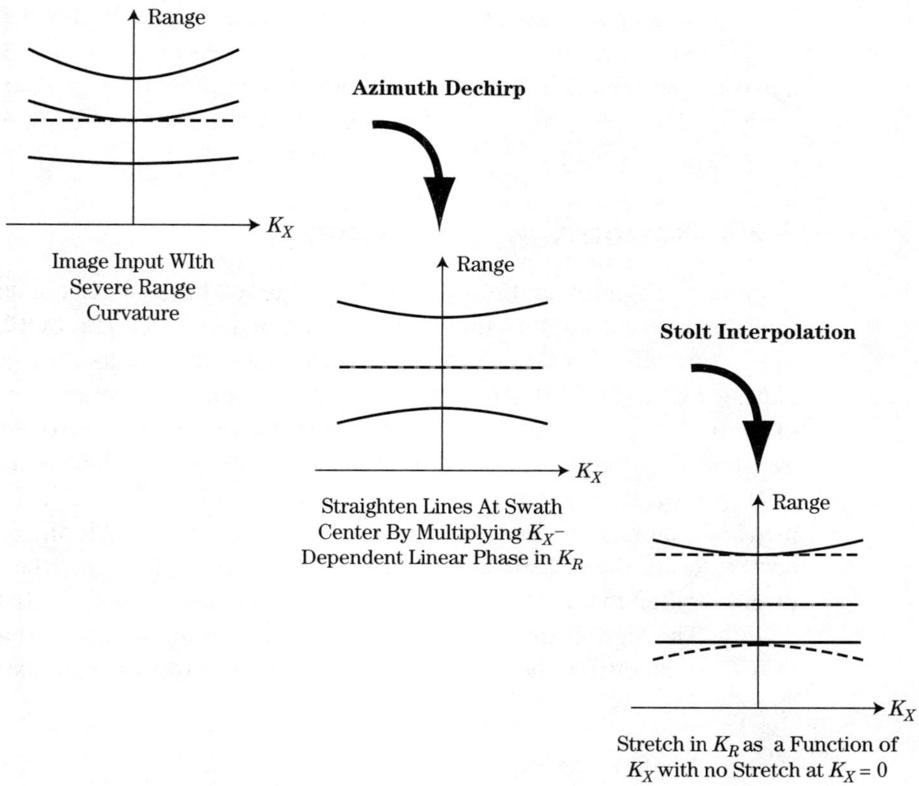

FIGURE 4–21

Range migration algorithm processing flow [1]

Along the midpoint of the image, the phase is constant on the $R_B = R_t$ line. However, at shorter and longer ranges, there is a residual phase curvature as shown in Figure 4–21.

The major difference between the RMA and the polar algorithm is that the RMA motion compensates to a straight line and polar algorithm compensates to a point at the center of the image. Furthermore, the Stolt interpolation in the RMA compensates for the differences in range curvature as a function of range. The Stolt interpolation straightens the lines of constant phase to form a linear phase in azimuth direction. The resampling provides the transformation

$$\sqrt{K_R^2 - K_X^2} \Rightarrow K_Y \tag{4.43}$$

Following the Stolt interpolation, the phase is now parallel to the K_X line for all ranges.

$$\Phi_{RMA}(K_X, K_Y) = K_X X_t + K_Y(R_B - R_s) \tag{4.44}$$

The signal from a point scatterer anywhere in the image has now been transformed to be linear in phase in K_X and K_Y space. As a result of this linearity, a two-dimensional IFFT can be performed to compress the signal simultaneously in range and cross-range. By using a two-dimensional FFT, the number of operations is reduced from N^3 to $N^2 \log_2 N$.

4.4 SAR Motion Compensation

Platform navigation and terrain height variations have direct effects on the focus and geolocation accuracy of the SAR images [18]. The FOPEN SAR image formation is carried out either to an image plane or to a digital terrain elevation data (DTED) height reference. As a result, the position and velocity of the RADAR platform need to be accurately known at the time of each pulse transmission along with the relative location on the image plane in the DTED reference of each resolution cell. An embedded GPS-INS (EGI) is used to provide accurate position, velocity, and orientation of the SAR antenna phase center during the SAR data collection. It is assumed here that the EGI has been specified to meet the short- and long-term accuracies of the SAR image quality. The significant unknowns then are due to the terrain variation and DTED accuracy and the departure of the sensor platform's flight path from a straight line.

4.4.1 Motion Measurement

The formation of the SAR image requires a coherent phase history be maintained during the time it takes the aircraft to fly the synthetic aperture. For FOPEN SAR, these apertures are very long; hence, the area for motion compensation is typically much larger than for microwave SAR. It has already been shown that UWB SAR requires significant processing to compensate for range curvature. If the motion measurement is not accurate, range curvature effects will cause severe image degradation to occur.

Figure 4–22 illustrates the types of errors that will result if there is an error in the location of the pixels in an image plane, either due to terrain height variation or to platform navigation errors. The estimated slant range to a scatterer location on the ground R_s (in latitude and longitude coordinates) is used to calculate the resultant phase for a path length. Over a synthetic aperture collection time, this phase is given by [17]

$$\varphi(t) = \frac{4\pi}{\lambda_c} R(t) = \frac{4\pi}{\lambda_c}\left[R_0 - \int V_L dt\right] \tag{4.45}$$

where V_L is the line-of-sight velocity between the antenna and the ground point, and R_0 is the initial range. The line-of-sight velocity is found from

$$V_L = \hat{t}_{SR}\overline{V} \tag{4.46}$$

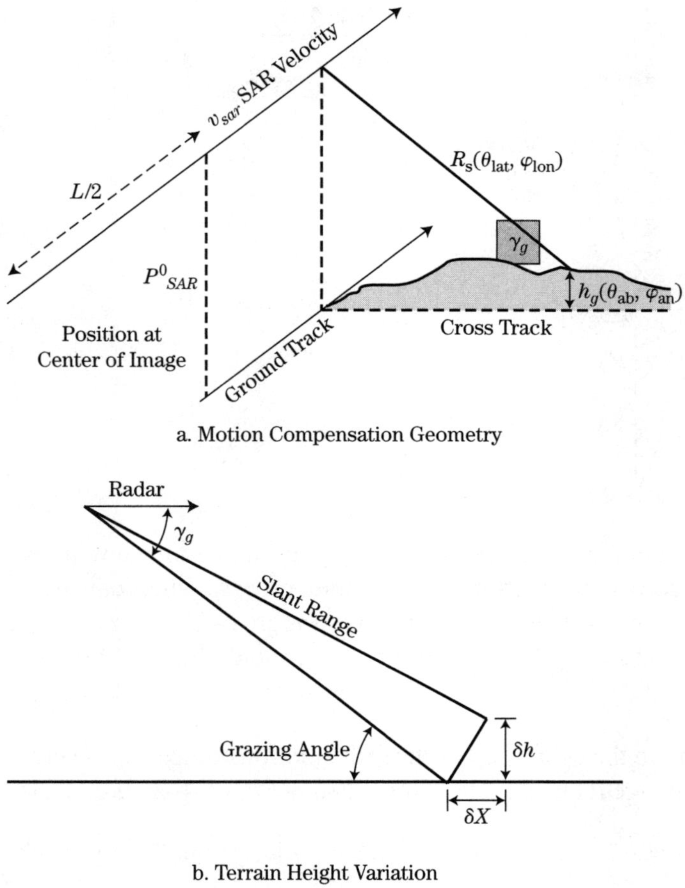

FIGURE 4–22

Motion measurement for SAR image formation © 1999 IEEE [20]

where, $\hat{t}_{SR}$ is the unit vector to the ground point, and $\bar{V}$ is the aircraft velocity vector. The accuracy of calculating $\hat{t}_{SR}$ and $\bar{V}$ is readily determined from the accuracy of position, velocity, and orientation specification on the GPS-IMU subsystem.

The actual slant range R_s will have an error δR in determining the MOCOMP processing correction, providing the estimated slant range $\tilde{R}_s$ as

$$\tilde{R}_s = R_s + \hat{t}_{SR}\delta R \tag{4.47}$$

If there is a velocity error, either or both of two effects can occur. If the velocity vector is incorrect, the azimuthal angle to the image point will be in error. Thus, the image point will either suffer error in focus or be rotated in geospatial coordinates. If the magnitude of the velocity is in error, the sample distance will be in error, and the Doppler frequency at each point in the scene will be scaled incorrectly. Both effects can give rise to focus and geospatial accuracy degradation [19].

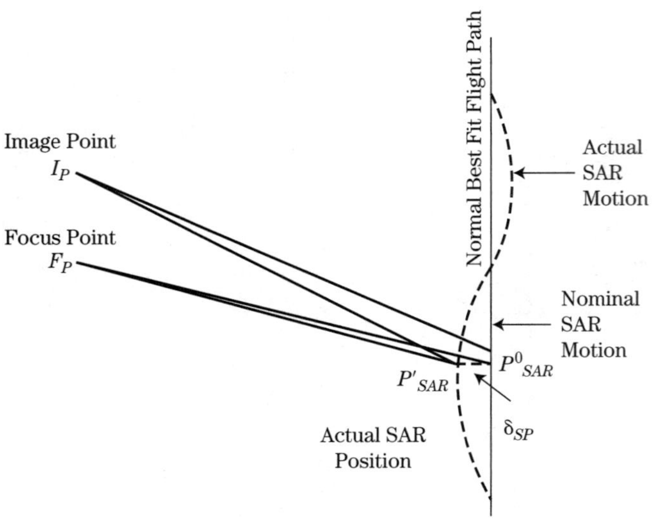

FIGURE 4–23

Motion measurement deviation from a straight line © 1999 IEEE [20]

For platform height errors, the phase to each point in the image will also be estimated incorrectly in the motion compensation process. This contributes to a layover effect in the SAR image as shown in Figure 4–22b. For a height error δh and grazing angle γ_g the position error in ground range δX will be

$$\delta X = \delta h \tan \gamma_g \tag{4.48}$$

For a given point in the synthetic aperture, the motion compensation error δR_{MC} for the SAR image focusing effect illustrated in Figure 4–23 is given by

$$\delta R_{MC} = |P^1_{SAR} - I_P| - |P^0_{SAR} - I_P| - |P^1_{SAR} - F_P| + |P^0_{SAR} + F_P| \tag{4.49}$$

where

I_P	Location of the scatterer
P^o_{SAR}	Nominal position of the SAR platform on the ground track line
P^1_{SAR}	Actual position of the SAR platform
F_P	Location of the focus point and

To examine the impact of motion effects on the RMA algorithm, the results of several data collection flights were examined in terms of the image quality. The P-3 UWB SAR was instrumented to collect data to analyze the impact of navigation errors on the image. Since the motion compensation for RMA is normally applied at broadside, the flight path was examined during a SAR integration period. This is illustrated in Figure 4–23 as a departure from the straight-line flight path.

The UWB P-3 operated at a slant range of approximately 5.5 km and with an integration angle of 35 degrees. In the image formation process, a best-fit straight line over the processing aperture is computed. Any deviation from this line will contribute to a degradation in the image quality [19].

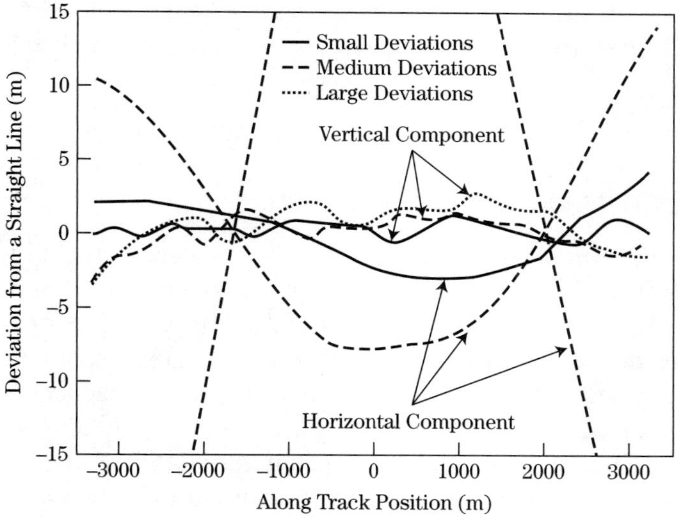

FIGURE 4–24

Motion measurement from P-3 image collection © 1999 IEEE [20]

Figure 4–24 illustrates typical motion errors that occur during in the UHF SAR image collection. Two types of deviations are illustrated: horizontal and vertical. The horizontal deviation for this platform is the larger and will have the greater impact on the image focus. These image degradation effects are due to the inability of the RMA algorithm to remove the cross-track variation of range curvature except at the motion compensation line at the center of the image. These data were collected from three classes of image collection runs. Figure 4–25 shows a histogram of the deviations from over 60 passes of the aircraft. In a significant number of collections the deviations were classified as being *small*, where the spatial errors were below 1.4 meters. Some collections

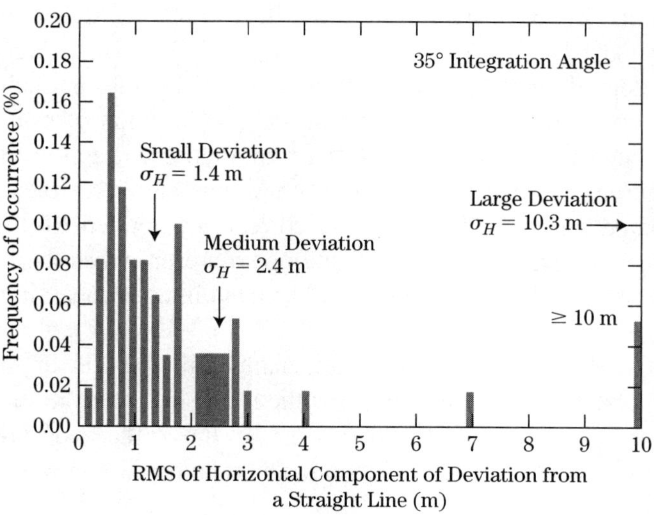

FIGURE 4–25

Histogram of P-3 UWB flight errors © 1999 IEEE [20]

had deviations on the order of 2.6 meters and were classified as being *medium*. One instance of a 10 meter deviation is shown and was classified as being *very large* [20].

4.4.2 Motion Effects on RMA Image Focus

SAR image focus is determined by two factors. The first, which is the ability to correct the phase delay from each of the image points to the scene center, is central to MOCOMP processing. If there is an error in the GPS-INS navigation, the variation in the phase is straightforward to calculate. The image defocus will be determined by the root mean square (RMS) error to each point in the image. The accepted MOCOMP processing standard is to measure the phase to better than 1/20 of a wavelength. Fortunately for FOPEN SAR, the wavelength is on the order of a meter; thus, the position needs to be measured only better than 50 centimeters. The use of Kalman filters and GPS-INS has demonstrated significantly better accuracy under normal SAR flight conditions. Therefore, the image degradation effect due to this source of error will be minimal.

Terrain height errors are the next most prevalent source of phase error. Knowledge of terrain height provided through the availability of very accurate DTED has been achieved with worldwide mapping from space and airborne instruments such as GeoSAR and the shuttle RADAR tomographic mapping (SRTM) systems. However, these errors will be on the order of a meter without the application of calibration techniques. The focus of the image will be improved by using multiple-channel measurements from interferometric SAR and autofocus techniques. The former will provide more accurate relative terrain height measurement and will improve the absolute geolocation accuracy. The latter will provide local improvements in image focus but will have degrading effects on multiple-channel operation, such as polarimetric detection and image characterization. The iteration of multiple techniques has been studied to improve the radiometric correction in images [19]. For example, in polarimetric FOPEN SAR the autofocus technique will be conducted on the HH image, and corrections applied to the other polarization channels. This approach will attempt to preserve the amplitude and phase relationships of the channels for subsequent image processing metrics.

Even if the MOCOMP is perfectly accurate, there will be an error in the RMA Stolt interpolation in compensating for the range curvature in the cross-range dimension. The MOCOMP difficulty arises due to the wide integration angles in FOPEN SAR image collection. The relationship between uncompensated motion and the phase error is given by

$$\Phi_{err} = \frac{4\pi\,\delta R}{\lambda} \tag{4.50}$$

When the motion is cross-track, there will be a residual phase error at the scene center due to a displacement of δy. At the edge of the beam β_a, a residual phase error for targets exists, given by

$$\Phi_{err}(y) = \frac{4\pi\,\delta y}{\lambda}[1 - \cos(\beta_a)] \tag{4.51}$$

This residual phase error due to the cross-track motion will be comparable for all pixels within the scene.

The effect of this platform motion was examined for P-3 UWB images, where two image formation processes were used. The first process employed a motion compensation that was broadside to the image formation, which is analogous to a strip map image formation. The second process applied the motion compensation to the center of the image, which is analogous to a spotlight image formation but using the RMA algorithm.

The impulse response (IPR) function for these two image formation and MOCOMP techniques are compared in Figure 4–26. The dotted line corresponds to MOCOMP at broadside to the image formation. The solid line corresponds to MOCOMP to a point in the center of the image collection. These two curves illustrate the IPR at the focal point. The strip-map MOCOMP processing exhibits a defocusing of the image during collection, resulting in an integrated sidelobe ratio (ISLR) of –7.7 dB. The spotlight MOCOMP

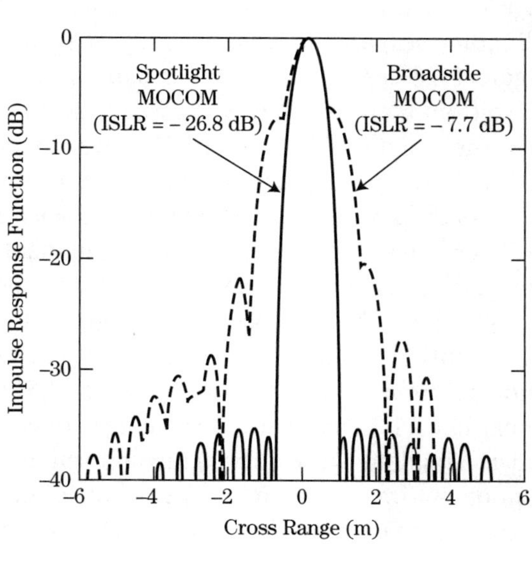

FIGURE 4–26
Measured impulse response with motion errors © 1999 IEEE [20]

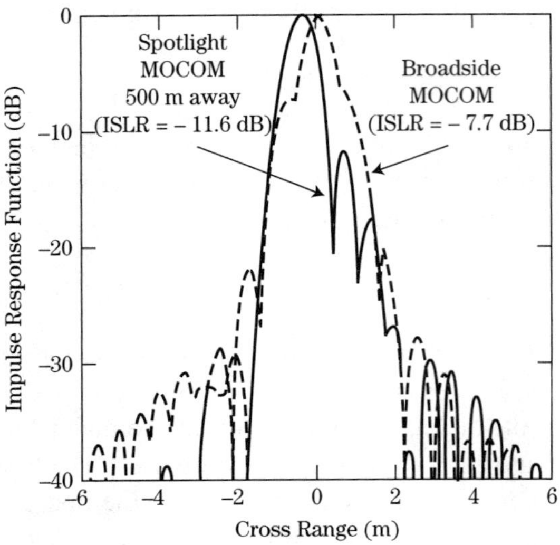

FIGURE 4–27

Measured impulse response 500 m along-track © 1999 IEEE [20]

processing had greatly improved IPR at the focal point of the image, achieving an ISLR of –26.8 dB.

Figure 4–27 illustrates the impact of image focus for points away from the center of the beam. Again, the dotted line is for the broadside compensation but with the image point 500 meters in cross-range from the focus point. This performance is comparable to an ISLR of – 7.7 dB when compared with the same condition in Figure 4–26. For the RMA image formation with MOCOMP performed to a central focus point, the displacement causes image quality degradation to increase the ISLR to –11.6 dB.

With the RMA spotlight approach, good image quality can always be achieved by limiting the output scene length L_{out} for a given ISLR performance objective. Figure 4–28 depicts the ISLR versus L_{out} for several cases of slant range deviation from the P-3 UWB SAR. Small deviations can obtain –20 dB ISLR for scene lengths of 500 meters. However for cross-track platform motion, the allowable scene length for acceptable performance is significantly reduced. These conditions require more processing since L_{out} is significantly less than the SAR integration length L. It is important to understand the maximum integration length that can be used as a function of the image quality and to plan the on-board processing to improve the overall efficiency.

The image quality (i.e., the amount of allowable defocus) can be related to the deviation δY from the correct path by using (4.51). The deviation δY will be different in each point of the SAR image collection. As such it can be considered as a random phase error in the uniform linear array formation. The maximum output image length can then be determined by specifying a maximum allowable phase error $\Phi_{\max}$ as a function of the integration angle

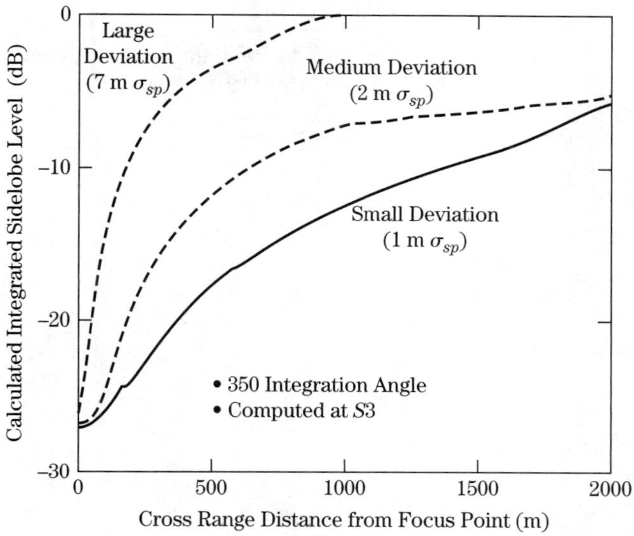

FIGURE 4–28

Motion deviation effects on integrated sidelobe level © 1999 IEEE [20]

$\Delta\theta$ by using the relation [20]

$$L_{out} = R_0 \sqrt{\frac{2\lambda_c \Phi_{\max}}{\pi \delta Y \cos^3(\Delta\theta/2)}} \tag{4.52}$$

The example illustrated in Figure 4–28 is for the 35 degree integration angle in the P-3 Image, with three examples of deviation shown (small $\sim$ 1 m; medium $\sim$ 2 m; and large $\sim$ 7 m).

The contribution of RMS deviation from a straight line to the extent of the focused scene is important. The data were fitted to the inverse square root of the deviation by two parameters. Values of these proportionality constants will serve as an analytic tool for specifying both the MOCOMP processing error and the size of the scene that will be appropriately compensated by the RMA algorithm.

If the deviation is known, then a modification of the RMA algorithm has been developed to compensate for the effects of systematic movement. However, the computations to estimate the error and recalculate the image are extensive. For nonreal-time operations, these computations have been successfully applied. The alternative would be to divide the image into subapertures and update the RMA image after forming the output scene length.

The points in Figure 4–28 fall along a curve that is approximately inversely proportional to the square root of the RMS motion deviation. The best-fit curve to the data was estimated from P-3 collection parameters as

$$L_{out} = \frac{\alpha}{\sqrt{\sigma_{SP}}} + \beta \tag{4.53}$$

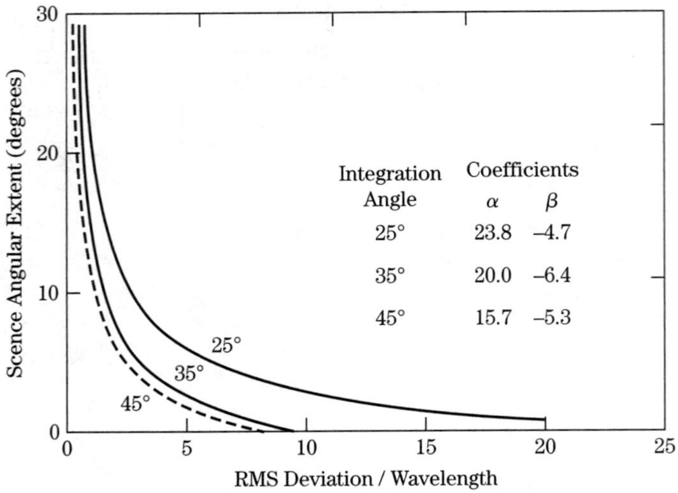

FIGURE 4–29

Curve fit to P-3 data for best image focus © 1999 IEEE [20]

The fit produces a good match to the data for σ_{SP} values of less than 5 meters. For σ_{SP} values that are larger than 5 meters, the best-fit curve significantly underestimates the achievable output scene length. When the output image length is below L_{out}, the focus will be within the design guidelines. This provides an estimate of how many subimages are required in the synthetic aperture collection to maintain the RMA algorithm focus over both range and cross-range.

To provide a general result that can be used in wide-angle FOPEN SAR design, the output scene length is expressed in terms of its angular extent:

$$\Delta\theta_{SL} = 2\tan^{-1}(L/2R_0) \tag{4.54}$$

The angular extent of the output scene was computed for integration angle of 2 degrees, 35 degrees, and 45 degrees. Based on 60 passes of P-3 UWB data, best-fit curves for each integration angle were generated. These data are summarized in Figure 4–29. This figure provides design data to illustrate how the angular extent of the output scene depends on the deviation of the platform motion from a straight line and on the SAR integration angle and wavelength.

4.5 References

[1] Carrara, W. G., Goodman, R. S., and Majewski, R. M., *Spotlight Synthetic Aperture RADAR*, Artech House, Norwood, MA, 1995.

[2] Soumekh, M., *Synthetic Aperture RADAR Signal Processing with MATLAB Algorithms*, Wiley, New York, 1999.

[3] Hellsten, H., "CARABAS—An UWB Low Frequency SAR," *IEEE MTT-S International* Vol. 3, June 1–5, 1992, pp. 1495–1498.

[4] Moyer, L. R., Technology Services Corporation, Silver Springs, MD, February 2010, private communication.

[5] Davis, M. E., Tomlinson, P. R., and Maloney, R. P., "Technical Challenges in Ultra-Wideband RADAR Development for Target Detection and Terrain Mapping," *Proc. 1999 IEEE RADAR Conference,* Boston, MA, pp. 1–6.

[6] Caputi, W. J., "Stretch: A Time-Transformation Technique," *IEEE Trans. AES* Vol. AES-7, No. 2, March 1971, pp. 269–278.

[7] Martinez, D.R., Bond, R. A., and Vai, M. M., *High Performance Embedded Computing Handbook- A Systems Perspective*, CRC Press, Boca Raton, FL, 2008, p. 157.

[8] Vickers, R., Gonzalez, V. H., and Ficklin, R. W., "Results from a VHF Impulse Synthetic Aperture RADAR," *Proc. SPIE* Vol. 1631, Bellingham, WA, 1992, pp. 219–226.

[9] Sheen, D. and Lewis, T. B., "P-3 Ultra-Wideband SAR," *Proc SPIE* Vol. 2747, Orlando, FL, April 1996.

[10] Hensley, S. and Wheeler, K., "The GEOSAR Mapping Instrument," Ultra Wide Band Mapping Conference, NASA Jet Propulsion Laboratory, Pasadena CA, September 28, 1999.

[11] Hellsten, H. and Andersson, L. E., "An Inverse Method for the Processing of Synthetic Aperture RADAR Data," *Inverse Problems* Vol. 3, IOP Publishing, London, UK, 1987, pp. 111–124.

[12] Skolnik, M. I., *RADAR Handbook* (2d ed.), McGraw Hill, New York, 1990, p. 10.31.

[13] Goldman, A., Werness, S. A., Stuff, M., DeGraaf, S., and Sullivan, R., "Radio Frequency Interference Removal in a VHF/UHF Deramp SAR," *Proc. SPIE* Vol. 2487, Orlando, FL, April 1995, pp. 84–95.

[14] Yegulalp, A. R., "Fast Backprojection Algorithm for Synthetic Aperture RADAR," *Proc 1999 IEEE RADAR Conference,* Boston, MA, April 1999, pp. 60–65.

[15] Hunter, A. J., Hayes, M. P., and Gough, P. T., "A Comparison of Fast Factorised Back-Projection and Wavenumber Algorithms for SAS Image Reconstruction," *Proc. of the World Congress on Ultrasonics,* Paris, France, September 2003.

[16] Cafforio, C., et al., "SAR Data Focusing Using Seismic Migration Techniques," *IEEE Trans on Aerospace and Electronic Systems* Vol. 27, No. 2, March 1991, pp. 194–206.

[17] Goodman, R., Tummala, S., and Carrara, W. F., "Issues in Ultra-Wideband, Widebeam SAR Image Formation," *Proc 1995 International RADAR Conference* Washington, DC, May 1995.

[18] Carrara, W., Tummala, S., and Goodman, R., "Motion Compensation Algorithms for Widebeam Stripmap SAR," *Proc. SPIE* Vol. 2487, Orlando, FL, April 1995, pp. 13–23.

[19] Kirk, J. C., "Motion Compensation for Synthetic Aperture RADAR," *IEEE Trans on Aerospace and Electronic Systems* Vol. AES-11, No. 3, May 1975, pp. 338–348.

[20] Kirk, D. R., Maloney, R. P., and Davis, M. E., "Impact of Platform Motion on Wide Angle Synthetic Aperture RADAR Image Quality," *Proc 1999 IEEE RADAR Conference,* Boston, MA, April 1999, pp. 41–46.

[21] Kirk, D. and Maloney, R. P., "Autofocus Techniques for Wide-Band Wide Angle Synthetic Aperture RADAR," *Proc SPIE* Vol. 3370, Orlando, FL, April 1998.

[22] Kirk, J. C., "A Discussion of Digital Processing in Synthetic Aperture RADAR," *IEEE Trans on Aerospace and Electronic Systems* Vol. AES-11, No. 3, May 1975, pp. 326–337.

CHAPTER 5

Radio Frequency Interference Suppression

Section 3.8 presented the problem of the radio frequency interference (RFI) environment in ultra wideband (UWB) RADAR operation from the standpoint of background clutter. This chapter will cover the significant design factors needed to operate a foliage penetration (FOPEN) RADAR, in either synthetic aperture RADAR (SAR) or ground moving target indicator (GMTI) modes, and under most operational conditions. Because of the worldwide regulation of the RF spectrum for telecommunications and active sensing operation, it is necessary to accommodate the limits that are imposed on power spectral density for transmission of the FOPEN UWB waveform. Equally as important is the need to remove the dense, high-power transmissions that will be intercepted by the wide beamwidth SAR antenna and that affect the dynamic range of the signal received by the SAR and the image formation processing. An example of this dense interference, along with the general sources of RF energy in the Adelphi, Maryland, area, is given in Figure 5–1 [1].

There are two parts to the RFI environment: (1) the bands that must be avoided due to federal regulations; and (2) the frequencies that represent strong interference to the FOPEN RADAR. It is evident that no contiguous part of the spectrum is allocated for the FOPEN SAR systems that have been characterized with the UWB designation. The proscribed transmit frequencies, which are regulated by the Federal Communications Commission (FCC) and the National Telecommunications and Information Administration (NTIA) in the United States, have strict requirements on the bandwidth and power spectral density of any UWB system operating in the environment. The International Telecommunications Union (ITU) imposes similar requirements in most of the world, affecting the geographic mobility of any FOPEN system. During the receive function, radio and television transmissions are particularly detrimental to the SAR image quality. As can be seen in Figure 5–1, the

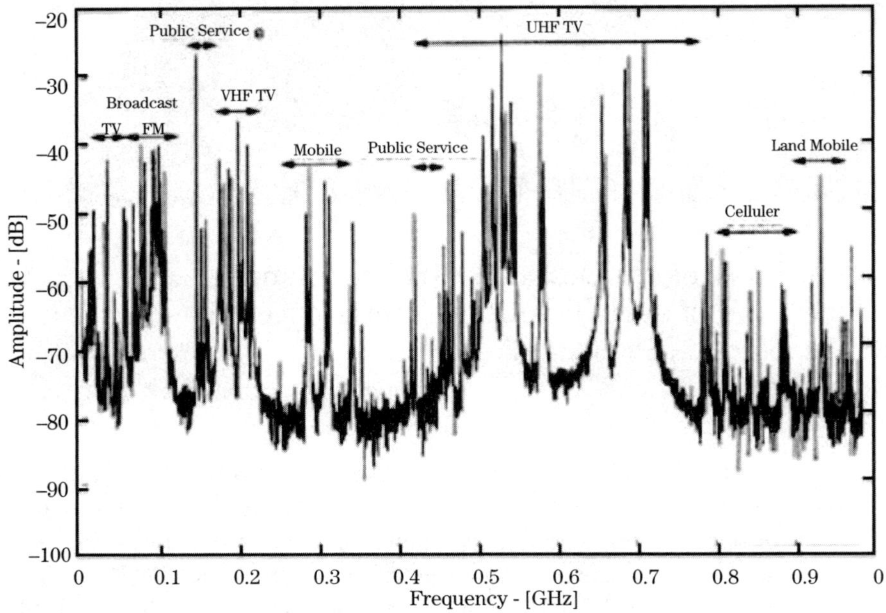

FIGURE 5–1

Typical radio frequency spectrum affecting FOPEN SAR design [1]

narrowband amplitude or frequency modulated (AM and FM) transmissions are from 20 to 60 dB above the noise level, as measured by the spectrum analyzer on the received RADAR signal. These interference sources must be removed to provide the necessary sensitivity for either the characterization of terrain features or the detection of targets under the foliage.

Chapter 2 covered the mixture of FOPEN SAR experimental sensors and platforms and illustrated the variation in designs for transmit waveforms. Most of the early experimental FOPEN RADARs either ignored the issue of interfering with sensitive or emergency receiver frequencies or avoided them by limiting bandwidth or segmenting the frequency coverage. However, there was significant scrutiny of all UWB sensor operation during the growth in WiFi system development in the late 1990s. As a result, the number of choices in UWB RADAR transmitter design was limited, and the spectral management design needs to be incorporated in any new system development. Section 5.1 will cover the details of these choices and their impact on the RADAR waveform and transmitter design [2].

The NTIA regulations on UWB transmission specify the power spectral density that can be transmitted (in the United States) by an unrestricted intentional radiator, within the fundamental frequencies, and for any out-of-band harmonics. Furthermore, the NTIA requires that emissions from any UWB

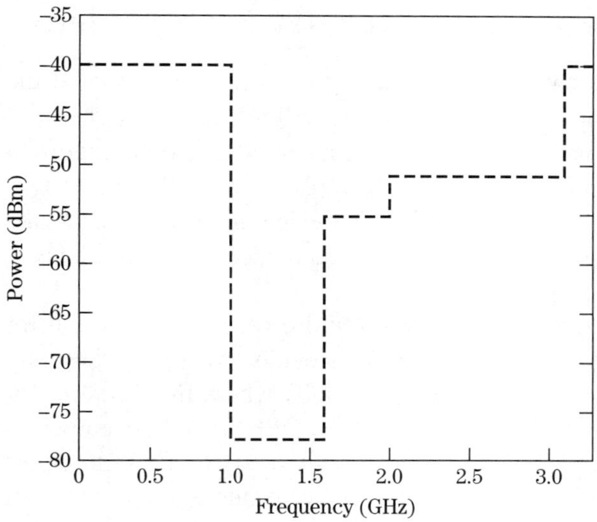

FIGURE 5–2

NTIA frequency mask for Part 15 compliance [2]

transmitter operating under the Part 15 provisions shall not exceed an equivalent isotropically radiated power (EIRP) density level over the frequency bands as shown in Figure 5–2. The power density, which is specified at a 3 meter distance and in a 1 MHz bandwidth, serves as the top-level requirement for qualifying the UWB device for unrestricted operation [2]. Since these power levels are significantly lower than required by any FOPEN RADAR, they should be used only as a first approximation for calculating the interference level at a remote receiver. Section 5.1 covers the measurement approaches for evaluation of a FOPEN RADAR and discusses the alternative mitigation techniques needed to minimize interference on a sensitive victim receiver. Although the regulations are different under ITU, this example of a signal power spectral density measurement will illustrate the approach for obtaining a license for commercial operation.

The preponderance of very high frequency (VHF) and ultra high frequency (UHF) television and radio stations has always limited the ability to communicate or sense in these RF spectrum bands. It is not possible to avoid these interference (i.e., they are actually jamming) sources because of the spatial and spectral density of the emissions. As a result, techniques were developed to remove the background interference by waveform design and adaptive processing techniques. Section 5.2 summarizes the approaches used in waveform design and illustrates the adaptive processing techniques used to remove the majority of the RFI energy from the SAR images. More recently, there has been a conversion to digital television, along with the emerging development of cognitive radios for cellular and personal communications. As a result, the efficient and real-time removal of RFI remains a continuing research topic in UWB SAR system design.

5.1 Transmit Waveform Design for RFI Environment

This section will treat the problems of avoiding transmitting signals at the proscribed frequencies, and at power levels that will affect the intercept receiver sensitivity. The avoidance of critical frequencies is a very important part of FOPEN RADAR development; since most areas in the world will not provide frequency allocation for any UWB device, unless it satisfies the NTIA guidelines on spectrum compliance or the equivalent ITU requirements in most developed nations [2].

Table 5–1 lists the frequency bands that require particular attention for transmission avoidance in UWB operation. It is evident that there are only a very few "empty" spaces in VHF or UHF bands where the freedom for complete operation can be achieved. Early in the FOPEN SAR development process, signal intercept collections and analyses were carried out to determine the extent of the problem. If the problem were one of removing RFI alone, there would not be a frequency allocation problem. However, to get a license to operate a FOPEN SAR in most of the developed world, there is a strict requirement to avoid sensitive frequencies that affect both civilian and government communications functions.

To satisfy the NTIA and obtain a license to operate the UWB SAR system, it is necessary to carry out the following analytic processes:

- Identify the geographic region of operation.

Table 5–1 VHF/UHF frequency bands requiring transmitter avoidance [2]

Function	Freq [MHz]	Function	Freq [MHz]
Radio Astronomy	37.5–38.25	AIS (Automatic Identification Service)	162.0125–167.17
Radio Astronomy	73–74.6	Fixed Mobile, Public Safety, Forest Fighting	167.72–173.2
Aeronautical Radionavigation	74.8–75.2	Fixed Mobile, Emergency	240–285
Aeronautical Radionavigation	108–121.94	Fixed Mobile, Radio Astronomy	322–335.4
Aeronautical Mobile Location	123–138	Radio Navigation Satellite	399.9–410
Mobile Satellite (earth to space)	149.9–150.05	Radio Astronomy	608–614
Maritime Mobile Safety	156.52475–156.52525	Aeronautical Radio Navigation Satellite	960–1240
Maritime Mobile–Distress	156.7–156.9	Radio Navigation Satellite	1300–1427

- Determine the location of all receivers that could be affected by the RADAR.
- Calculate the line-of-sight range from the FOPEN SAR to the intercept receiver along with an assessment of the intercept power.
- If the intercept power exceeds the levels indicated in Figure 5–2 or the receiver has a lower intercept power density, the waveform must be modified to reduce the power spectral density to the receiver.

There are two fundamental waveform approaches to frequency avoidance:

- *Frequency jump burst (FJB) waveforms:* A UWB RADAR employing FJB uses narrowband subpulses and jumps from one frequency subband to the next to avoid the proscribed frequencies. This approach is preferred for systems that can operate in limited spectral coverage or are designed with limited component instantaneous bandwidth. The coherent all radio band sensing (CARABAS) RADAR is an excellent example of an early UWB system employing this approach [3].
- *Notched linear FM (LFM):* To provide fine range resolution, a single LFM pulse is transmitted with the required wide bandwidth. However, narrow notches are required to be inserted into the spectrum. These notches are controlled to meet the power spectral density at protected frequencies. This approach provides significant advantage in operation where the controlled spectrum usage can be updated with rapid digital control, and software controls the waveform generation and signal processing. Both the Environmental Research Institute Michigan (ERIM) UWB P-3 and JPL GeoSAR employed this frequency avoidance technique [4,5].

The system impact to each FOPEN SAR waveform, from the percentage of the spectrum omitted, will result in an increase in the waveform sidelobe levels. From simple array theory, if 10% of the spectrum is randomly missing the integrated sidelobe ratio (ISLR) will be degraded to only –10 dB [6]. However, if the distribution of the missing frequencies is not random, the impact on the SAR performance can be significant. Alternatively, techniques of matched signal processing combined with adaptive processing can restore some of the spectral content, albeit with a decrease in the signal-to-noise ratio (SNR). Specific analysis of these effects will be needed for each system application.

5.1.1 Intercept of FOPEN Signal by Receiver

To obtain an operating license for any UWB RADAR system, a system developer is required to have the NTIA assess the amount of interference the RADAR will impart on any existing system, operating within the same frequency band. The NTIA requirement on received signal levels is expressed in terms of the frequency mask shown in Figure 5–2. This is predominantly a case of RADAR intercept analysis by a sensitive receiver. The key parameter to be considered is the signal-to-noise ratio SNR_I at a victim (intercept) receiver [7]:

$$SNR_I = \frac{\hat{P}_T G_{TI} G_I \lambda_T^2 L_t}{k_B T_0 N F_I B_I (4\pi R_{TI}^2)} \left(\frac{\Gamma_{IP}}{\Gamma_{IPN}} \right) \tag{5.1}$$

where

$\hat{P}_T$	Peak RADAR emitted power
G_{TI}	Gain from the transmitter toward the intercept receiver
G_I	Gain on the victim receiver
λ_T	Wavelength transmitted
L_I	Loss between the RADAR and the victim receiver
$k_B T_0 NF_I B_I$	Noise bandwidth at the receiver
R_{TI}	Range from the transmitter to the receiver
$(\Gamma_{IP}/\Gamma_{IPN})$	Receiver processing gain with respect to noise, plus losses

The key factor in (5.1) that affects the ability to meet the NTIA Part 15 specification is the victim receiver processing gain $(\Gamma_{IP}/\Gamma_{IPN})$. Without detailed knowledge of the receiver spectral and signal processing characteristics, the intercepted signal can be estimated only by the ratio of the bandwidth between the transmit signal and the published receiver bandwidth. The most conservative approach would be to give the victim receiver a coherent processing gain within its bandwidth for the UWB transmitted signal. The victim receiver's intercept signal could then be related to the received power P_I and the antenna gain G_I through the transmitted power-gain $\hat{P}_T G_{TI}$ and the propagation loss between the two antennas:

$$S_I = \frac{P_I}{G_I} = \frac{\hat{P}_T G_{TI} \lambda_T^2 L_I}{(4\pi)^2 R_{TI}^2} \left(\frac{B_I}{B_T} \right) \tag{5.2}$$

This system sensitivity represents the power received by a victim receiver at its location, such that the received power density results in the required detection SNR. Since the SAR signal is swept across a bandwidth $B_T > B_I$, the intercepted receiver power density is approximately reduced by the ratio (B_I/B_T). This is only an approximation since the filter time constants in the intercept receive have a specific, but unknown, impact on the signal reaching the receiver detector.

If the system sensitivity levels from the NTIA Part 15 frequency mask are used, a first-order assessment of the interference ranges from a candidate SAR system can be calculated [2]. Given a FOPEN RADAR with 1,000 watt peak power, 200 MHz bandwidth, and a 10% duty factor, the sensitivity at a receiver can be calculated. This is shown in Figure 5–3, where the antenna

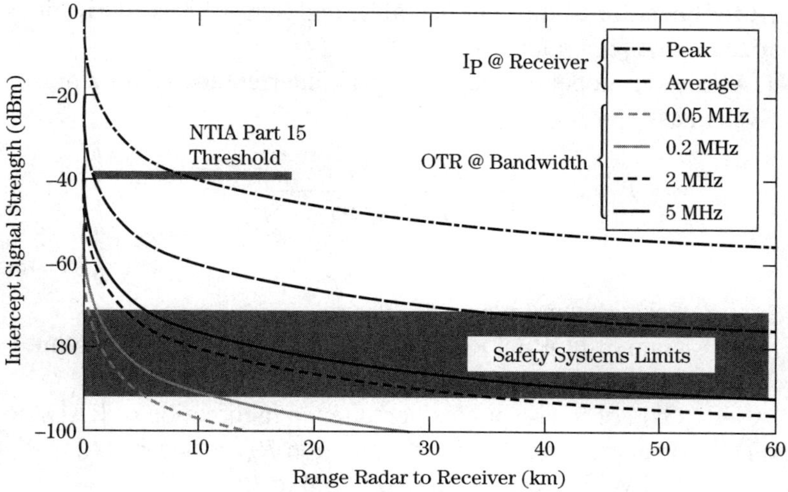

FIGURE 5-3 Intercept power of RADAR pulse transmission as function of range © 2010 IEEE [8]

gain is assumed to be 6 dBi and is pointed directly at the victim receiver. Various bandwidths of intercept receivers between 50 KHz and 5 MHz are evaluated.

The power received is calculated based on the ratio of the victim receiver and the total bandwidth of the SAR LFM signal bandwidth. If the SAR collection plan is closer than 2.0 Km for the 5 MHz receiver, the RADAR will violate the FCC Part 15 specification. Moreover, it is important to note that many of the sensitive receivers have EIRP avoidance thresholds lower than Part 15—often as low as –90 dBm. Because of these receiver intercept requirements, a different model of the RADAR waveform is required.

There are major differences between a high time bandwidth (BT) product waveform's effect on the intercept receiver signal as given in (5.2) and the classic NTIA receiver model. These differences are due to:

1. The pulsed nature of the FOPEN waveform—the signal is changing frequency rapidly to satisfy the high BT product, within the pulsewidth τ.
2. The very small time that the FOPEN SAR signal is in the passband of the victim receiver, which can limit the signal interference.
3. Deviations between a narrowband receiver model and the actual receiver passband [9].

The next section will present the NTIA receiver model that is used to evaluate licensing requirements for a FOPEN SAR.

5.1.2 NTIA Receiver Model

The derivation of temporal and spectral effects of a UWB RADAR on a victim receiver was carried out by NASA Jet Propulsion Laboratory, while qualifying the GeoSAR for operation in the continental United States. These derivations are summarized in the paragraphs below [9].

The NTIA receiver model evaluates the peak interference I_p for a pulsed waveform system as

$$I_P = \frac{\hat{P}_T G_{TI} G_I \eta_I}{L_b(R_{RI}^2) FDR(\Delta f) \eta_0(\Delta f)} \tag{5.3}$$

where

$\hat{P}_T$	Peak transmit power
G_T	Gain of FOPEN RADAR antenna in direction of victim receiver
G_I	Gain of intercept receiver antenna in direction of RADAR
$L_b(R_{\mathrm{RI}})$	Transmission loss for a separation R_{RI} between RADAR and victim receiver, proportional to R_{RI}^2
η_I	Ratio of average peak power of the transmitter $\overline{P}_I / \hat{P}_I$, where $\overline{P}_I$ is the average transmit power
$\eta_0(\Delta f)$	Duty cycle of the response waveform induced by the interfering signal in the receiver bandwidth
$FDR(\Delta f)$	Frequency-dependent rejection measured by the NTIA standard receiver

The FDR is calculated by either the on-tune rejection (OTR) or the off-frequency rejection (OFR). The appropriate measure depends on whether there is an overlap with the RADAR bandwidth B_T centered at frequency f_c, with the victim receiver centered at f_{cI}. This relationship is given by

$$FDR(\Delta f) = \begin{cases} OTR & |f_c - f_{cI}| \leq \frac{B_T}{2} + B_I \\ OFR(\Delta f) & otherwise \end{cases} \tag{5.4}$$

Thus, OTR is used when there is an overlap in bandwidths and OFR when there is no overlap as shown in Figure 5–4.

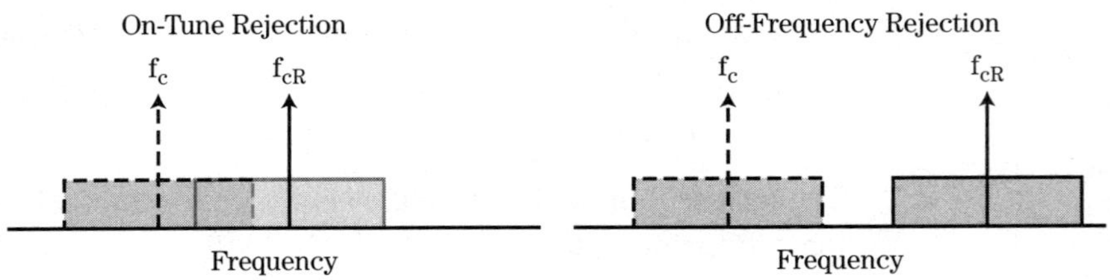

FIGURE 5–4

Definition of On-Tune Rejection and Off-Tune Rejection
Source: NASA Jet Propulsion Laboratory [9]

It is often assumed that average interference level is more applicable to continuous interfering signals; whereas the peak interference level is appropriate for pulsed signals. This assumption is not necessarily valid for a large class of intercept receiver systems without some modification to the standard peak interference model. More importantly for UWB chirp signals, the rapid change in frequency of B_T/τ will have a reduced impact on the victim receiver depending on its transfer function.

The average and peak power from the FOPEN RADAR, expressed as an input to the victim receiver, are defined respectively by

$$\bar{P} = \frac{1}{2T}\int s(t)^* s(t)\,dt = \frac{1}{2T}\int S(\omega)^* S(\omega) d\omega \tag{5.5}$$

$$\hat{P}_I = \max(P(t)) \quad t \in [-T, T] \tag{5.6}$$

where T is the duration of the transmit signal, $s(t)$ is the RADAR temporal waveform, and $S(\omega)$ is the Fourier transform of $s(t)$.

To analyze the extent that the transmit signal affects the intercept receiver, one needs to characterize the receiver transfer function $h(\omega)$ in terms of the time domain output of the victim receiver. The average and peak power output of the victim receiver are given respectively by

$$\overline{P_0} = \frac{1}{2T}\int_{-\infty}^{\infty} S(\omega)h(\omega)(S^*(\omega)h^*(\omega))d\omega = \frac{1}{2T}\int_{-\infty}^{\infty} P(\omega)H(\omega)d\omega \tag{5.7}$$

$$\hat{P}_0 = \max(y(t)y^*(t)) \quad \text{where} \quad y(t) = \int_{-\infty}^{\infty} S(\omega)h(\omega)e^{j2\pi\omega t} \tag{5.8}$$

From the previous definition of OTR combined with (5.7) and (5.8), we have the resulting relationship

$$\frac{\eta_I}{FDR\eta_0} = \frac{\hat{P}_0}{\hat{P}_I} = \max_{t\in[-T,T]}\left(\left|\int_{-\infty}^{\infty} S(f)h(f)e^{2\pi ift}\,df\right|^2\right) \tag{5.9}$$

Thus, the peak response is a function of both the victim receiver transfer function and the RADAR waveform. Consequently, any evaluation of the peak interference for a given system amounts to analyzing the peak of the convolution of the receiver transfer function with the RADAR waveform.

The relationship of pulse width τ and the PRF, defines the duty cycle of the RADAR waveform as

$$\eta_I = \frac{\tau}{2}PRF = \frac{\tau}{2T} \tag{5.10}$$

Based on Figure 5–4, the victim receiver frequency response (or transfer function) $h(\omega)$ can be approximated by

$$h(\omega) = rect_{BI}(\omega - \omega_I) + rect_{BI}(\omega + \omega_I) \tag{5.11}$$

where ω_I and B_I are the receiver center frequency and bandwidth, respectively. Furthermore, the total power of a single pulse in the receiver's output can be expressed from the pulse width and the ratio of the transmit and receiver bandwidths as:

The total power of a single pulse in the receiver's output is

$$P_0 = \frac{B_I}{B_T}\frac{\tau}{2} \tag{5.12}$$

In analyzing impact of an UWB signal on a victim receiver, there are two cases of interest: (1) the victim receiver has a large bandwidth compared with the RADAR bandwidth ($B_I \gg B_T$); and (2) the victim receiver has a small bandwidth compared with the RADAR bandwidth ($B_I \ll B_T$).

Using the above definitions of peak and average power, the output duty cycle can be expressed as

$$\eta_0 = \frac{\overline{P}_0}{P_0} = \begin{cases} \dfrac{B_I}{B_T}\dfrac{\tau}{2}PRF & \text{for } \dfrac{B_I^2\tau}{B_T} \gg 1 \\ \dfrac{PRF}{2B_I} & \text{for } \dfrac{B_I^2\tau}{B_T} \ll 1 \end{cases} \tag{5.13}$$

Since the center frequency of the receiver is most likely contained within the bandwidth of the UWB transmitting chirp, FDR will be determined by the on-tune rejection. OTR is the ratio of the input average power to the output average power and is expressed as

$$OTR = \frac{\overline{P}_I}{\overline{P}_0} = \frac{B_T}{B_R} \tag{5.14}$$

The comparison of peak output power can similarly be expressed for the two cases as

$$\frac{\eta_I}{FDR\,\eta_0} = \overline{P}_0 = \begin{cases} 1 & \text{for } \dfrac{B_I^2\tau}{B_T} \gg 1 \\ \dfrac{B_I^2\tau}{B_T} & \text{for } \dfrac{B_I^2\tau}{B_T} \ll 1 \end{cases} \tag{5.15}$$

During the process of obtaining a license to operate GeoSAR, NASA Jet Propulsion Laboratory ran a series of tests at the Van Nuys California Airport [9]. These tests were witnessed by Federal Aviation Administration (FAA) personnel from the Los Angeles airport and the frequency spectrum manager from Edwards Air Force Base. An FAA standard receiver was used as the victim receiver, with bandwidth of 23 KHz (representative of the FAA air-to-ground communicator) centered at 317.5 MHz and a threshold (squelch break) of –98 dBm. The GeoSAR waveform used consisted of a 40 μsec pulse width at 350 MHz center frequency and 160 MHz bandwidth.

It was observed that the FAA receiver would not reach its detection threshold until –24 dBm of received power, corresponding to 74 dB of signal attenuation. The attenuation of the GeoSAR signal, as seen by the FAA standard receiver, was due to a combination of the short pulse width attenuation (38 dB) and the wide bandwidth mismatch compared with the victim receiver (36 dB). These test conditions are representative of the OTR from (5.13) and the signal attenuation corresponding to $B_I^2\tau/B_T \ll 1$ in (5.15).

The analysis of the RADAR specified UWB waveforms in terms of several classes of victim receivers is conducted by NTIA to obtain a license to operate. The policy to date has been to allow only an experimental license for FOPEN SAR systems to operate in the US and to require that the collection plan and frequency avoidance be filed prior to operation. The measurement of UWB transmitter operation against the majority of commercial receivers has been successful (i.e. no measurable interference), largely because of the extremely fast LFM sweep and the narrowband receivers. However, for the sensitive receivers associated with safety of flight and emergency operations, the NTIA requires that the waveforms be modified to limit the power spectral density for discrete frequencies and bandwidths. The following two sections details two approaches successfully used to avoid these frequencies.

5.1.3 Frequency Jump Burst

FJB scheduling of transmit waveforms has been used on many systems to achieve higher range resolution than the hardware would normally allow. The approach is to transmit a subpulse at one frequency and bandwidth, receive the returns, and then transmit on the next frequency, and so forth. By digitizing the subpulse returns, a full UWB signal bandwidth can be coherently constructed at the expense of increased signal processing over LFM. For example, the Swedish CARABAS system uses up to 37 frequency bursts of 1.875 MHz each to cover the 20 MHz to 90 MHz frequency range. The added benefit of the narrowband subpulses is the use of much wider dynamic range analog-to-digital converters (ADC), up to 14 bits in the case of CARABAS [10].

The problem that FJB presents for a wide swath SAR system is the limitation on slant-range extent for wide-swath system operation and the potential for range ambiguities affecting the image quality. If the pulse repetition interval (PRI) is too short for the maximum range to the swath, the receiver must be "opened" at the appropriate time for scheduled returns. If the system requires wide swath operation, the desired range extent may be longer than the sub-pulse interval, causing an eclipsing loss in returned signal. Consequently, a requirement on simultaneous transmit and receive is imposed, which greatly complicates the design of the RADAR receiver and associated receiver input protection.

When there is a requirement for avoiding proscribed frequencies, those frequencies can be eliminated in the FJB waveform formation. However, that process leaves "holes" in the spectrum that will raise the range sidelobes. More importantly, it is often difficult to control the spectral purity near the pulse edges; and hence a wider avoidance of spectra is needed.

If there are no critical frequencies to be avoided within the total bandwidth B, equal subpulse bandwidths can be used. A generalized sequence of pulses can be described, where each subsequent subpulse is a regular sequence of both center frequency and time delay. The bandwidth B_M for each of M_{SB} subpulses is given by [11]

$$B_M = \frac{B}{M_{SB}} \tag{5.16}$$

The frequency and timing conditions for a generalized FJB sequence is illustrated in Figure 5–5 as

- PRI_{ij} Transmit at the i-th PRI window for the FJB waveform and receive on the j-th window
- f_{cm} Center frequency at the m-th frequency burst
- B_M Bandwidth of the m-th frequency burst

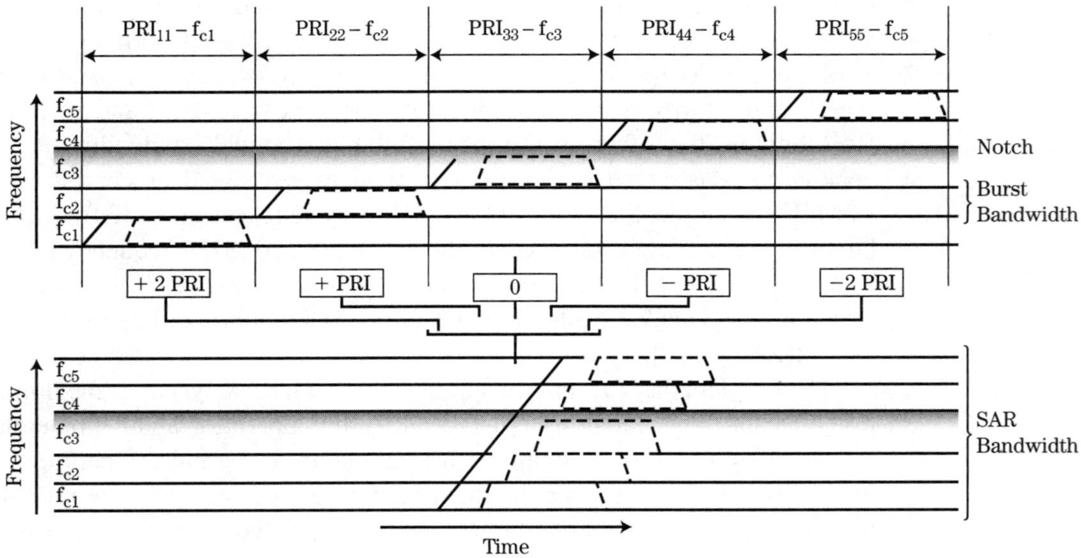

FIGURE 5–5

FJB waveform sub-pulse temporal alignment

For ease of analyzing the FJB waveform, each subpulse is assumed to transmit an equal bandwidth B_M over an equal subpulse width T_{PM}. Furthermore, the LFM chirp rate γ is preserved on each subpulse. The chirp rate is given by

$$\gamma_M = \frac{B_M}{T_{PM}} = \frac{B}{T_p} = \gamma \tag{5.17}$$

For the case of a short-range SAR, as shown in Figure 5–5, the receiver can be opened up for a subpulse period of τ_{SP} and can receive the imaging pulse over a range extent of

$$R_{SP} = c\tau_{SP}/2 \tag{5.18}$$

This receive window is scheduled with a short delay from the transmit pulse to allow for transmit-receive switching. This also provides receiver blanking, covering the ground bounce and covering the minimum slant range from the RADAR to the near range of the imaging swath. If the image window has a short-range delay after the transmit pulse, the control and timing of the waveform is simplified.

Figure 5–5 also illustrates the timing for a five-frequency FJB waveform; along with the associated receive windows for collecting SAR signal returns. For simplicity only five sub-pulses are shown within the bandwidth B required for the range resolution.

The FJB waveform signal characteristics can be developed based on the LFM processing presented in Section 4.2.1. The LFM pulse width T_p is divided into M subpulses of width T_{PM}, where

$$T_{PM} = \frac{T_p}{M} \tag{5.19}$$

and transmitted at a fraction of the bandwidth B_M per subpulse. The return from each subpulse is received during fractional PRI interval [11]

$$T_M = \frac{T}{M} \tag{5.20}$$

Based on these bandwidth and timing conditions, the LFM pulsed waveform phase function, from (4.15), has the modified form

$$s_{FJB}(x, m) = rect\left(\frac{t}{T_{Pm}}\right) \exp\left[j\omega_c(m)t + j\frac{2\pi B_M}{T_{PM}}t^2\right] \tag{5.21}$$

The modulation frequency for each subpulse will have a separate center frequency, which is given by [8]:

$$\omega_{CM}(m) = \omega_C + 2\pi B_M\left(m - \frac{M+1}{2}\right) \tag{5.22}$$

In order to correctly align the sub-pulses, each subpulse must be time delayed by an incremental amount given by

$$\Delta t(m) = T_{PM}\left(m - \frac{M+1}{2}\right) \tag{5.23}$$

before it can be compressed. The time delays corresponding to (5.23) are also illustrated in Figure 5–5, and represent integral multiples of sub-PRI.

In addition to time delay, there is a phase delay for each subpulse required to compensate for the quadratic phase change in the chirp waveform, as presented in (5.5) [8]:

$$\Phi_{SP}(m) = \exp\left[j\pi\gamma_p\left(T_{PM}\left[m - \frac{M+1}{2}\right]\right)^2\right] \tag{5.24}$$

If these time delays and phases are properly applied and the phase due to platform motion between subpulses is correctly compensated, the compressed FJB waveform will correspond to that of LFM.

The systems advantage of this approach lies in the use of narrow bandwidths within the subbands B_M compared with the UWB LFM bandwidth of B. Significantly higher dynamic range exists for ADC at these lower sampling frequencies, as was shown in Figure 4-5. However, the data collected during a subpulse repetition interval will be correspondingly reduced by the same factor M. This reduction in collection window will reduce the available range swath. In a dense RFI environment, these trades have been rewarding with improved SAR image quality [12].

The subpulses are aligned on receive when compensating for the individual pulse repetition intervals and the phase advance between subpulses. The spectral content of each subpulse is shown in Figure 5–6 along with the time alignment shown in Figure 5–5. The frequency and phase compensation between the subpulses have effectively reduced the phase discontinuity of

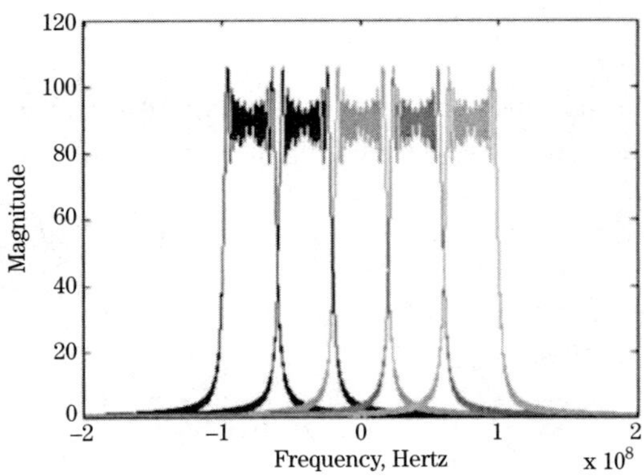

FIGURE 5–6

Spectral content of aligned FJB waveform with no frequency notching

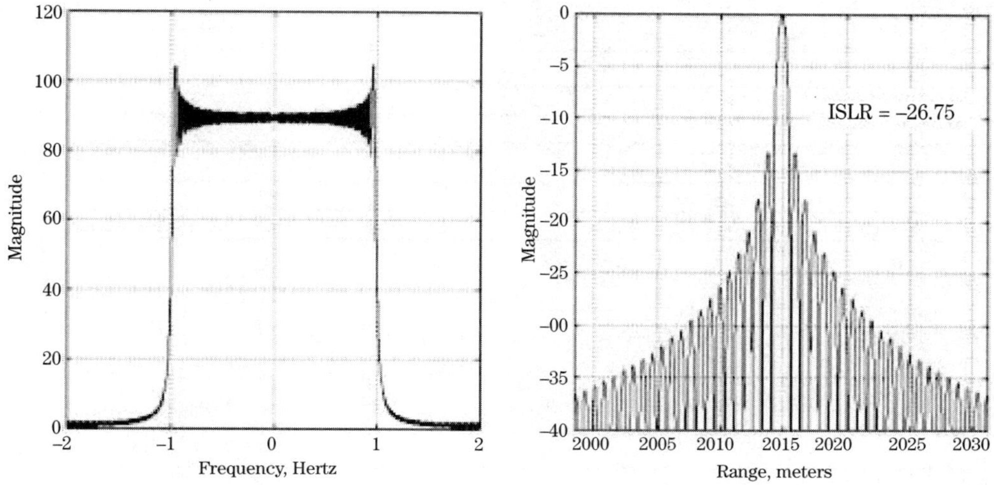

a. Pulse Spectrum b. Range Return

FIGURE 5–7

Combined spectral content of aligned FJB waveform with no frequency notching

combining the separate returns. Both the spectral content and the compressed range return from the five subpulses are shown in Figure 5–7. The waveform sidelobes are consistent with an unweighted pulse at the range resolution of the combined time bandwidth.

It is important to anticipate that strong clutter will exist beyond the short PRI window. The time delay from returns at each sub-pulse frequency can interfere with the system sensitivity if appropriate filtering is not included in either the IF or digital receiver designs. One design approach is to chose subsequent pulses with a wider separation of the frequencies, as illustrated in Figure 5–8a. Depending on the intermediate frequency (IF) band pass filtering and any time constants in the receiver, the center frequency of the next pulse should be separated from the previous pulse by multiples of B_{SP}. If this approach is taken, the representative time delay must be applied by choosing a different modulus in (5.23) for time delay and (5.24) for phase compensation.

For a long-range standoff SAR system, the range delay to the near edge of the swath may be more than a PRI of the m-th sub-pulse. Figure 5–8b illustrates the case where the receive window is opened on the second PRI of the collection timeline. Now the center frequency of both the subsequent PRI and the current PRI must be separated from the transmit pulse by B_{SP}. This approach will complicate time, frequency and phase compensation, and any subsequent image formation processing. However, it will significantly reduce clutter returns from adjacent sub-pulses.

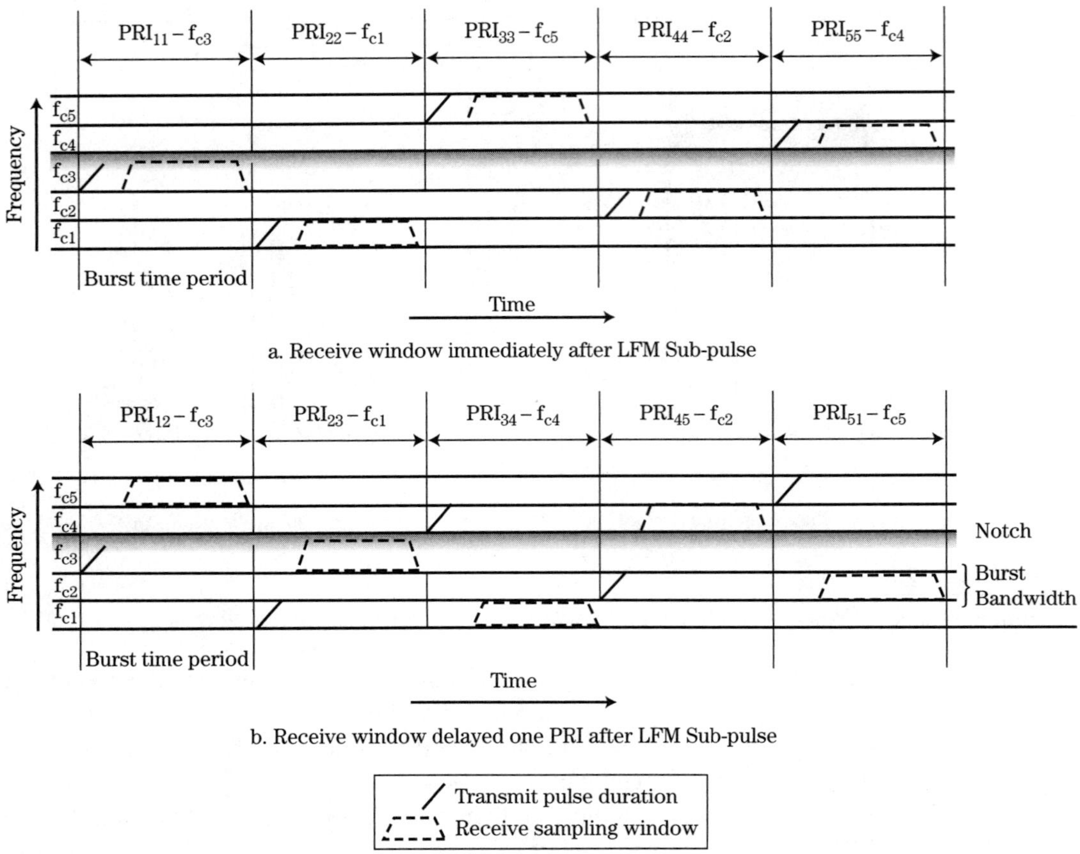

FIGURE 5–8

Alternative FJB temporal scheduling

The notching to avoid sensitive receivers has been used on a few systems [10, 12]. One technique for notching is to select the subpulses and leave gaps in the frequency by reducing the coverage of individual pulses. This is illustrated in Figure 5–9 for two gaps between five subpulses and a combined 8 percent reduction in the time bandwidth coverage The compressed pulse is shown in Figure 5–9b, after matched filter processing of the return and accounting for the missing frequencies in the waveform. There is a modest increase in ISLR due to these notches.

It should be noted that the Fresnel effects at the edges of the pulses directly limits the depth of notches. A wider separation between the frequencies in the subpulse scheduling is required than for a notched LFM waveform. Techniques for smoothing the amplitude and phase discontinuities across these

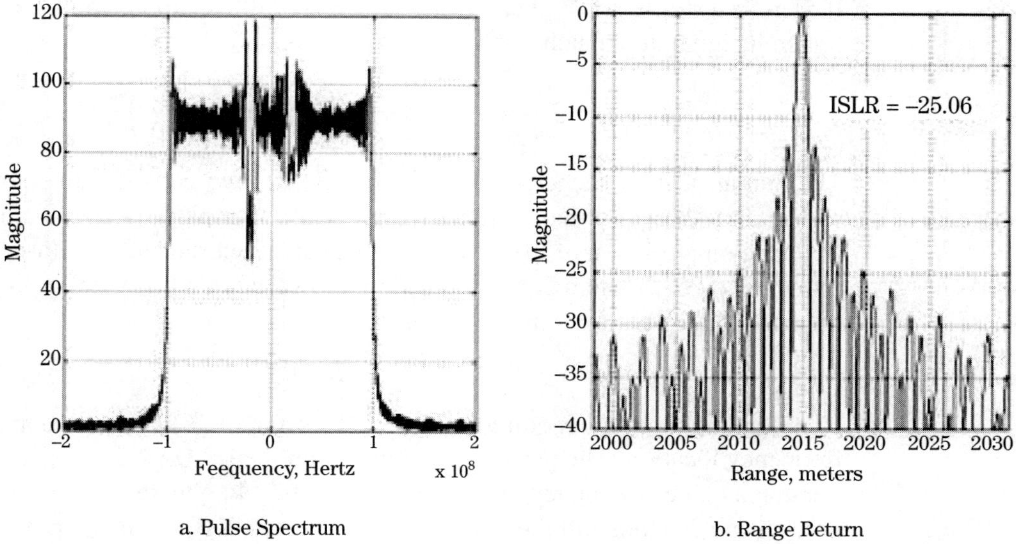

FIGURE 5–9

Combined FJB waveform with 8 percent frequency notching

gaps are required, which in turn make significant demands on the linearity and turn-on characteristics of solidstate transmitter amplifiers. For real operation in FOPEN SAR, the number and widths of these notches will be significantly larger, requiring improvements in range match filtering to restore ISLR characteristics.

5.1.4 Notched Linear FM Waveform

A number of approaches have been considered to obtain a notch in the transmit spectrum and to avoid interfering with existing spectral users. These approaches are implemented using either an analog waveform exciter or digital waveform synthesis. The ERIM P-3 UWB SAR, which was the first system that confronted the need for notching on transmit, used an analog waveform synthesis approach [13].

Because the frequencies are known a priori, a notch in the transmit spectrum can be constructed deterministically. The performance impact on the waveform is analyzed in terms of the LFM transmit waveform and the spectrum of the notch. The LFM waveform has a spectral response given by [14]

$$S(\omega) = rect\left[\frac{\omega}{2\pi B}\right] \exp\left(-j\frac{\omega^2}{4\pi\gamma}\right) \tag{5.25}$$

If there is a single notch at carrier ω_1 and bandwidth B_1 to be applied to the transmit signal, the matched filter in spectral domain will be

$$H(\omega) = S^*(\omega) - rect\left[\frac{\omega - \omega_1}{2\pi B_1}\right] \exp\left(j\frac{\omega^2}{4\pi\gamma}\right) \tag{5.26}$$

The output of the matched filter with a notch has two parts: (1) the desired signal; and (2) a perturbation to the signal. This latter part represents a high-frequency modulation of the waveform that can affect the clutter return. The influence on the system can be considered to be *clutter-like* with an impact on the signal-to-clutter ratio (SCR) of

$$SCR = 20 \log (B_1/B) \tag{5.27}$$

If multiple notches are required, the second term in (5.26) is repeated at each frequency location. The performance impact on the RADAR is a function of the notch center frequency, bandwidth, and depth. In general, as long as the notches are random and the sum of all the notches $\Sigma B_i \ll B$, there will be little impact on the range resolution. However, these notches will have a major impact on the waveform sidelobes, as will be illustrated below.

Figure 5–10 illustrates the generic process for waveform generation of an LFM signal and notch removal. The top of the diagram shows the LFM waveform synthesis in both the time and frequency domains. A desired notch is synthesized as a time-varying signal at the appropriate frequency within the pulse duration. The transformation of both LFM signal and the frequency notch to the frequency domain is performed, and the two signals are then subtracted. After transformation back to the time domain, the resultant output is the chirp signal with a narrow notch within the spectrum.

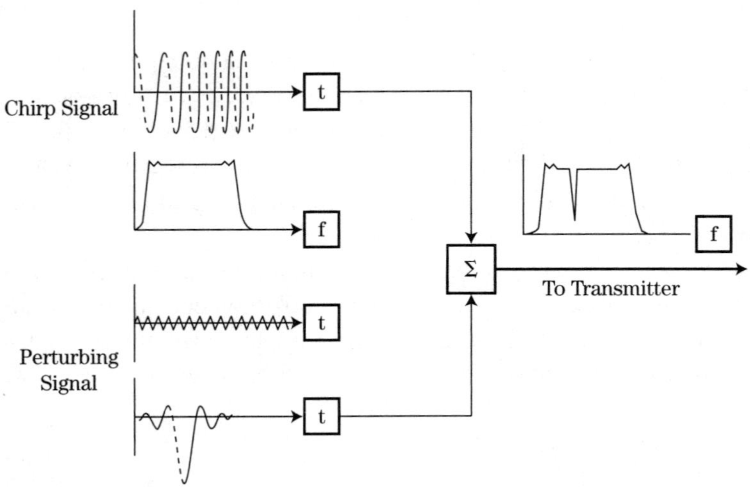

FIGURE 5–10

Transmitter notching with perturbed signal

This approach requires the use of a waveform generator capable of producing arbitrary waveforms. The waveform modulator must be sufficiently stable that unwanted sidebands are not created. Furthermore, all devices after the waveform generator must be accurately modeled, especially nonlinear devices such as frequency doublers. Given today's high-speed logic and direct waveform synthesis technology, the equivalent parallel synthesis of the notches can be calculated and applied in one processing chain.

As an illustration of the notching process, a simple example is given to examine the process and impact on the waveform IPT. First, the specific frequency must be avoided in programming the LFM frequencies within the waveform generator. The narrowband waveform is represented by a single tone at ω_j, such as [15]

$$S_{ISLR} = b_j \, rect\left(\frac{t}{T_r}\right) \sin(\omega_j t) \tag{5.28}$$

where $\omega_j = 2\pi f_i$ is the frequency to be avoided.

When this function is subtracted from the LFM waveform in the spectral domain, finite notches are formed, as shown in Figure 5–10. If multiple notches are required, one notching signal function from (5.28) is derived for each frequency. The summation of these notching signals is subtracted from the LFM waveform before transmission.

The LFM waveform in Figure 5–11 has five notches that comprise less than 0.6% of the signal bandwidth. As a result, the range integrated sidelobes shown in Figure 5–11b are not significantly different from the uniformly weighted sinc function. The ISLR under errored conditions is an important measure of the SAR image quality and is covered in more detail in Section 5.2.3. For the example of 0.6 percent of bandwidth notching, the ISLR is raised by 0.46 dB, as seen by comparing Figure 5–7 to Figure 5–11.

Notching an UWB SAR signal is effective in reducing the interference from the FOPEN SAR to most sensitive receivers, such as emergency communications and airport glide slope indicators. Because the frequency is being swept at a rate significantly faster than the receiver time-bandwidth can detect, the victim receiver will not measure the full UWB waveform spectrum. In addition, the power spectral density at any one subband is reduced by the typically high time bandwidth ($BT \gg 100$). If the FOPEN SAR is also going to be used for terrain mapping or characterization, it is important that both the spectral and polarimetric characteristics be maintained. This may be accomplished by digital preselection or emphasis on transmit or by compensation in the digital processing on receive.

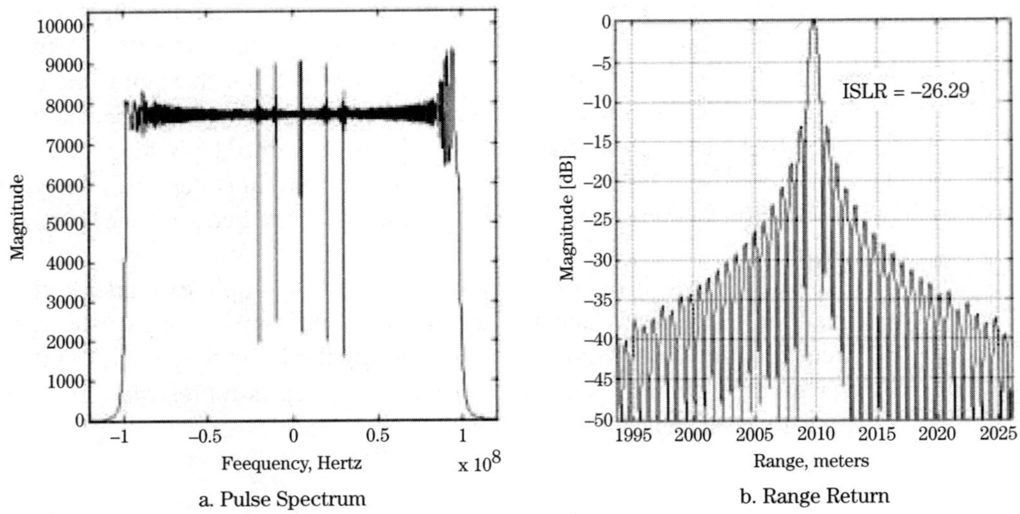

FIGURE 5–11

Digitally notched SAR LFM spectrum

The effectiveness of notching was evaluated by the GeoSAR system at the same test location as covered in Section 5.1.2. A 20 dB deep notch was placed in the 160 MHz bandwidth LFM pulse at 317.5 MHz. To verify the transmit waveform; a spectrum analyzer measured the pulse characteristics directly out of the RADAR. The pulse exhibited the same characteristics as the original LFM waveform, with the exception of an 18 dB notch at 317.5 MHz. The FAA receiver was tested with the 40 msec LFM pulse along with its notch, and the threshold was measured at –3 dBm compared with the −21 dBm without the notch. This verifies the signal interference attenuation of 18 dB and the benefit of notching a transmit waveform to satisfy the frequency avoidance requirements [9].

5.1.5 Impact of Notching on IPR

The requirement for removing the frequency content in a transmit waveform has a direct impact on the impulse response (IPR) of the SAR system. Linear array theory provides a first-order assessment on the waveform sidelobes. It is first assumed that the notches are randomly placed and are a small percentage of the spatial illumination function.

The effect on IPR sidelobes was given in (3.6) for an N-element linear array, where the total mean square error σ_T^2 is equal to the sum of the amplitude and phase errors. When treating an array with small random errors, the mean squared sidelobe level (MSSL) that bounds the SAR resolution sidelobe levels was shown in (3.7) to be the ratio of the total errors to the number of elements. As a result, the MSSL for random and small perturbations to the waveform elements can be expressed by [16]

$$MSSL_{rand} = \frac{\sigma_T^2}{\eta_a N(1-\sigma_T^2)} \tag{5.29}$$

where η_a is the sidelobe weighting efficiency. When the frequencies (or elements) in the array are missing, this is equivalent to a failed element. The number of missing or failed elements can be characterized by a probability P, such that

$$MSSL = \frac{(1-P)+\sigma_a^2+P\sigma_\phi^2}{\eta_a PN} \tag{5.30}$$

The factor $(1 - P)$ represents the percentage of frequencies in the LFM waveform that are zeroed. The range compression aperture efficiency due to weighting for sidelobe reduction is given by η_a.

When the number of elements zeroed is a large percentage of the number of elements (over 10%), the amplitude distribution of the errors is directly affected. As a result, (5.30) can be modified by

$$\sigma_{a,thin} = \frac{1}{N}[((1-P)N)^2+(PN\sigma_a)^2]^{1/2} \tag{5.31}$$

The assumption for this case is that the number of elements removed $(1 - P)N$ have an amplitude error of $\sigma_a = 1$, and their effects are directly added to the error density. The remaining elements have a normal distribution with mean value of σ_a.

Figure 5–12 shows parametrically how the number of frequency notches K_N affects the MSSL given in decibels. This is an extreme case where the number of frequencies are all zeroed, and $\sigma_a = \sigma_\varphi = 0.1$. When the notching perturbation amplitude is less than the LFM waveform component, the effect on the range sidelobes will be reduced. However, it is important to understand the power spectral density of each frequency and the frequency impact of interference on the specific location of the sensitive victim receivers.

5.1.6 Notched LFM Integrated Side Lobe Response

One of the most important figures of performance in evaluating SAR images is the Integrated Side Lobe Ratio (ISLR). This metric evaluates the return in a SAR image from distributed clutter in the sidelobes, that competes competing with a point scatterer in the image resolution cell. This measure is especially

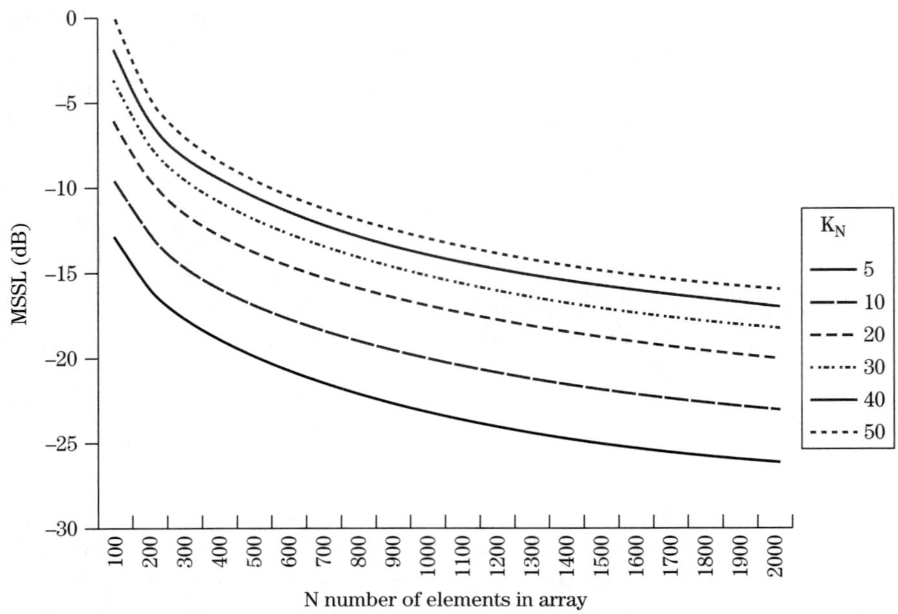

FIGURE 5–12

Mean sidelobe level with missing frequency elements

indicative of effects from strong scatterers in the neighborhood of a target of interest. The metric is expressed by:

$$ISLR = \frac{\text{energy in sidelobes}}{\text{energy in mainlobe}} \tag{5.32}$$

ISLR can be similarly determined from linear array theory by examining the response of an array to a point scatterer in the far field of the linear array. If $s_p(t)$ is the incident plane wave on a field of elements having amplitude $a(t)$, the array response is [17]

$$S_P(\omega) = \int a(t) \exp[-j\omega t + \delta\varphi(t)]d\omega \tag{5.33}$$

The function $S_P(\omega)$ is the Fourier transform of the plane wave and provides the impulse response of a point scatterer at broadside in the presence of a phase error of $\delta\varphi(t)$. A Taylor series expansion of the response of the function $S_p(\omega)$, when $\delta\varphi(t)$ is small yields

$$S_P(\omega) = S_0(\omega) + S_1(\omega) \tag{5.34}$$

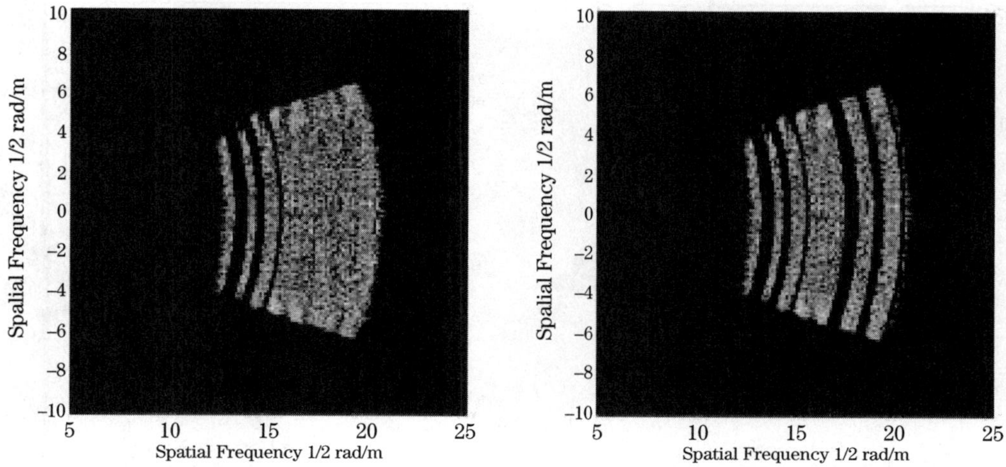

a. 7 Percent Nulled

b. 15 Percent Nulled

FIGURE 5–13

LFM waveform with frequencies removed

where $S_0(\omega)$ defines the mainlobe return, and $S_1(\omega)$ is the sidelobe return of the point function. Using these factors, the ISLR can be determined from (5.32) as

$$ISLR = \frac{\int |S_1(\omega)|^2 d\omega}{\int |S_0(\omega)|^2 d\omega} \tag{5.35}$$

The scene described in Figure 4-10 was evaluated with two waveforms where each had frequencies zeroed or nulled during its formation. Figure 5–13 shows the waveform support characteristics in the fast- and slow-time domain. The left image represents a 7% elimination or nulling of the transmitted frequencies during the LFM formation. The right image represents a 15% nulling of the frequencies.

The missing frequencies can clearly be seen as stripes in the fast-time spectral dimension. These are maintained throughout the imaging to avoid the sensitive frequency bands. However, the abrupt zeroing of frequencies will induce spectral sidelobes that do not satisfy the NTIA requirements. This is the principal reason that approaches for frequency avoidance maintaining the out-of-band spectral content of the waveform have been developed.

After SAR image formation, the image can be examined. Figure 5–14 shows the range and cross-range impulse function from the frequency nulled waveforms in Figure 5–13. The waveform has not been weighted, and the two-dimensional characteristics of an LFM signal are evident. As the frequencies in

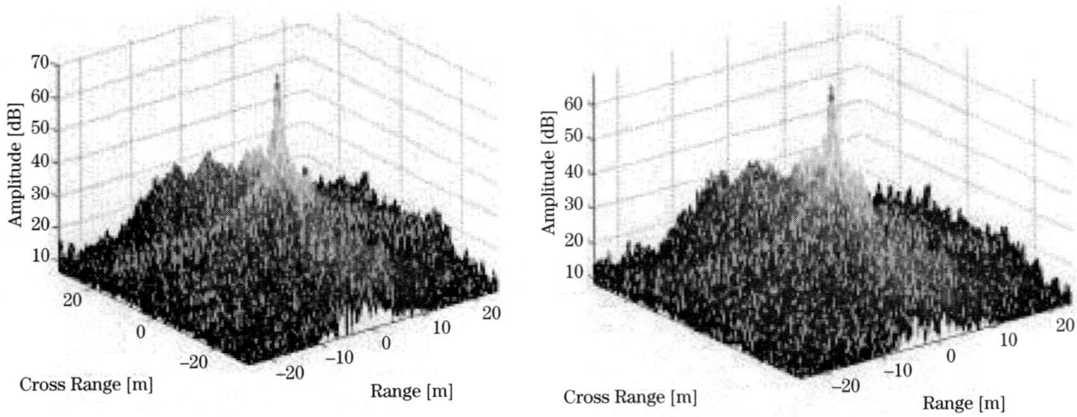

a. 7 Percent Nulled

b. 15 Percent Nulled

FIGURE 5–14

Three dimensional waveform IPR with frequencies notched

Table 5–2 Comparison of ISLR for frequency nulled waveforms

	No Notching	7% Nulling	15% Nulling
Range	−31.73	−30.38	−29.01
Cross Range	−25.78	−24.44	−23.06

the support waveform function are nulled, the sidelobes are randomly raised. A summary of the resulting ISLR for targets under conditions of no nulling, 7% nulling, and 15% nulling is given in Table 5–2.

5.2 Cancellation of Radio Frequency Interference

It has been noted several times that RFI will directly influence the background noise in the image, adversely affecting the ability to discern features of the SAR image and any target detection. There has been significant development carried out to determine the most effective RFI removal technique. Two competing objectives were used to determine the most appropriate algorithm: (1) processing efficiency for real-time image formation; and (2) RFI removal effectiveness to restore the image quality.

The primary approach is to estimate the interference spectrum and remove it by digital filtering prior to image formation. Depending on the number of interfering signals and their bandwidth, the RFI removal can be effective in increasing the signal to interference characteristics. However, it is important to understand both the time-varying nature of the environment and the motion of these interferers in the SAR collection aperture. The range and angle to the interferers will be constantly changing (slow-time effect), and the signal characteristics of the interferers will vary during a single PRI (fast-time effect).

One early method of removing the interference due to narrowband radio and television transmissions was to sample the environment without the clutter return from the RADAR transmitted waveform. A spectrum analysis function in the processing path measures the interference spectrum and determines the location and strength of each transmission. Based on a straightforward constant false alarm rate (CFAR) technique, the interference peaks are identified and removed from the RADAR returns. This approach, although simple to implement, does not preserve the full characteristics of the image. Either the spectral lines are completely removed, or the RADAR signal energy is clipped prior to image formation. In the first case, the waveform spectral content is highly thinned, and the image loses valuable resolution. In the second approach, residual RFI energy remains in the return, and the image noise level is adversely raised.

Significant progress was made in the late 1990s to understand the effects of strong RFI and providing efficient removal to enhance the FOPEN SAR image quality.

The objectives of the *RF suppression in UWB receivers* project was to define and develop a subsystem design in which a UWB RADAR receiver can operate in the presence of nonhostile in-beam interference, narrowband interference, and communication-type interference. Algorithms have been developed such that the RFI is sufficiently suppressed, negligible loss in target detection performance is achieved, target resolution performance is maintained, and SAR image integrity is preserved.

The simplest approach is to first collect receiver data over several pulses and then to estimate the interference in the environment. Two approaches that have been documented with varying impact on the design and operation of the RADAR are illustrated in Figure 5–15.The RF environment signal must be sampled over sufficient time that the spectral characteristics can be measured. In the first approach, samples are obtained during the RADAR dead time as shown in Figure 5–15a. In this case, there must be sufficient isolation between the RFI sampling receiver and the transmitter that leakage energy leakage does not corrupt the interference estimation. This normally is accomplished using a separate antenna and receiver to meet the isolation requirements. The second approach is to collect data during a quiet time in the receiver operation as shown in Figure 5–15b. This would be done with the same receiver and signal

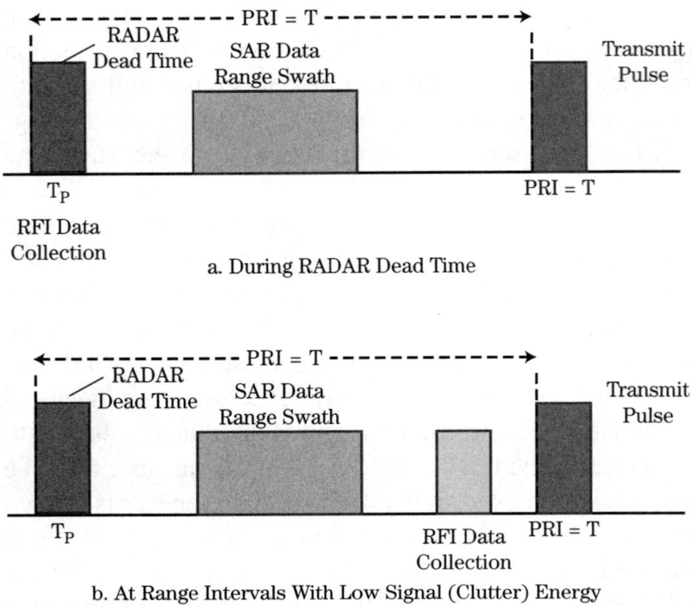

FIGURE 5–15

RFI sampling during RADAR operation [18]

processing chain as the SAR operation. The challenge is to sense and identify the interference sources for real-time image formation processing.

The following sections illustrate several approaches for removing RFI from the SAR signal, along with an assessment of their processing complexity.

5.2.1 Adaptive Transverse Filter

An early approach that showed promise in stationary environments used adaptive transverse filtering (ATF), similar to signal processing approaches used for wideband channel matching and digital signal conditioning. The processing flow for RFI removal is shown in Figure 5–16. In the ATF process, three primary sources of interference data are considered for RFI removal [18]:

- Off-line data collection to characterize the spectral characteristics of the system—used for calibration of the signal across the RFI interference span
- Collection of the RFI spectrum from the environment—used to calculate the finite impulse response (FIR) filter weights for cleaning the UWB SAR image
- SAR signal time records—with RFI and targets included

First, the reference signal Z_S within the receiver chain is formed by convolving the weighted reference signal with a synthetic target signal x_S, which

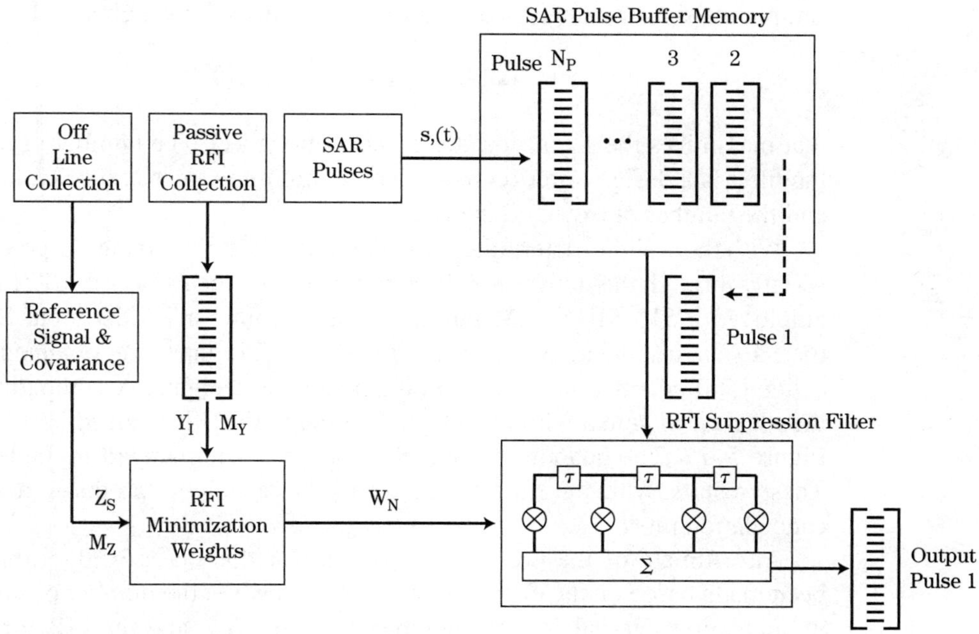

FIGURE 5–16

Transverse filter RFI removal processing flow [17]

is used to determine the receiver characteristics. The reference signal is given by

$$Z_S = w_Q^T s_R \otimes x_S \tag{5.36}$$

where w_Q^T is the sidelobe weighting of the reference signal s_R. A covariance of Z_s is then formed for use in the RFI minimization process, given by

$$M_Z = E(Z_S^* Z_S) = \frac{1}{N_A} \sum_{i=1}^{N_A} Z_S^* Z_S \tag{5.37}$$

Next, the system collects time samples of the interference environment Y_I, without the effects of targets and clutter. The covariance of this record M_Y is formed similar to the receiver reference function in (5.37).

The correlation, r_D, between the interference signal and the receiver function can then be obtained given by

$$r_D = E[Z_S(Y_I + Z_S)^*] \tag{5.38}$$

This function acts as a "steering vector" for calculating the transversal filter weights W_N for processing the input data. The approach, is a form of the sample matrix inversion (SMI) algorithm, produces the weights

$$W_N = [M_Z + M_Y]^{-1} r_D \tag{5.39}$$

that maximizes the signal-to-interference ratio (SIR). The number of taps in the filter is a design criterion based on the minimum interference bandwidth and the number of expected interferers.

With the weights determined, each record of the SAR image is processed serially through the transversal filter. For example, the P-3 UWB SAR signal employed a 515 MHz LFM pulse, which exhibited a 7.6 dB input SIR in the RFI environment. A transverse filter with 251 taps was obtained after using 1500 adaptive iterations to obtain suitable weights. A comparison of the ideal LFM signal with the adapted output with RFI removed is shown in Figure 5–17. The quantitative signal response is summarized in Table 5–3. These results, which are representative of the technique, produced a 25 dB cancellation ratio.

An estimate of the processing required for this class of algorithm has been made based on the choice of number of taps N_T, the number of samples in adaptation interval N_A, the number of pulses N_P, and the PRI for SAR collection of T seconds. The computation process and the total number of real operations are summarized in Table 5–4 [18].

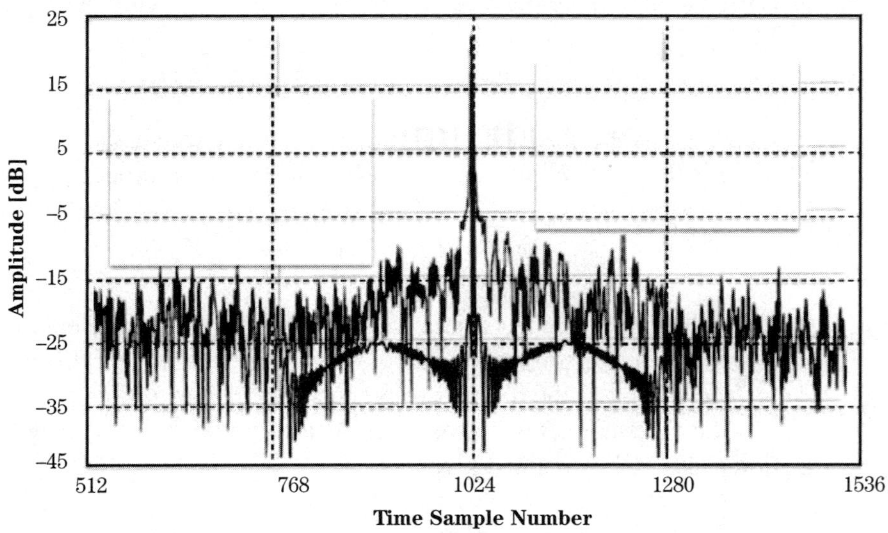

FIGURE 5–17

P-3 reconstructed range resolution using adaptive transversal filter [17]

Table 5–3 Comparison of adaptive transverse filter removal of P-3 RFI energy [17]

	Ideal Response	Adaptive Response
Peak Signal Amplitude	22.7 dB	20.0 dB
Peak Sidelobe Level	43.8 dB	22.9 dB
RMS Sidelobe Level	56.5 dB	38.9 dB

Table 5–4 Operations count for adaptive transverse filter algorithm [17]

Operation	Symbol	Count	Operation	Symbol	Count
Weighted Match Filter Signal	Z_S	$8N_P^2 + 8N_S N_P$	Interference Correlation	r_D	$8N_A N_T$
Reference Convolution Vector	M_Z	$4N_T^2 N_A + N_T N_A$	Matrix Addition	$M_Y + M_Z$	$2N_T^2$
Reference Covariance	$W_Q^T S_R$	$6\ N_P$	Correlation Vector Add	$r_Y + r_D$	$2N_T$
In Line Deskew		$15N_S \log_2 N_S + 12N_S$	Matrix Inversion	$(M_Y + M_Z)^{-1}$	$4/3N_T^3 + 8N_T^2 - 4N_T$
Interference Covariance	M_Y	$4N_A^2 N_T + 4N_T N_A + N_T^2 - N_T$	Filter Coefficient Determination	$W = M^{-1} r_D$	$8N_T^2 – 8N_T$

The adaptive transversal filter algorithm has shown to be an effect method of removing significant RFI while providing equalization of the signal and maintaining signal resolution. However, it can require significant adaptation cycles (e.g., 20–50) depending on the environment's RFI source density and the need to handle broadband interference. As a consequence, the results have been seen to degrade in a dynamically changing environment, as would be anticipated for a high-altitude unmanned air vehicle.

5.2.2 Chirp Least Squares Adaptive Processing

An alternative adaptive algorithm that has been developed to provide signal quality improvements in a dynamically changing environment is referred to as the *chirp-least-squares algorithm with clipping* (CLSC) [19]. The following advantages are provided by this RFI removal processing technique:

- As an estimate-and-subtract algorithm, it provides the narrowest possible stopband for a given data length and therefore minimizes time sidelobes.

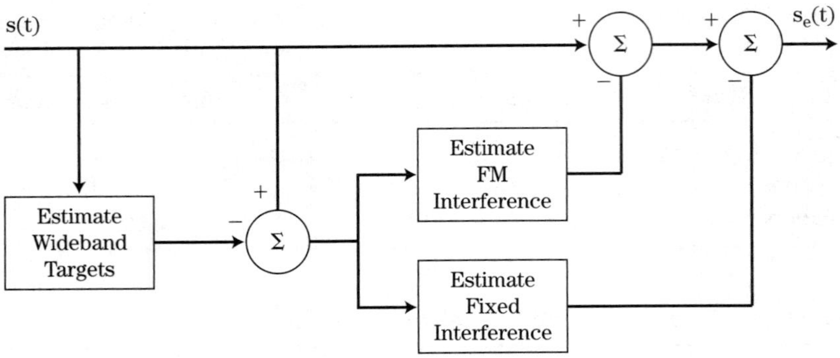

FIGURE 5–18

RFI removal process flow [18]

- It allows iterative, nonlinear signal (target) excision, which further reduces time sidelobes and signal loss.
- There are no filter edge effects.

Similar to the ATF algorithm, the objectives of CLSC processing are to attain (1) maximum reduction in RFI energy, (2) minimum loss and distortion of the wideband target response, and (3) real-time implementation. With this algorithm, knowledge of the characteristics of the interference is used to assist in the reduction in processing. Two assumptions are made. First, no pulse compression occurs that modifies the characteristics of the interfering signals, i.e. all the interfering signals and the short time response of the targets are captured in the time domain. Second, the bandwidth of any interfering signal is small compared with the bandwidth of target return signal. Based on these assumptions, that the signals can be modeled as a sum of sinusoids, over the short duration of the target return. There are two classes of interference considered: (1) those whose frequencies are fixed, but whose amplitude changes slowly; and (2) those whose frequency and amplitude change slowly.

The general flow of the processing algorithm is shown in Figure 5–18 and can be summarized as follows. Broadly speaking, the algorithm has two paths, an interference estimation path, and a signal path where the output is the input signal minus the estimated interference. The processing steps can be summarized as follows:

Step 1 *Target Signal Excision:* Since large targets cause large sidelobes, while sidelobes from small targets are small enough to be ignored,target signal excision is used to largely eliminate the sidelobes that would be caused by large targets. Because target signals are characterized by wideband (short-time) signatures and are removed from the processing, their features are preserved in the SAR image and are useful for recognition. Strong returns are identified early in the processing (e.g., after-pulse compression) and and are removed (clipped) so that the interference is estimated more accurately –i.e. with less perturbation from the large target returns. Clipping of large targets takes place only in the interference estimation path and therefore does not introduce any artifacts in the output signal, but reduces them instead.

The threshold and clip approach substitutes signal $s_c(n)$ for the time series $s(n)$ by

$$s_C(n) = \begin{cases} s(n) & s(n) < \Theta_C \text{ or } > \Theta_C \\ 0 & \text{otherwise} \end{cases} \tag{5.40}$$

where the clipping levels $\pm\theta$ are adaptively set to three times the root mean square (RMS) of the average interference signal as

$$\Theta_C = 3\left(\frac{1}{N}\sum_{n=0}^{N-1} |s(n)|^2\right)^{1/2} \tag{5.41}$$

After the largest interfering signals are removed, the target clipping step is reiterated to remove any additional targets that exceed a new threshold.

Step 2 *Fixed frequency signals:* Many interference signals have very little deviation from the know center frequency and bandwidth. These signals are due to AM and narrowband FM (NBFM) transmitters, which are well regulated by the FCC. The interfering signal spectrum consists of L_1 known frequencies that can be written as

$$r_1(n) = \sum_{i=1}^{L_1} A_i \sin(\omega_i k) = [A]z \qquad n = 0, \ldots, N-1 \tag{5.42}$$

The standard least square estimate of the solution to (5.42) is given by

$$\hat{r}_1 = [A]\hat{z}, \qquad \hat{z} = (A^T A)^{-1} A^T s \tag{5.43}$$

Since the frequencies are known, the matrix $[A]$ is known. The estimate of the $2L_1 < N$ unknowns can be estimated to minimize the mean square error between $s(n)$ and $r_1(n)$. Using sinusoids enables a set of orthonormal vectors, thereby avoiding the need for inverting the matrix $[A]$. The approach uses a Gram–Schmidt process to precompute the orthonormal vectors

$$O = [o_1(n)\ o_2(n)\ o_3(n) \ldots o_{2L_1}(n)] \tag{5.44}$$

Given O, the solution to (5.38) is simplified to

$$p = O^T x, \qquad r_1 = Op \tag{5.45}$$

Step 3 *Frequency-modulated interference cancellation:* Commercial FM radio signals have wide bandwidths, and thus the spectral spread is not well controlled. CLSC exploits the fact that FM signals are contained in the 88–108 MHz band, and the sampling frequency of the SAR receiver allows for decimation of the data over the band. The incoming signal is bandpass filtered to isolate the FM band and to permit decimation without aliasing. An FIR filter of N_T taps is used on the data, along with a 4:1 decimation. With the bandpass filter h(n), the filtered signal with record length of $S = (N - N_T)/4$ is given by

$$g(n) = \sum_{n=0}^{N_T - 1} h(N_T - n)\, s(2k + n), \qquad k = 0, \ldots, S - 1 \tag{5.46}$$

Next, an oversampled frequency representation of the interference is obtained with a chirp transform. This provides equally spaced sample of the Fourier transform of $g(n)$ over the FM band

$$G(n) = c(n) \sum_{n=0}^{S-1} g(n) e^{j\omega_k n}, \qquad \omega_k = \omega_0 + k\Delta\omega \tag{5.47}$$

The largest magnitude of the signal in the frequency domain is used to estimate the amplitude, frequency, and phase of the interference. Similar to the fixed-frequency RFI extraction, interference sinusoids are constructed and subtracted from the incoming signal.

The processing for CLSC was exercised with the interference spectrum at Army Research Laboratory in Adelphi, Maryland, previously shown in Figure 5–1. Performance of the adaptive processing was evaluated with and without a strong (injected) target, as shown in Figure 5–19. This figure illustrates the point that removal of RFI alone is not sufficient for obtaining quality images. Any targets present in the data records will cause the misestimation of the RFI, amplitude, and phase. Moreover, it causes removal of target energy at the frequencies removed, directly affecting the IPR of the targets. Figure 5–19 compares the sidelobe energy of s_C with and without a target present. The dashed lines provide the average for each of the three curves presented. Since averaging of s_C is equivalent to SAR image formation, it can be seen that coherent misestimation of the RFI excision processing raises the effective noise floor and decreases the sidelobes of s_C and increases the required dynamic range. With a target present during the RFI process, the sidelobe energy in the lower plot is 7 dB higher than the processing in the upper plot.

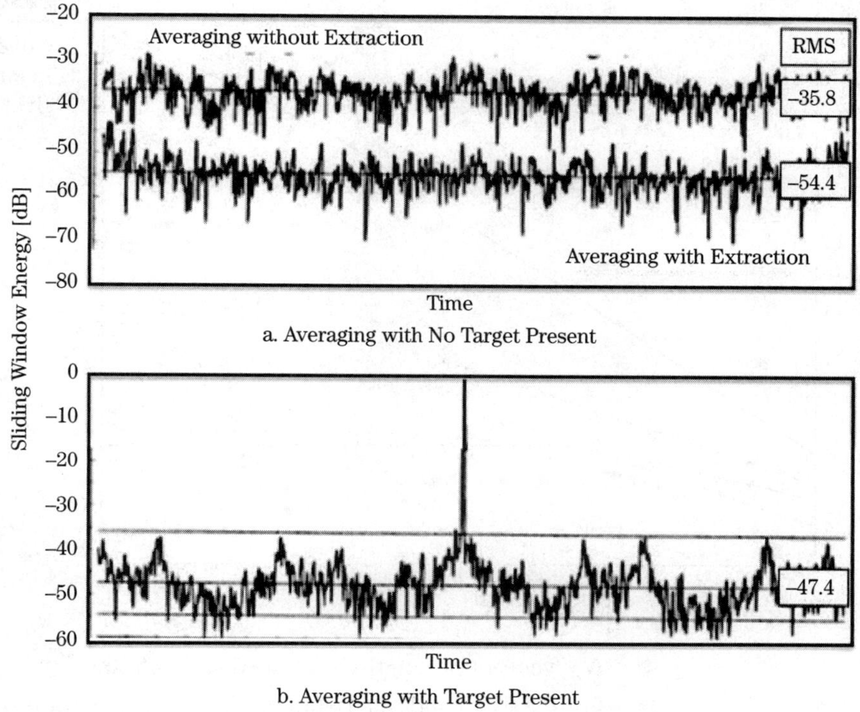

FIGURE 5–19

Target sidelobe levels after RFI processing © 1997 IEEE [19]

The effect of processing with target extraction can be better quantified in Figure 5–20. Three alternative algorithms that were considered in the literature are compared: the maximum likelihood estimation algorithm (ML); the coherent least squares algorithm (CLS); and CLSC. The CLSC algorithm exhibited improved signal to interference ratio over the competing algorithms. When no targets exist in the image record, the range sidelobes are significantly improved, especially, even when a large number of sinusoids are extracted. When weak targets were present, the performance is adequate, since the strong RFI effects that mask target features have been significantly reduced. For the competitive approaches, the performance did not improve even with an increase of the number of RFI sinusoids [19].

An estimate of the processing operations count can be obtained by cataloging the various steps in the algorithm, as summarized in Table 5–5. Key parameters are as follows:

- N_I iterations are required to excise all of the targets and interference frequencies.

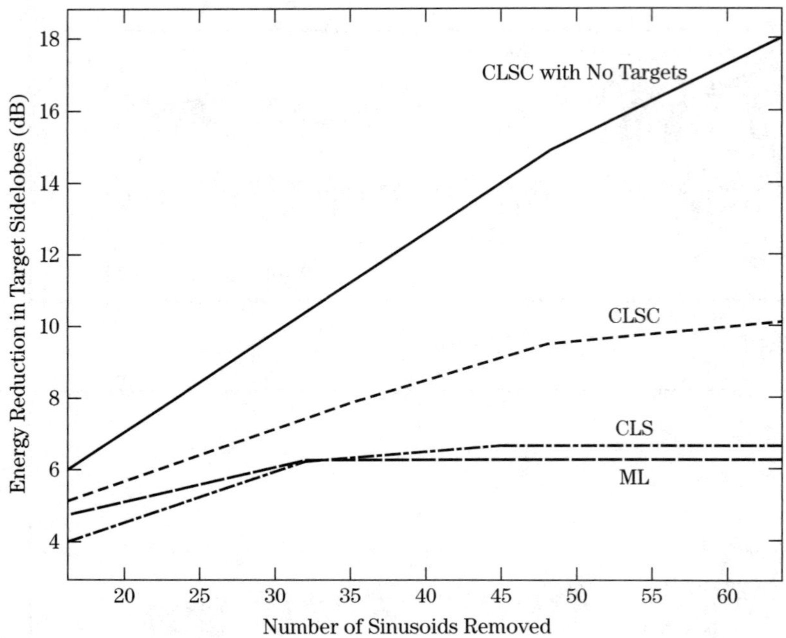

FIGURE 5–20

Coherent energy reduction in target sidelobes © 1997 IEEE [19]

- N_K vectors of length N are used on each iteration.
- For the FM band, the data length after decimation is N_S, and the number of frequency chip frequency bins is N_B.
- Each iteration includes several sinusoidal vectors:
  - N_{M1} for the known frequencies
  - N_{M2} for the FM interference signals
  - N_{M3} for unknown spurious interference signals

The total processing is $O(N \log_2 N)$ operations count. When the interference in Figure 5–1 was processed, $N_{M1} = 10$, $N_{M2} = 5$, $N_B = 512$, and $N_K = 15$. The RFI processing required 1.5 million operations to remove 64 RFI signals from a data record of 2,048 samples. Assuming a PRI of 1 millisecond, the sustained throughput would be 1.5 GOPS [19].

Table 5–5 Processing count for chirp least squares with clipping algorithm © 1997 IEEE [19]

Operation	Symbol	Count
Estimate Known Frequencies	$x_C(n)$	$4\, N_K\, N_{M1}$
Estimate Known FM Parameters	$r_1(n)$	$N_I(55N_S + 4N_S \log_2 N_S + 6N_B + 20N_BN_{M2} + 4N_KN)$
Generate and Subtract Estimated RFI	g(n)	$N_I(4N_{M1}N + 3N_{M3}N)$
Chirp Z Unknown FM	G(n)	$N \log_2 N + 3N + 4N_{M3}N$

5.2.3 Deramp RFI Removal

The P-3 UWB SAR was used in stretch (or deramp-on-receive) processing to manage the signal conversion to baseband. During the SAR system development it was necessary to develop an RFI mitigation algorithm that could correctly down-convert the SAR wideband signal, along with performing of the RFI removal. Equally as important to the reduction of RFI was the need to limit the number of computational operations. The parametric maximum likelihood (PML) algorithm was developed to estimate the RFI and to enable subtraction from the SAR image [20].

Because the SAR signals are similar to wideband noise in their nature, it was important to estimate the RFI during the period when receivers were not receiving SAR signals. Even more important was the determination that the RFI from radio and TV transmissions were stable in angle and frequency during P-3 SAR signal collection. Hence, the RFI tones could be estimated early in the image collection and therefore did not need to be reestimated. Specifically, for the P-3 standoff range and platform velocity, the RFI was stable for a period on the order of 200 pulses.

The PML algorithm is shown in Figure 5–21 and consists of several stages. First the received samples are bandpass filtered, and then the tones for the strongest interference sources are estimated. For the P-3, this was accomplished on a single pulse, using the N_K range samples collected in the dead time of the first pulse.

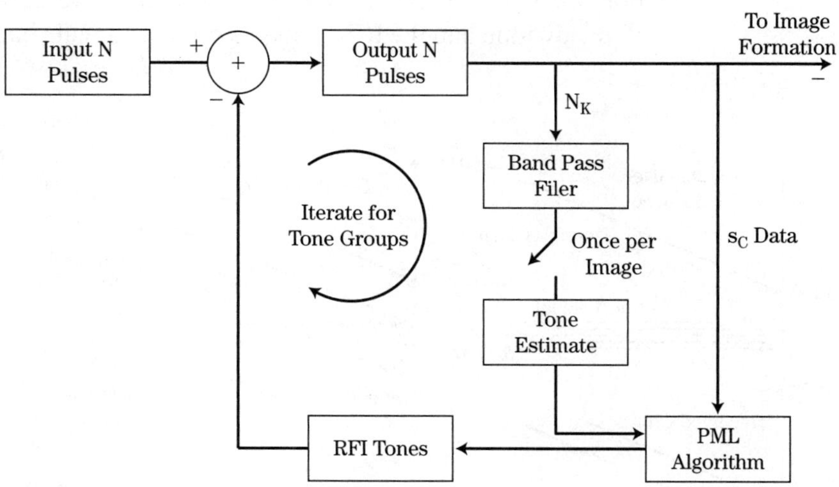

FIGURE 5–21

Parametric maximum likelihood algorithm processing [20]

The incoming signal is then bandpass filtered into several subbands to isolate the RFI tones, with an objective of one tone per subband. The amplitude, phase, and frequency of the interferer in each subband are estimated, and those signals are subtracted from the incoming data. This step was important to provide a signal estimation on the smaller tones; which are normally masked by the stronger tone energy. An iterative estimation of RFI tones is conducted with multiple thresholds, each threshold covering success levels of signal strengths. For the P-3 data, six thresholds were used with up to 28 bandpass filters.

The frequency estimation is provided using a forward–backward linear prediction model, based on the single pulse time series data of order N. Data records of length L are constructed next, then forming a correlation matrix of rank $2(N - L)$. The prediction estimate is greatly improved the approaching the maximum likelihood estimate capability but with a reduction in processing complexity. This improvement is assured by limiting the number of interfering sinusoids in each subband filter [21].

Because the SAR collection uses stretch processing, it is important to have an RFI removal technique that can be effective with the deskew process. In a deramp SAR, the receive signal is mixed with the frequency shifted replica of the transmitted chirp waveform. Any RFI becomes a chirped signal when passing through the intermediate frequency amplifier. However, the receiver has a bandpass filter after the deskew mixing, and not all of the RFI passes through the analog filter.

Figure 5–22 shows the stretch processing, previously presented in Section 4.2.3. The SAR signal will be deramped into constant tones within the IF bandwidth, and the RFI is spread out over a wide bandwidth. For the SAR

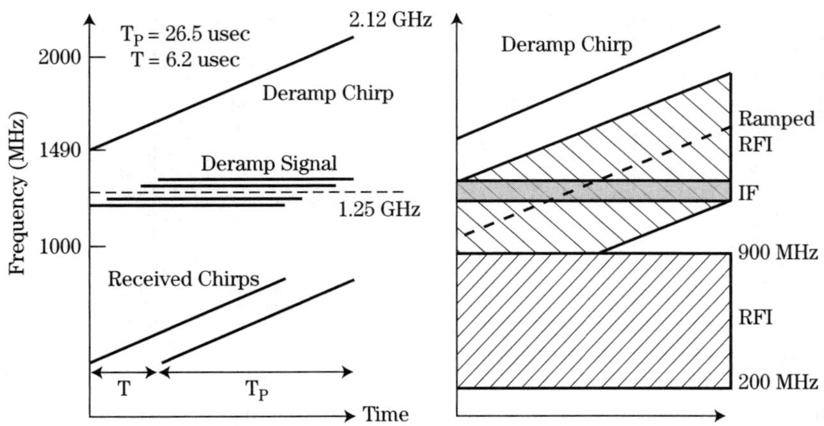

FIGURE 5–22

Deramp process during UWB SAR image formation [20]

signal, the signal after deramp and IF filtering has the following form [20]:

$$s(t) = A\ \exp\left[j\pi(t - t_0)^2\right]\ rect\left[\frac{t - t_0}{T}\right] \tag{5.48}$$

where the parameters in Figure 5-22 and (5.48) are:

T_P	Duration of the SAR pulse
A	Complex amplitude of the tone,
T	Delay of the chirped tone at the IF output
$B_{IF} = \gamma T$	Bandwidth of the SAR IF
t_0	Delay from the start of the SAR chirp until the deramped tone frequency equals the center IF

After the deramp process, the signal history is digitally processed with a deskew followed by removal of the residual video phase (RVP) history. The deramped signal s_0 is Fourier processed, multiplied by the residual phase function, and inverse Fourier transformed. The inverse Fourier transform recovers the SAR signal with frequency tones proportional to the range extent, as given by

$$\Phi_{RPV} = \exp\left[-j\frac{\omega^2}{4\pi\gamma}\right] \tag{5.49}$$

The same sequence of operations are applied to the RFI, causing them to appear as pulse modulated signals, whose parameters are a function of the product of a pulse waveform and a chirp waveform. The first FFT on the product of these two signals results in a convolution of the transforms of the two individual functions as

$$\begin{aligned} S(\omega) &= \left(A\sqrt{\frac{j}{\gamma}}\exp\left[j\left(\frac{\omega^2}{4\pi\gamma}\right)\right]\exp\left(-j\omega t_0\right)\right) \\ &\quad \otimes \left(T\,\mathrm{sinc}\left(\frac{\omega T}{2}\right)\exp\left(-j\omega t_0\right)\right) \\ &\approx A\sqrt{\frac{j}{\gamma}}\exp\left[j\frac{\omega^2}{4\pi\gamma}\right]\exp\left(-j\omega t_0\right) \end{aligned} \tag{5.50}$$

The multiplication by Φ_{RPV} and the final IFFT results in a sinc function output in time:

$$s_0(t) = A\sqrt{\frac{j}{\gamma}}\frac{\sin[\pi B_{IF}(t - t_0)]}{\pi B_{IF}(t - t_0)} \tag{5.51}$$

After the first two steps of the deskew/RVP removal process, a single tone at ω remains; which is the time that the i-th tone is swept through the center of the IF

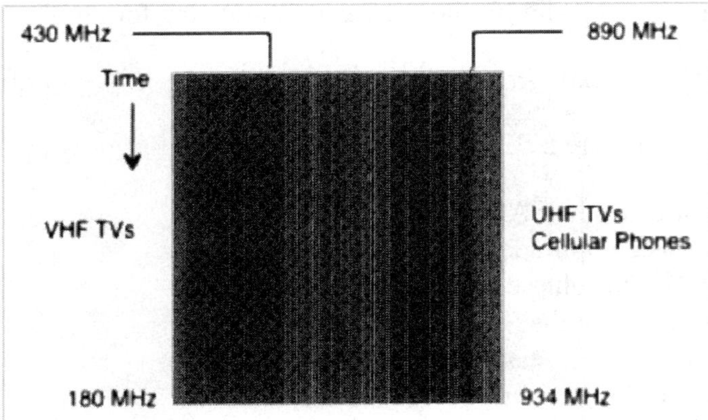

FIGURE 5–23

RFI during deramp processing [20]

bandwidth. This tone is the same form as a single complex sinusoidal of radian frequency ω over the time period. Thus, the PML algorithm can estimate the time of each of the tones through the IF over the period $-\pi B_{IF} < \omega < \pi B_{IF}$. At this point of the deskew/RVP removal output, the compression of the RFI in the processor can either clip the RFI impulse peaks or detect their presence and remove them later in the PML process. Figure 5–23 shows the FFT of the output of the deskew process and illustrates the single frequency tones of the RFI. The stability of the RFI tones over time is verified by the straight lines as a function of time.

Next, the deramp RFI removal technique is applied in the time domain, based on the assumption that a significant portion of the RFI can be eliminated by the deskew process. However, the technique counts on the use of stretch processing in the signal conversion and image formation.

Figure 5–24 illustrates the recorded spectrum from a P-3 UWB collection in Grayling, Michigan. The original image shown at the upper left was formed without removing the RFI from the image processing. It is apparent that strong interference from radio and television transmissions severely degrade the image noise level. The image on the right was obtained when the RFI spectrum was removed using the PML algorithm. For this example, the noise equivalent σ_0 was improved from -6 dB with RFI present to -18 dB after RFI removal.

The processing complexity of the PML algorithm can be estimated in a straightforward manner. Because of the iterative nature of the algorithm, one must specify the number of samples used to estimate the RFI tones N_K, and the number of tones to be removed M. Moreover, the estimation process needs to accounted for the number of bandpass filters, along with the number of iterations N_I. A summary of the PML algorithm operations count is provided in Table 5–6 . The P-3 UWB SAR operated with a PRF of 300 Hz with the

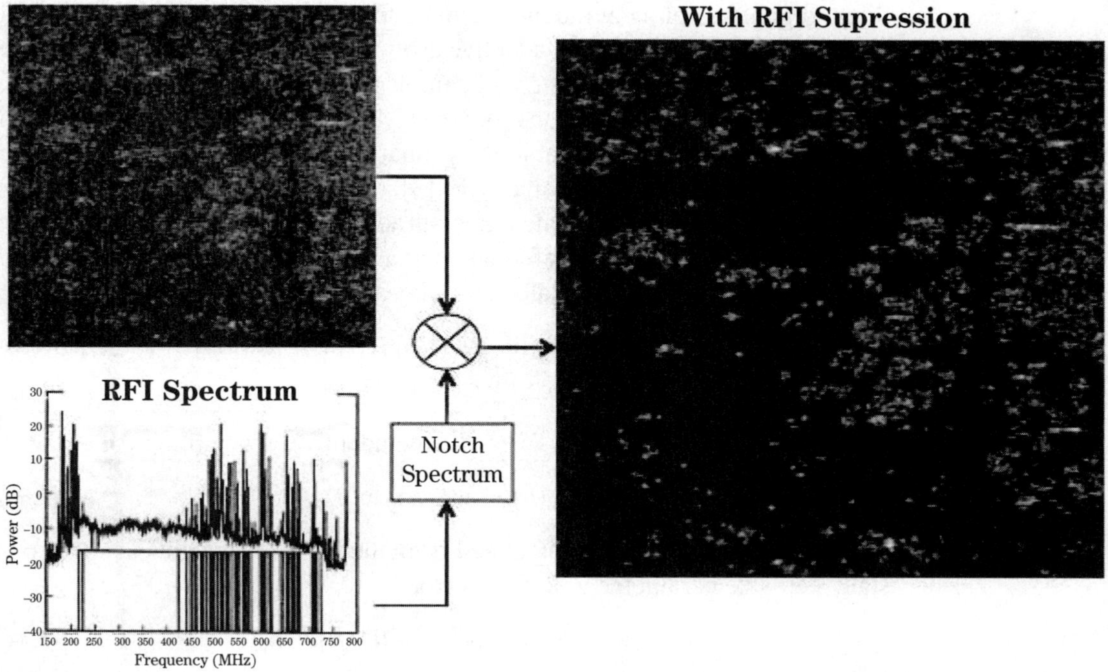

FIGURE 5–24

Radio frequency removal improvement in P-3 SAR
Source: MIT Lincoln Laboratory [22]

Table 5–6 Operations count estimation of parametric maximum likelihood algorithm [2019]

Operation	Count
PML Iteration	$(7/3)M^3 + 34M^2 + 6MN$
Total Processing	$N_F\ N_T\ M$ (PML)

number of samples per pulse being 4,096. The number of tones per pass per threshold was on the order of eight. As a result the real-time operations for PML would be 80 GFlops.

5.2.4 Adaptive RFI Removal

The GeoSAR system developed an adaptive processing technique to provide RFI removal in the frequency domain, rather than in the time domain. The bandwidth of GeoSAR was smaller than that of the UWB P-3; and as a result, direct digital sampling at the intermediate frequency was used [23]. The RFI removal technique resembles adaptive array processing. By using a reference channel, the RFI environment was sampled and transformed into the SAR fast-time domain. A least means square (LMS) adaptive algorithm could then correlated the reference signal to the primary input and iteratively determine a weighted signal for reduction of RFI [3].

The following process calculates an adaptive weight for each iteration of the pulse processing. A reference signal $x(n)$ is an L-dimension vector, constructed from the decorrelated, Δ-delayed version of the input signal [24]:

$$\begin{aligned} x(n) &= [x(n), x(n-1), \ldots, x(n-L+1)]^T \\ &= d(n-\Delta) \end{aligned} \tag{5.52}$$

Then an output signal $y(n)$ is formed from adaptively weighting the data:

$$y(n) = w^T(n) \cdot x(n) \tag{5.53}$$

The weights are iteratively calculated from the previous weights using a constant step-size parameter μ, and the error vector $e(n)$, by

$$w(n+1) = w(n) + \mu x(n) e^*(n) \tag{5.54}$$

where

$d(n)$	Input signal
$w(n)$	Complex filter weights of n-th iteration
$x(n)$	Reference signal
$y(n)$	Filter output
$e(n)$	Error vector from adaptive process
μ	Constant step size parameter
Δ	Constant decorrelation parameter
L	Filter length

Several techniques were explored for the interference subtraction in either the temporal or frequency domains. Figure 5–26 illustrates LMS adaptive interference subtraction from the pulse compression filter in the frequency domain. The first spectrum plot shows the received signal including the RFI and the transmitted waveform. It is impossible to detect the target or terrain feature in this representation. The adaptive filter helps to reduce the interference and enhancing the target visibility. The bottom graph in Figure 5–26 represents the ideal performance of the pulse compression process without RFI. By comparing the middle and lower spectra, it is evident that system sensitivity within 3 dB of ideal was achieved.

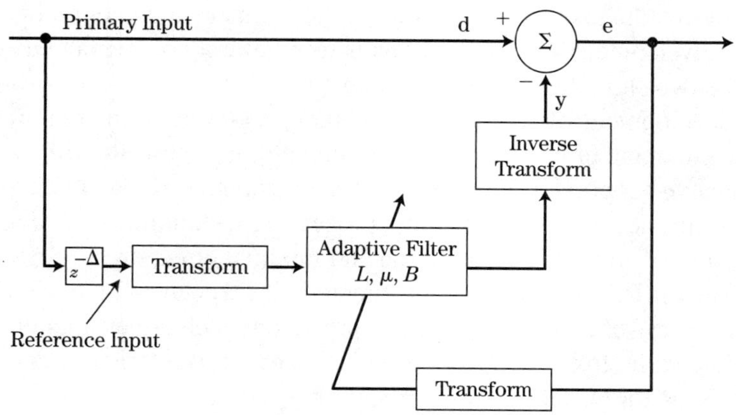

FIGURE 5–25

Adaptive RFI removal process [24]

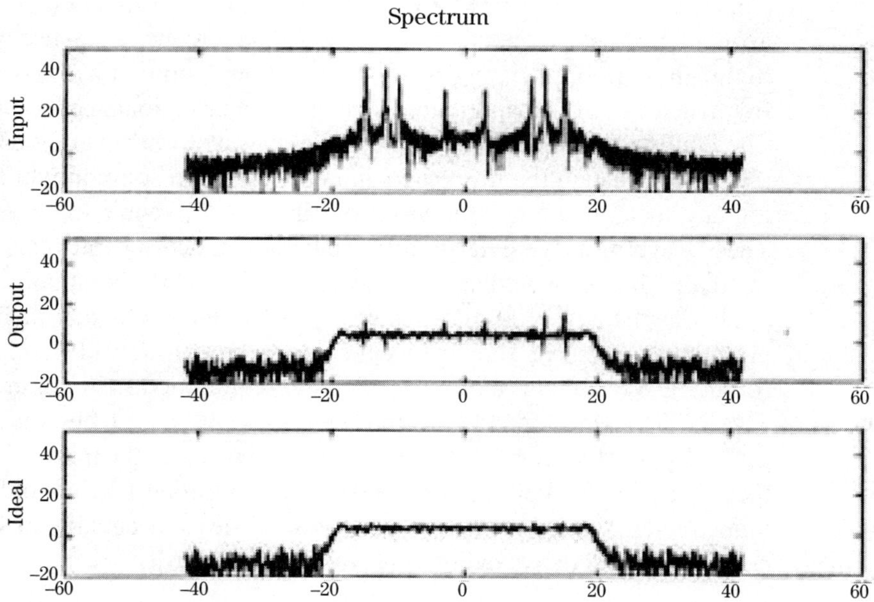

FIGURE 5–26

RFI removal from GeoSAR waveform [24]

5.3 RFI Suppression Summary

The subject of RFI suppression for FOPEN SAR systems has been a major challenge to developing and operating these important systems in any civilian environment. This chapter has covered the two major challenges: (1) adaptive processing for RFI removal; and (2) waveform design to satisfy the requirements for avoiding sensitive receivers in the RF band and geographic location. Each development was and continues to require a complementary awareness of the two signal processing requirements.

The early development of RFI removal of radio and television sources required the development of wideband adaptive processing. These algorithms relied on the relative geographic and stationary characteristics of the interference sources. Major developments in signal processing techniques were developed to enable real-time image formation and minimal effect on the SAR impulse response, ISLR, and polarimetric characteristics. The techniques also demonstrated a variety of innovative adaptive processing techniques, including compatibility with stretch processing, polarimetric characteristics, and interferometric SAR for terrain elevation mapping.

The avoidance of proscribed frequencies was both a technical and a sociopolitical challenge to FOPEN SAR systems. Worldwide there is a strict avoidance of any frequency associated with emergency response or safety of flight. In addition, it is necessary to avoid any communications or sensing frequencies that are important to a geographic or regional economic process. The FOPEN SAR system needed to develop waveform generation techniques that would sustain the fine range and cross-range resolution while eliminating any spectral energy that was proscribed by the regulating agencies [25]. These waveforms were developed and demonstrated several years before the regulating bodies would agree to license the RADAR operation.

It was the advent of ultra wideband communications that facilitated this eventual acceptance. The NTIA developed a regulation and testing approach for any UWB emitter and the verification requirements for commercial use. The RADAR developments were forced to apply these techniques and have generally been successful at getting (limited) operational approval. However, there is still a compelling need to develop more robust RFI avoidance techniques and to work with the regulating authorities to accept their use in a very competitive and dense radiofrequency spectrum [26].

5.4 References

[1] Miller T., McCorkle J., and Potter L., "Near Least Square Radio Frequency Interference Suppression," *Proc. SPIE,* Vol. 2487, Orlando, FL, April 1995, pp. 72–83.

[2] Federal Communications Commission (FCC), "Part 15 Regulations," July 10, 2008, http://www.fcc.gov/oet/info/rules/part15/PART15_07-10-08.pdf

[3] Hellsten, H., "CARABAS—An UWB low frequency SAR," Pr*oc. 1992 IEEE MTT International Symposium* Vol. 3, June 1992, pp. 1495–1498.

[4] Sheen, D. R., Vandenberg, N. L., Shackman, S. J., Wiseman, D. L., Elenbogen, L. P., and Rawson, R. F., "P-3 Ultra-Wideband SAR: Description and Examples," *IEEE AES Systems Magazine,* November 1996, pp. 25–30.

[5] Wheeler, K. and Hensley, S., "The GeoSAR Airborne Mapping System," *Proc. 2000 IEEE International Radar Conference,* Washington, DC, May 2000, pp. 831–835.

[6] Ruze, J., "Physical Limitations on Antennas," MIT Research Laboratory of Electronics, Technical Report 248, October 1952, Cambridge MA, p. 20.

[7] Lynch, D. Jr., *Introduction to RF Stealth*, Scitech Publishing, Raleigh, NC, 2004, Chapter 2.

[8] Davis M.E., "Design Alternatives for Foliage Penetration SAR Ultra Wideband Waveforms", Proc. 2010 IEEE International Radar Conference, Washington DC, May 2010

[9] Hensley, S., Le, C. T. C., and Gurrola, E., "A Rigorous Expression for the Amount of Interference from a Chirped Waveform Using Ideal Band Pass and Realistic Receive Models," Unpublished NASA Jet Propulsion Laboratory GeoSAR Memo, September). Sept 20, 1999.

[10] Ulander, L., "Precision Processing of CARABAS HF/VHF-Band SAR Data," *Proc. 1999 IEEE RADAR Conference,* Boston, MA, May 1999, pp. 47–53.

[11] Lord, R. T. and Inggs, M. R., "High Resolution SAR Processing Using Stepped-Frequencies," *Proc. IGARSS 1997,* Vol. 1, August 1997, pp. 490–492.

[12] Lord, R. T. and Ingss, M. R., "Approaches to RF Interference Suppression for VHF/UHF Synthetic Aperture Radar,", *Proc. 1998 South African Symposium on Communications and Signal Processing (COMSIG '98),* September 1998, pp. 95–100.

[13] Goodman, R., Tummala, S., and Carrara, W., "Issues in Ultra-Wideband, Widebeam SAR Image Formation," *Proc. IEEE 1995 International RADAR Conference,* Arlington, VA, May 1995, pp. 479–485.

[14] Chang, W., Cherniakov, M., Li, X., and Li, J., "Performance Analysis of the Notch Filter for RF Interference Suppression in Ultra-Wideband SAR," *Proc 9th International Conference on Signal Processing (ICSP200),* 2008, pp. 2446–2451.

[15] Le, C. T. C. and Hensley, S., "Removal of RFI in Wideband RADARs," *Proc 1998 IGARSS Symposium,* Seattle, WA, July 1998, p. 2032.

[16] Cheston, T. C. and Frank, J., "Phased Array Radar Antennas," in *RADAR Handbook* (2d ed.), Ed. Skolnik, M. I., McGraw Hill, New York, 1990, .pp. 7.38–43.

[17] Bogler ,P. L., "Motion-Compensated SAR Image ISLR," IEEE T*rans. on* G*eoscience and* R*emote Sensing* vol. GE-25, no. 6, November 1987, pp. 871–878.

[18] Koutsoudis, T. and Lovas, L., "RF Interference Suppression in Ultra Wideband Radar Receivers," *Proc. SPIE* Vol. 2487, Orlando, FL, April 1995, pp. 107–118.

[19] Miller, T., Potter, L. C., and McCorkle, J., "RFI Suppression for Ultra Wideband RADAR," *IEEE Trans. on Aerospace and Electronic Systems* Vol. 33, No. 4, October 1997, pp. 1142–56.

[20] Golden, A., Werness, S. A., Stuff, M., DeGraaf, S., and Sullivan, R., "Radio Frequency Interference Removal in a VHF/UHF Deramp SAR," *Proc. SPIE,* Vol. 2487, Orlando, FL, April 1995, pp. 84–95.

[21] Tufts, D. W. and Kumaresan, R. "Estimation of Frequencies of Multiple Sinusoids: Making Linear Prediction Perform like Maximum Likelihood," *Proc. IEEE,* Vol. 70, No. 9, September 1982, pp. 975–989.

[22] Toups, M., unpublished images provided by MIT Lincoln Laboratory as part of DARPA/Navy UWB P-3 Collection and Verification Program, 1997.

[23] Hensley, S. and Madsen, S. N., "Interferometric Radar Waveform Design and the Effective Interferometric Wavelength," *Proc. 2007 Waveform Diversity & Design Conference,* Kaui, HI, April 2007, pp. 287–291.

[24] Le, C. T. C., Hensley, S., and Chapin, E., "Adaptive Filtering Of RFI In Wideband RADARs Using The LMS Algorithms. Part I: The TDLMS Adaptive Filter," NASA Jet Propulsion Lab unpublished report, 2000,), http://trs-new.jpl.nasa.gov/dspace/bitstream/2014/15603/1/00-1309.pdf

[25] Lord, R. T., "Radio Frequency Interference Suppression Applied To Synthetic Aperture RADAR Data," *XXVIIIth General Assembly of International Union of Radio Science, URSI 2005,* New Delhi, India, October 2005.

[26] Lindenfeld M.J., "Sparse Frequency Transmit and Receive Waveform Design", IEEE Trans. Aerospace and Electronic Systems, Vol. 40, No. 3, July 2004, pp.851-860

CHAPTER 6

FOPEN Target Detection and Characterization

This chapter provides a summary of the major advances in the ability of RADAR to detect objects under foliage with sufficient characterization for scientific and tactical applications. Because of the spiky nature of foliage penetration clutter and the propagation losses through foliage, the concept of probability of detection and false alarm from microwave radars needs to be reexamined [1]. Specific areas of RADAR research to improve foliage penetration (FOPEN) synthetic aperture RADAR (SAR) target detection include the uses of polarization diversity and change detection along with their impact on the image formation processing. Target characterization includes techniques to discriminate man-made from natural objects by using polarization and image morphological filtering. The chapter starts with a short discussion of the image formation processing chain for efficient target detection. The details of image formation were covered in Chapter 4, and a consideration of radiofrequency interference (RFI) mitigation processing was given in Chapter 5. However, it is instructive to look at the whole image formation, target detection, and feature characterization chain to understand the importance of polarimetry and change detection.

The results of FOPEN SAR collection campaigns and subsequent image processing analysis have shown three important advantages of a polarization diverse SAR system. The first is the ability to counter the speckle obtained from wide collection angle SAR image processing through the use of polarization whitening. The second is the effect of polarization on differentiating several target features from clutter and other object classes. The third point is the importance of polarization on image change detection will be presented.

Several benefits of polarization discrimination have developed in the SAR literature over the past two decades. Specific to foliage penetration SAR, it is evident that the statistical distribution of clutter returns is very spiky in nature. In addition, the image segment transitions from one type of clutter to the next will provide strong specular returns at discrete aspect angles. Image polarization diversity provides the potential for reducing strong returns that can often resemble tactical targets [2].

Man-made objects exhibit returns that are strongly correlated at discrete angles. Thus, a physical object will provide similar returns to a RADAR when viewed at angles that provide strong correlation in magnitude and polarization returns. These features are usually attributed to the large planar structures and the junctions between these structures that will reflect incident energy with enhanced radar cross section (RCS). By knowing the characteristics of man-made objects and applying spatial filtering over the SAR image, it is possible to reduce the false alarms due to strong clutter discretes [3].

It is also known that the polarization of the scattering from various terrain types or slopes will have characteristic returns. If the terrain is flat, the principal polarization components as well as single and double bounce phase changes will be predictably returned. As a result, polarization has been proposed as a method for characterizing terrain type and slope [4].

6.1 Target Detection Processing

To detect targets under foliage, signal processing in ultra wideband (UWB) SAR includes several important image processing steps, as shown in Figure 6–1. The processing steps and images in the figure are from a P-3 UWB collection in Grayling, Michigan, where the three polarization images are HH, VV, and HV. Image formation algorithm processing counts were covered in Section 4.2, and Figure 6–1 summarizes the appropriate number of operations per pixel for each step in the process. RFI removal was covered in Chapter 5 and is typically the first step in the image processing procedure. The illustration of signal processing steps and their representative operations count include the following:

- The range migration algorithm (RMA) used on the P-3 FOPEN SAR images required an operations count of approximately 2,000 operations per pixel per polarization. The effects of RFI are clearly shown in the first image segment, motivating the need to first remove RFI.
- The second image segment visually illustrates the improvement in noise equivalent sigma-0 (i.e., σ^0_{ne}). The RFI removal carried out during the deskew image formation processing required 100 operations per pixel.

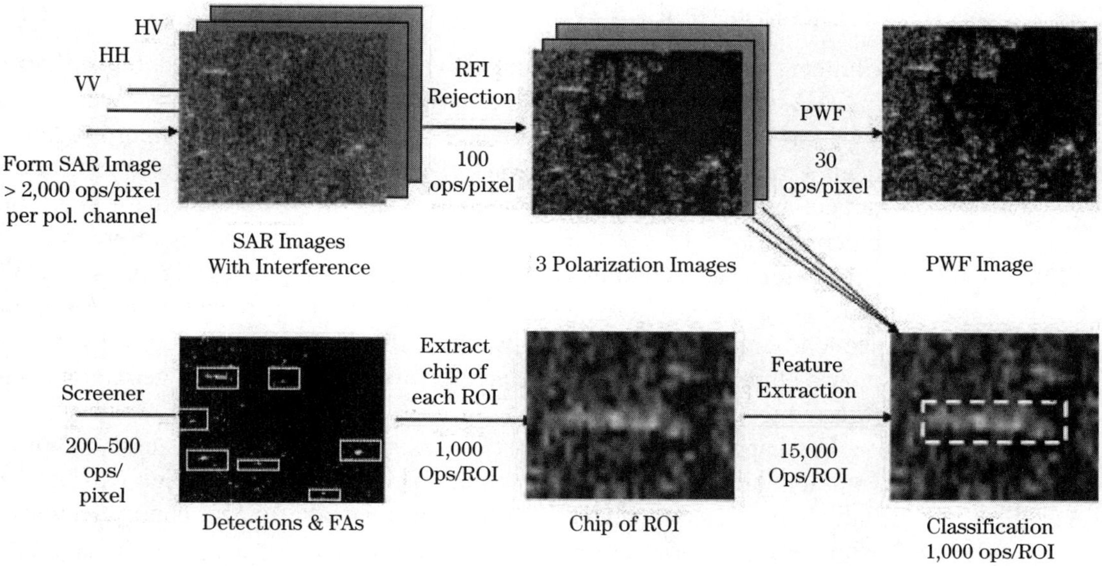

FIGURE 6–1

FOPEN SAR target detection and characterization processing flow [5]

- The polarization whitening filter (PWF) step provides a significant improvement in the target-to-clutter ratio, at a cost of 30 operations per pixel. PWF enhances the ability to nominate areas of the image that are likely to contain man-made targets, with lower false alarm rates.
- Target detection is accomplished through an area constant false alarm rate (CFAR) on the image, providing a nomination of regions of interest (ROI) for subsequent feature examination. Between 200 and 500 operations per pixel produce small areas (i.e., chips) around potential target along with the background clutter to be passed on for subsequent feature extraction.
- These "chips" are next extracted from the image along with auxiliary data of location and orientation, with a cost of approximately 1,000 operations per ROI. It is important that the pixels for each of the polarizations are located correctly to ensure correct pixel association and to enable subsequent target characterization.
- These ROI data will be examined finally for features that include size, shape, and polarization characteristics for final classification as a man-made target versus terrain artifact. The operations count is large, typically 15,000 operations per ROI. As a result the processing is likely to be in a ground processor after being sent down a data link. Hence, the management of false alarm rate is important in the previous processing step.

6.1.1 Polarization Whitening

Clutter speckle has been recognized to be the cause of significant false alarms in SAR images [1]. As a result, polarization diversity was examined extensively as a technique to reduce image speckle, primarily for microwave frequency radars. The early analysis of polarimetric returns from man-made targets and foliage clutter provided verification of the benefit of using the independent degrees of signal polarization characteristics. Therefore, it was a straightforward technology development to investigate applications of PWF processing to assist in reducing false alarms in foliage penetration SAR. Consequently, the P-3 UWB SAR was specifically built to collect fully polarimetric clutter and target characteristics over a wide range of geographic and forested regions.

In Chapter 3, the foliage clutter was shown to be non-Gaussian. With a Gaussian clutter model, each resolution cell of the SAR image will be spatially homogeneous and have the same average polarimetric energy. However, with the foliage log-normal clutter model, this uniform clutter characteristic cannot be assumed. In fact, there is a significant inhomogeneity of the clutter distribution over a typical SAR scene. Some of the inhomogeneity is due to differences in clutter type, some is due to boundaries between clutter such as tree lines, and much of the variation is due to speckle within normally homogeneous clutter types. PWF processing has been found to be beneficial for the later application.

A polarimetric FOPEN SAR system typically collects three polarization components by using two receiver channels. On successive pulses, horizontal and vertical transmit pulses illuminate the scene, and the copolarization and cross-polarization returns are recorded. Because the two cross-polarization channels HV and VH have been observed and analyzed as being reciprocal, only three channels are typically recorded: HH, HV, and VV.

Polarization whitening is the process that combines the three input channels into a complex vector to equalize the intensity in three polarization vector quantities and to decorrelate the three polarizations. The polarization measurements of the signal return are collected into a complex vector [1]

$$Y = \begin{bmatrix} HH \\ HV \\ VV \end{bmatrix} = \begin{bmatrix} HH_I + jHH_Q \\ HV_I + jHV_Q \\ VV_I + jVV_Q \end{bmatrix} \tag{6.1}$$

Note that the polarization covariance is a 3×3 unit matrix. The vector Y is assumed to be the product of a complex Gaussian vector X (representing the speckle) and a spatially varying, gamma-distributed texture variable g:

$$Y = \sqrt{g}X \tag{6.2}$$

The probability density function of the complex speckle vector X is defined b

$$f(X) = \frac{1}{\pi^3|\Sigma|}\exp(-X^*\Sigma^{-1}X) \tag{6.3}$$

where Σ is the polarization covariance matrix. It is common to express the clutter covariance matrices in terms of normalized linear-polarization bases, in the form

$$\sum = \sigma_{HH} \cdot \begin{pmatrix} 1 & 0 & \rho\sqrt{\gamma} \\ 0 & \varepsilon & 0 \\ \rho^*\sqrt{\gamma} & 0 & \gamma \end{pmatrix} \tag{6.4}$$

In (6.4), the parameter γ is the ratio of the expectation value of the VV intensity return to the HH intensity return

$$\gamma = \frac{E(|VV|^2)}{E(|HH|^2)} \tag{6.5}$$

Similarly, ε is the ratio of the HV cross-polarization intensity return to the HH intensity return

$$\varepsilon = \frac{E(|HV|^2)}{E(|HH|^2)} \tag{6.6}$$

Finally, ρ is the complex cross-correlation between the HH and VV returns

$$\rho = \frac{E(HH \cdot VV^*)}{[E(|HH|^2)E(|VV|^2)]^{1/2}} \tag{6.7}$$

From the polarimetric measurements at each pixel, the objective is to construct a new image that minimizes the variance of speckle among the three channels. This is carried out by a transformation of the complex vector Y in the quadratic form

$$y = Y^\dagger AY = gX^\dagger AX \tag{6.8}$$

The measure of speckle in the image is expressed as the ratio of the standard deviation of the image pixel intensities to the mean of the intensities (σ/μ)

$$\frac{\sigma_y}{\mu_y} = \frac{\text{std.dev}(y)}{\text{mean}(y)} \tag{6.9}$$

A is the desired weighting matrix that minimizes the (σ/μ) in the output SAR intensity image, y. This has been shown to equal the inverse of the polarization covariance matrix. So the polarization vector Y from each pixel in the image is passed through the whitening filter $\Sigma^{-1/2}$ to obtain a new image function Y_W:

$$Y_W = \Sigma^{-1/2}Y = \sqrt{g}\Sigma^{-1/2}X \tag{6.10}$$

The whitened vector Y_W then forms a single SAR (intensity) image with the requisite reduction in background clutter speckle, given by $1/sqrt(3)$.

$$Y_W = \left[HH, \frac{HV}{\sqrt{\varepsilon}}, \frac{VV - \rho^* \sqrt{\gamma} HH}{\sqrt{\gamma(1 - |\rho|^2)}} \right] \tag{6.11}$$

The scale factors in (6.11) come from the three ratios of image polarization channel data given by γ, ε, ρ in (6.5) to (6.7).

Figure 6–2 shows the three P-3 FOPEN SAR polarimetric input channels from the Grayling, Michigan, collection, where the clutter and targets from this collection have been extensively analyzed [3]. The figure clearly shows that the HH and VV channels exhibit a spiky distribution of clutter discretes,

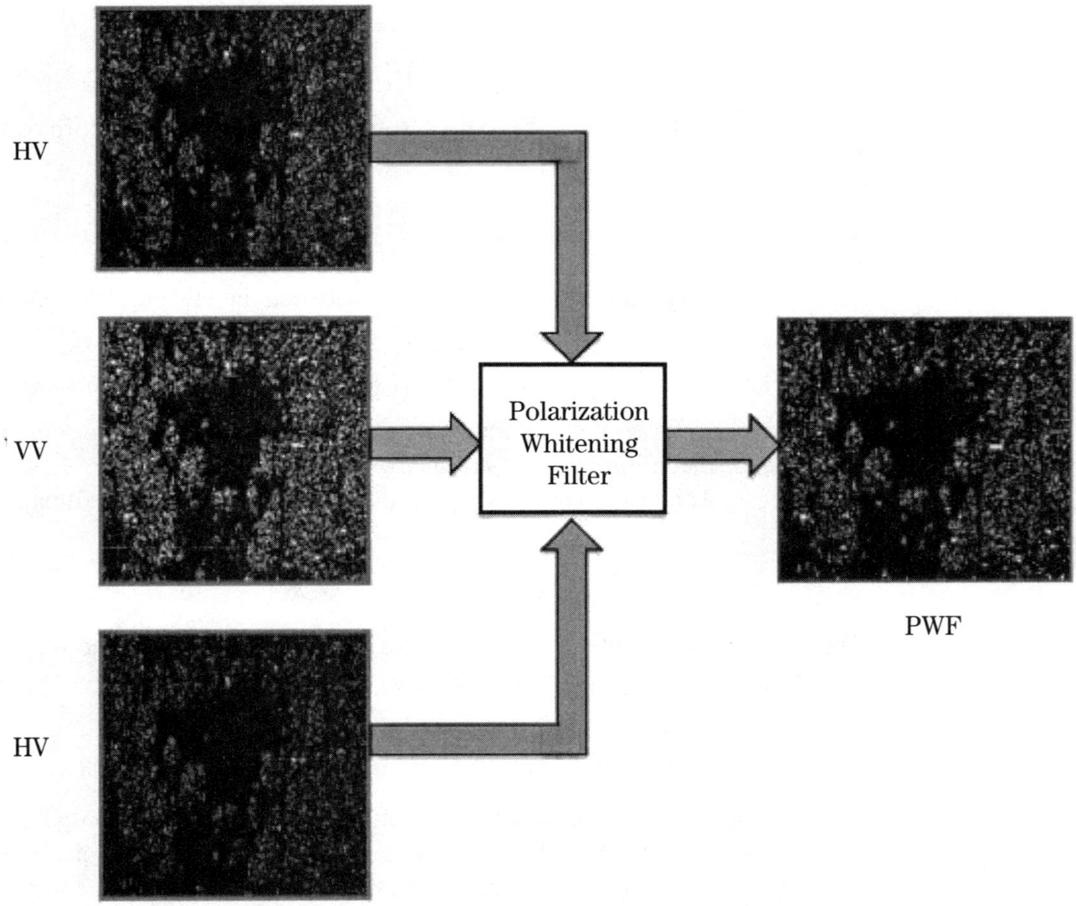

FIGURE 6–2

Polarization whitening of Grayling P-3 UWB data. *Source:* MIT Lincoln Laboratory representative of a log-normal distribution as summarized in Chapter 3. To characterize this clutter distribution, a set of 500 clutter chips were extracted from the images, each containing 200 pixels in range and 100 pixels in cross-range. Then, the values of the parameters of a polarization covariance matrix were estimated for homogeneous stands of tree clutter. For the northern latitude tree clutter, the parameters are found to be approximately [2]

$$\begin{aligned} \sigma_{HH} &= 0.08 \\ \varepsilon &= 0.25 \\ \gamma &= 0.5 \\ \rho\sqrt{\gamma} &= -j0.125 \end{aligned} \tag{6.12}$$

When the three channels were combined with the PWF, the clutter distribution of scatterers is modified significantly. This is evidenced in Figure 6–2 by the narrower distribution of return amplitudes in the forested area and the reduction in speckle noise in the open field.

The signal-to-clutter ratio (SCR) level for targets under the foliage was also improved from 9 dB to 19 dB. More importantly, the targets suffered no loss in resolution. Thus, it is expected that after PWF processing the target detection process will be greatly improved.

6.1.2 Constant False Alarm Rate Processing

It is important to recognize that FOPEN SAR returns are volumetric due to the foliage clutter surrounding the target. Therefore, targets concealed beneath foliage clutter need be detected in the presence of both the scattering and attenuating propagation through the volumetric clutter. Moreover, the amplitude of the signal return will be reduced by the foliage loss, and target detection cannot generally be improved with finer resolution because of competing volumetric clutter. Hence, image processing techniques are needed to discriminate the man-made objects under foliage from the surrounding natural clutter.

This foliage volumetric clutter phenomenon contrasts with imaging from microwave SAR, which cannot penetrate the foliage and detects targets only above the clutter. For microwave SAR, the only "volumetric" returns are from clutter layover, when objects that are higher than the target appear in the same image resolution cell.

The detection process starts by defining the ROI though a screening process. The clutter statistics are sampled around a particular pixel or group of pixels, as shown in Figure 6–3. It is important that the statistics are obtained with sufficient separation from the expected target area; so that returns from the target pixels will not bias the estimate of the clutter. A mean μ_Z and

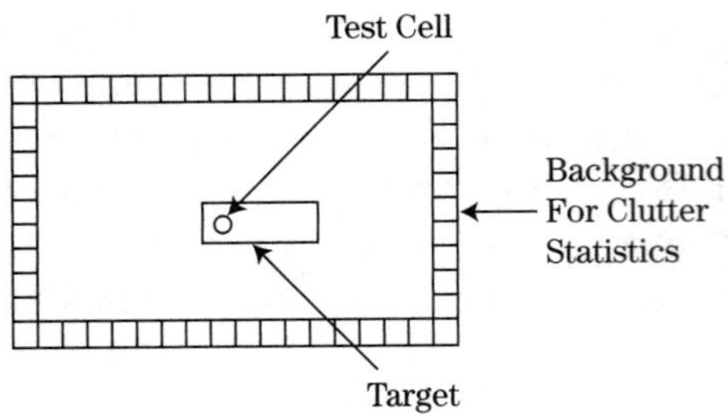

FIGURE 6–3

CFAR window for target detection [1]

standard deviation σ_Z of the clutter background is calculated, and a decision concerning a potential target is determined from the threshold.

The CFAR test consists of the following steps:

- A test cell is examined, within a small region of pixels representative of the shape of the expected object.
- An exclusion zone around the test cell is used to calculate the statistics of the background clutter. The test statistic includes calculation of the average clutter and the logarithm of the standard deviation of the clutter.
- Several cells associated with the target and within the boundary will typically pass the CFAR test. These cells are grouped within a ROI for subsequent examination.

A modified two-parameter CFAR detector has been found to be adequate for defining the FOPEN ROIs. This detector is defined by the rule [1]

$$y \underset{clutter}{\overset{target}{\underset{<}{\geq}}} \mu_z + K_{CFAR}\,\sigma_z' \tag{6.13}$$

The definition of CFAR is used in general terms in this text because high-resolution FOPEN SAR does not follow Gaussian statistics. Nevertheless, the detector defined by the constant K_{CFAR} can be chosen to limit the number of false alarms in the image and to provide a mechanism to reduce the number of detections for further processing. It should also be noted that a modification of the standard deviation σ_Z is used, where [1]

$$\sigma_z' = \max\{\sigma_z, 4dB\} \tag{6.14}$$

The modification is needed to protect against too low of a standard deviation where the false alarm levels will be raised. A limit on the standard deviation to being greater than 4 dB has been used primarily for microwave SAR. This hard limit needs to be reexamined in various foliage environments, depending on the clutter characteristics and signal propagation loss.

The prescreening eliminates large areas of clutter, where it is unlikely that an object of interest would exist. In the CFAR process, the objective is to maintain a high probability that all of the valid targets have been selected. However, with the nature of FOPEN SAR, many false alarms will pass through this process. As previously illustrated in Figure 6–1, the next step in the target nomination procedure is to examine each ROI and to classify the image as to whether it is a man-made object or cultural clutter. This characterization process has been the source of significant development and remains largely an art. Several powerful techniques are employed in the effort, including the following:

- Using morphological filters to determine whether the ROI has sufficiently strong detections to represent the expected size and shape of the object.
- Using characteristics of the detected pixels to determine if the grouping has the traits of an expected object.
- Examining the polarization statistics of the pixels to determine if the scattering has the polarization characteristics of man-made objects of interest.

These techniques will be developed in the next sections.

6.2 Polarimetric Scattering

Polarimetry plays a significant role in the characterization of both natural and man-made objects. Scattering from various terrain types exhibit significant differences, especially when the scattering sources are random in size and orientation. This condition is particularly true for foliage penetration radars with long wavelengths. The scale of the electromagnetic interaction with natural terrain features is such that the returns become Rayleigh in general. There are exceptions, when the objects are on the same scale as the wavelength and when the objects are regular in shape. The return from a large tree trunk over a ground plane is a classic example of an exception to the rule, since it provides a return very similar to a top-hat reflector.

A polarimetric RADAR return, for a system having two linearly independent transmit and receive polarizations, is characterized by the separate backscatter of the electromagnetic fields from a target or element of clutter. For horizontal and vertical linear polarization (H-V) coordinate basis, the matrix of coefficients is given by [6]

$$S(f,\theta) = \begin{bmatrix} S_{HH}(f,\psi) & S_{HV}(f,\psi) \\ S_{VH}(f,\psi) & S_{VV}(f,\psi) \end{bmatrix} \tag{6.15}$$

The monostatic scattering coefficients of $S(f, \psi)$ are a function of the frequency f and the incidence angle ψ, as obtained from the complex SAR imagery. It is important that, when obtaining the scattering coefficients, the multiple polarization receiver channels are coherent and matched in amplitude and phase. Each image resolution cell is assumed to contain a dominant scattering center, whose scattering response is averaged over the bandwidth and aperture angle of the SAR. However, many more scattering centers in the image cell can often contribute to the composite polarization of the electromagnetic return.

6.2.1 Scattering Entropy

A key attribute of the FOPEN SAR images is the entropy of the pixels and the contributions of individual scatterers to the intensity distribution. In a two-dimensional image of pixels x_{nm} the entropy H of pixel scattering is defined as [7]

$$H(x_{nm}) = -\sum p_{nm} \log(p_{nm}) \tag{6.16}$$

where p_{nm} is the proportional contribution of the pixel to the total power P:

$$p_{nm} = \frac{|x_{nm}|^2}{P}, \quad P = \sum_{nm} |x_{nm}|^2 \tag{6.17}$$

Entropy has the important property of a lower bound of 0 when there is a single scatterer and an upper bound of $\log(N)$ when $p_n = 1/N$. Thus, smaller values of entropy correspond to a more localized concentration of scattering. The SAR image entropy has a scale and shift invariant property and thus is insensitive to the global image gain, phase or translational shift, assuming identical distributions of clutter in the scene [8].

Entropy is also a very useful characteristic of SAR images, especially foliage penetration SAR returns. Because entropy can isolate and characterize principal scattering centers, it has been used for object characterization [8], image focusing [9], and image registration [10]. To establish the relative polarimetric scattering mechanisms, an entropic approach will be taken in explaining the SAR returns from several classes of scatterers, both natural and man-made.

Foliage penetration RADAR scattering has several important mechanisms that are based on the geometry and features of the foliage and surrounding terrain. These effects, which are illustrated in Figure 6–4, show five classes of primary scattering, and their attendant impact on the received polarization. These scattering effects are summarized by the following characteristics [11]:

Type I: For a surface roughness, which has a root mean square (RMS) scattering coefficient σ that is much less than a wavelength, the scattering will

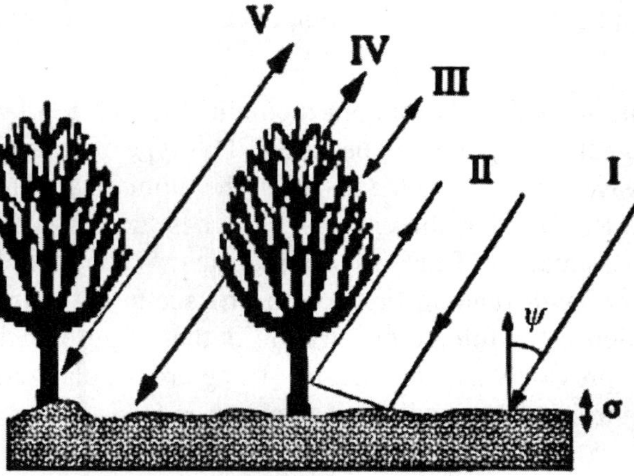

FIGURE 6–4

Five important polarization scattering types [11]

be highly specular. As a result, the entropy is almost zero, and hence there is little or no cross-polarization component. This scattering is characterized by either a dominant HH or a VV scattering cross section and a variation with the grazing angle of the incident wave.

The scattering matrix of the surface, which is usually characterized as Bragg scattering, has the general form of

$$[S_I] = A \begin{bmatrix} \alpha_{HH} & 0 \\ 0 & \alpha_{VV} \end{bmatrix} \tag{6.18}$$

The scattering coefficients are a function of the average incidence angle θ and complex relative permittivity ε_r of the surface such that

$$\alpha_{HH} = \frac{\cos\psi - \sqrt{\varepsilon_r - \sin^2\psi}}{\cos\psi + \sqrt{\varepsilon_r - \sin^2\psi}} \tag{6.19}$$

$$\alpha_{VV} = \frac{(\varepsilon_r - 1)[\sin^2\psi - \varepsilon_r(1 + \sin^2\psi)]}{(\varepsilon_r\cos\psi + \sqrt{\varepsilon_r - \sin^2\psi})^2} \tag{6.20}$$

Type II: A double bounce scattering between the ground and a vertical object represents a low-order scattering, as occurs for dihedrals reflections in forests and urban settings. This will be the product of two Type I scattering processes:

$$[S_{II}] = \begin{bmatrix} \alpha_{HH1}\alpha_{HH2} & 0 \\ 0 & -\alpha_{vv1}\alpha_{vv2} \end{bmatrix} \tag{6.21}$$

The values of $\alpha_{P\varepsilon}$ $[\alpha_{HH}, \alpha_{VV}]$ are given by (6.19) and (6.20) for horizontal and vertical polarization, respectively, with the appropriate permittivity for the scattering material on each bounce. The most important aspect of this scattering type is that there is little entropy in the process, so little or no cross-polarization components are created. Furthermore, there is a π radian phase difference between the HH and the VV components over a wide range of incidence angles.

Type III: The tops of the trees form a random medium, for which a large portion of the incident signal energy does not penetrate. This type of scattering is common with microwave SAR, where the losses in the biomass are large. For long wavelength SAR, however, the energy penetrates the tops of the trees, and a volumetric summation of the scattering occurs. When there are large number of scatterers with random orientation, the scattering entropy will be 0.95. The experience for foliage penetration is that large branches with random orientation provide characteristically strong cross-polarization returns. The Type III scattering matrix will have the form of [11]

$$[S_{III}] = A \begin{bmatrix} 3 & 0 & 1 \\ 0 & 2 & 0 \\ 1 & 0 & 3 \end{bmatrix} \tag{6.22}$$

The third dimension in (6.22) comes from a third eigenvalue of the entropy due to HV cross-polarization scattering. As a result, Type III scattering is orthogonal to the Type II and Type I scattering. Thus, the combination of the three scattering types makes a useful set of bases functions for characterizing the foliage.

Type IV: For long-wavelength RADARs, Type IV scattering is characteristic of surface reflections but after propagation trough the tops of the trees has occurred. The scattering from the surface is a Type I low-entropy process but only after it undergoes random scattering and attenuation by the foliage. Depending on the nature of the foliage scattering, the cross-polarization component could be small or negligible.

Type V: This model represents single-path scattering from anisotropic structures such as tree trunks, which can be modeled as a rough dielectric cylinder. The scattering from the tree trunk is another low-entropy process, represented by the return from a cylinder over a ground plane. Depending on the orientation of the cylinder, there may be a change in the polarization, analogous to a tilted dihedral. For normal incidence the $[S_V]$ matrix is diagonal in the HV base. For oblique angles, the scattering matrix is no longer diagonal, as components of HH and VV polarizations are returned.

6.2.2 Polarization Entropy

Polarization entropy is an important approach to characterizing the scattering centers and mechanisms. The development of the polarization entropy starts with the coherent scattering matrix expressed in (6.15). The following treatment will focus on the linear polarization components of [HH HV VH VV]. Analogous treatment of the coherency and polarization processes can be derived for other polarimetric sensor implementations. The matrix is converted to a vector, with the form [10]

$$\vec{k} = [S_{HH}\ \ S_{HV}\ \ S_{VH}\ \ S_{VV}]^T \tag{6.23}$$

Under the assumption of reciprocity of the cross-polarization component, a reduced target vector can be used and has the form

$$\vec{k}_3 = [S_{HH}\ \ \sqrt{2}S_{HV}\ \ S_{VV}]^T \tag{6.24}$$

Note that the $\sqrt{2}$ multiplier for the HV term is used to maintain the norm of the scattering matrix. The coherency matrix is formed from the outer product of the k vector with its conjugate transpose:

$$[T_3] = k_3 g k_3^{*T} = [T_3]^{\dagger} \tag{6.25}$$

This leads to an important concept in the target decomposition into eigenvalues of the coherency matrix. Since the eigenvalue problem is basis invariant, they can be used to generate a diagonal form of the coherency matrix. This is physically interpreted as statistical independence between a set of target vectors. As such, they are expected to yield a general decomposition of the independent scattering processes. The coherency matrix is written in the form

$$[T_3] = [U_3]\,[\Sigma]\,[U_3]^{-1} \tag{6.26}$$

where $[\Sigma]$ is a 3×3 diagonal matrix with nonnegative real elements:

$$[\Sigma] = \begin{bmatrix} \lambda_1 & 0 & 0 \\ 0 & \lambda_2 & 0 \\ 0 & 0 & \lambda_3 \end{bmatrix} \tag{6.27}$$

where the eigenvalues are in order, $\lambda_1 > \lambda_2 > \lambda_3$. [U3] is a unitary transformation that diagonalizes the polarization matrix.

From this decomposition of the monostatic RADAR image pixel scattering, the polarimetric entropy H_P is defined as

$$H_P = \sum_{i=1}^{3} -P_i \log_3 P_i, \quad P_i = \frac{\lambda_i}{\sum_{j=1}^{3} \lambda_j} \tag{6.28}$$

This approach considers the entropy of the scattering medium and not the electromagnetic wave itself. If the entropy H_P is low, the environment is considered to be weakly depolarizing and the dominant target scattering component is the largest eigenvalue. If the entropy is high, then the target is depolarized, and there is no longer a dominant scattering matrix. As such, the full eigenvalue spectrum must be considered.

6.2.3 Polarization Basis Vectors

Polarization diversity in SAR measurements has been shown to yield improved detection and characterization of man-made targets in clutter [11,12]. Just as the characteristics of clutter can be characterized by basis functions, man-made objects are made up of scattering centers with unique polarization signatures. When the resolution of the SAR is sufficient to isolate these scattering centers, the return will readily provide discrimination. Even in the case of multiple scattering centers in a resolution cell, the characteristics can be separated, provided that sufficient degrees of freedom exist and the SCR is sufficiently large. The separation of these features has been the subject of significant research, which includes techniques of maximum likelihood, Bayesian maximum a posterior (MAP) probability and minimum entropy or eigenvalue processing. The common theme in these target characterization techniques is obtaining a set of orthogonal basis vectors from which the processing can be trained [12].

Man-made objects can be classified into canonical, primitive shapes based on their geometry and orientation in the electromagnetic scattering. The principal scattering return and RCS of many targets were developed in Section 4.2. In this section, those targets and other general shapes will be presented from a polarization scattering perspective. When the primitive shapes summarized in Figure 6–5 are representative of man-made object scattering centers as well as a few objects in nature, polarization can provide a strong discrimination. Polarimetric scattering from the basic objects is characterized by a consideration of symmetry, where the return is either linear or rotational. Linear symmetry primarily affects the linear component of the polarization (horizontal or vertical), whereas the rotational symmetry will affect the component of circular polarization (right hand or left hand).

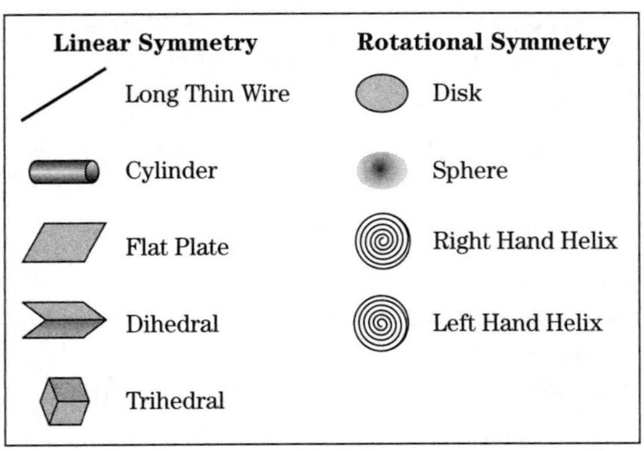

FIGURE 6–5

Primitive polarimetric scattering elements

In addition to symmetry, the number of bounces in the wave propagation from a shape is important in terms of the phase relationship of a coherency matrix. Thin wires, cylinders, and flat plates exhibit single bounce reflections of the RADAR signal radiation. A dihedral has double bounce, whereas trihedral will exhibit a triple bounce.

Polarization scattering has been treated in terms of basis functions, which have been analyzed in the literature. The dihedral–trihedral basis functions have been used in a Bayesian decision process for classifying man-made objects. By training the classifier, it is computationally convenient to examine each pixel and to determine the principal type of scatterer.

Under the assumption that the scattering is reciprocal, the HV and VH contributions are the same. As such, the polarization vector is treated as a three-component vector, as previously described in (6.26):

$$S = [S_{HH} \;\; \sqrt{2}S_{HV} \;\; S_{VV}]^T \tag{6.29}$$

These primitive scattering types, illustrated in Figure 6–5, can now be determined, assuming monostatic measurements in free space, a tilt angle ψ, and backscattering field measurements in a HV basis. The scattering vector S is expressed as a normalized signature vector $b(\tau, \rho, \psi)$, and scaled by a complex amplitude

$$A_c = A\exp[j\rho] \tag{6.30}$$

In this formulation, the parameters (A, ρ, ψ) describe the amplitude, absolute phase, and orientation angle of the scattering center, respectively.

The scattering representation of the primitive objects is given by

$$S = A_c R_{HV}(\psi) b(\tau, \mu, \gamma) \tag{6.31}$$

$R_{HV}(\psi)$ is a 3×3 matrix in the HV coordinate system. To provide a simpler interpretation in other scattering frames, the scattering function is transformed by a rotation matrix B into an alternate frame, for example, horizontal-vertical basis (H-V), trihedral-dihedral (T-D) basis, or a left-right (L-R) circular-basis.

A rotational transformation matrix $R_B(\psi)$ for scattering basis B can be expressed by

$$R_B(\psi) = B^H R_{HV}(\psi) B \tag{6.32}$$

Three common bases and their corresponding rotation matrices are given in Table 6–1. Many of the features on man-made objects such as vehicles are either trihedrals or dihedral scatterers. Using the previous notation, the polarimetric signatures for scattering centers in the T-D coordinate system can be defined, as summarized in Table 6–2.

Table 6–1 Different bases for canonical scattering centers [6]

Basic	B	$R_B(\psi)$
Horizontal-Vertical (H-V) Basis	$\begin{bmatrix} 1 & 0 & 0 \\ 0 & 1 & 0 \\ 0 & 0 & 1 \end{bmatrix}$	$\begin{bmatrix} \cos^2\psi & -\sqrt{2}\sin\psi\ \cos\psi & \sin^2\psi \\ \sqrt{2}\sin\psi\ \cos\psi & \cos^2\psi - \sin^2\psi & -\sqrt{2}\sin\psi\ \cos\psi \\ \sin^2\psi & \sqrt{2}\sin\psi\ \cos\psi & \cos^2\psi \end{bmatrix}$
Trihedral-Dihedral (T-D) Basis	$\begin{bmatrix} \frac{1}{\sqrt{2}} & \frac{1}{\sqrt{2}} & 0 \\ 0 & 0 & 1 \\ \frac{1}{\sqrt{2}} & -\frac{1}{\sqrt{2}} & 0 \end{bmatrix}$	$\begin{bmatrix} 1 & 0 & 0 \\ 0 & \cos(2\psi) & -\sin(2\psi) \\ 0 & \sin(2\psi) & \cos(2\psi) \end{bmatrix}$
Left-Right (L-R) Circular Basis	$\begin{bmatrix} \frac{1}{2} & \frac{\sqrt{2}}{2} & \frac{1}{2} \\ \frac{j}{2} & 0 & -\frac{j}{2} \\ \frac{1}{2} & \frac{\sqrt{2}}{2} & -\frac{1}{2} \end{bmatrix}$	$\begin{bmatrix} \exp(-j2\psi) & 0 & 0 \\ 0 & 1 & 0 \\ 0 & 0 & \exp(j2\psi) \end{bmatrix}$

Table 6–2 Polarization scattering matrices for TD coordinate system [10]

i	Scattering Center	$R_{TD}(\psi)b(\tau, \omega, \gamma)$
1	Trihedral	$\begin{bmatrix} 1 \\ 0 \\ 0 \end{bmatrix}$
2	Dihedral ($a = 0$)	$\begin{bmatrix} a \\ \cos(2\psi) \\ \sin(2\psi) \end{bmatrix}$
3	Dipole ($a = 1$)	
4	Cylinder ($a = 3$)	
5	Narrow Diplane ($a = 1/3$)	
6	Quarter Wave ($a =$ j)	
7(8)	Left (Right) Helix	$\exp(j2\psi)\begin{bmatrix} 0 \\ 1 \\ (-)j \end{bmatrix}$

To exploit these basic target characteristics, the FOPEN RADAR must be designed to operate and maintain orthogonality and phase-coherency between the polarization senses. Two alternative approaches have been used in the remote sensing designs, as illustrated in Figure 6–6. Use of the linear polarization in RADAR design is most common for UWB SAR, because of the limitations in antenna technology over the very large instantaneous bandwidths. However, circular polarization can be synthesized by a full set of polarization channels. It should be noted that the major challenge in using

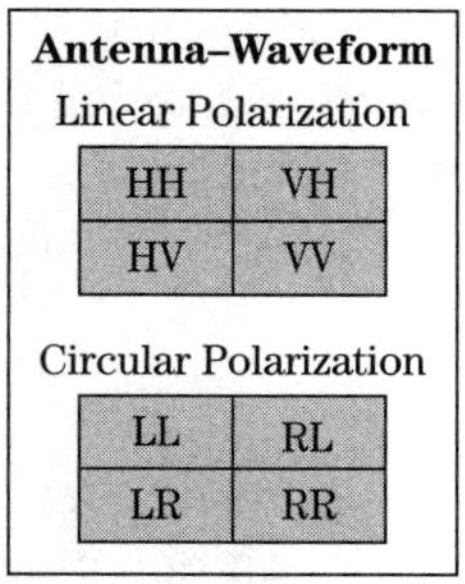

FIGURE 6–6

Alternative full polarization channels for FOPEN imaging

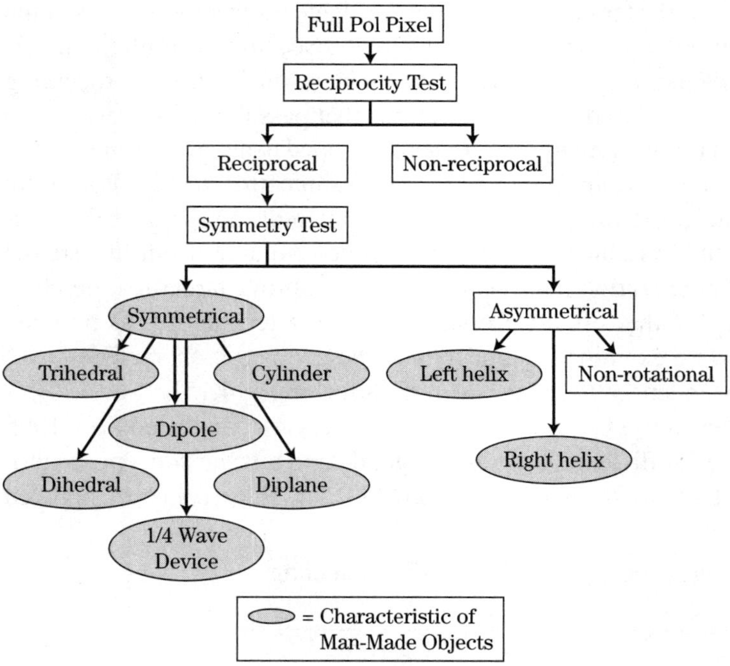

FIGURE 6–7

Polarization symmetry of man-made and natural objects [10]

polarization for target characterization is the ability to maintain channel characteristics of amplitude and phase, along with orthogonality of the polarization vectors. This will be illustrated in Chapter 7.

Recognizing that polarization has been a strong discriminate for man-made objects is a strong motivation to look at scattering classes. Many systems will test the return for reciprocity and symmetry prior to a decision on the scattering type.

As indicated in Figure 6–7, canonical man-made objects will show symmetry. However, it is also possible to get strong asymmetric scattering from fundamental shapes. If the polarization return does not exhibit reciprocal or rotational symmetry, it is unlikely to be man-made. But it should be noted that some cultural objects will exhibit reciprocal and symmetric scattering. Thus, it is important to take the polarimetric scattering data in context with the surrounding scattering environment.

6.3 Target Characterization

After efficient image formation and RFI mitigation have been carried out, image processing of FOPEN SAR provides an opportunity to improve the detection of tactical targets. However, the false alarm rate may remain high due to reduction of the target signature due to foliage propagation losses or the similarity of natural objects to the desired targets. Because of similarity in target strength and feature size to many natural objects, further image processing techniques are needed to discriminate returns that pass the CFAR test.

Polarization is one of many techniques developed to determine whether an individual object is man-made or natural. At the same time, there was a strong emphasis on increased image resolution in an attempt to segregate the scattering centers and to reduce the competing clutter volume. Both fine spatial resolution and polarization have been shown to improve target feature characterization [13]. Other effective image processing techniques, as outlined in Table 6–3, were examined to understand the geometric characteristics of tactical objects under the forest versus terrain features [14,15].

The sequence of P-3 UWB FOPEN data collections, conducted from 1995 to 1999, provided sufficient clutter and target data to test the concept of automated target detection and characterization (ATD/C). These concepts included

Table 6–3 Image processing techniques for detecting man-made objects [13]

Features Types	Algorithm
Textural	Standard Deviation Fractal Dimension Ranked Fill Ration
Size	Mass Diameter Normalized Square Rotational Intertia
Contrast Features	Peak CFAR Mean CFAR Percent Bright CFAR
Polarimetric Features (Fully Polarimetric Only)	Percent Pure (odd or even) Percent Pure Even Percent Bright Even

techniques of image segmentation and target strength for both false alarm reduction and classification of man-made targets. It is not anticipated that FOPEN SAR should provide any useful automatic target recognition due principally to two factors. First, there are insufficient pixels on a target at the long wavelengths of UHF and VHF to expect a satisfactory target recognition performance. These target recognition objectives have been tried for several decades with limited success for microwave targets in the open. But, more importantly, the few features that exist on man-made and tactical objects will be occluded by the propagation of signal energy through large trees. These trees essentially mask many of the pixels and cause amplitude and phase perturbation of the RADAR returns. So the most that can be reasonably expected from FOPEN SAR ATD/C is to place objects in broad classes, based on the features that robustly exist in the measurements.

6.3.1 FOPEN Target Features

Four types of features have been employed in classification analyses, as summarized in Table 6–3 [13]:

- *Texture:* The spatial variation of the returns in the neighborhood or as part of the area being characterized. The returns for clutter and target within the CFAR box from Figure 6–3 are examined for their particular statistical characteristics. The *standard deviation* of the returns in the area is a measure of the fluctuation of the intensity in an image. The *fractal dimension* measures the N brightest scatterers in the region and characterizes it in terms of the geometric shape. A single bright pixel will have a dimension of zero, a line a dimension of 1, and solid rectangle a dimension of 2. Finally, the *rank fill* feature measures the percentage of total energy contained in the N largest pixels. N is typically chosen as being 5.
- *Size:* The grouping of strong scatterers and association with an object via a morphological filter for determination of length and width, as illustrated in Figure 6–8. The *mass* feature is obtained by counting the number of pixels in this morphological shape. The *normalized square rotational inertia* feature is the second mechanical moment of the shape around its center of mass; normalize by the inertia of an equal mass.

  The *center of mass* of the object within the image chip can be calculated by the weighted position of each pixel that exceeds the CFAR threshold. Given (x_i, y_i) as the position of the pixel, the centers of mass M_x and M_y are calculated b

$$M_x = \frac{\sum_i x_i}{N} \qquad M_y = \frac{\sum_i y_i}{N} \tag{6.33}$$

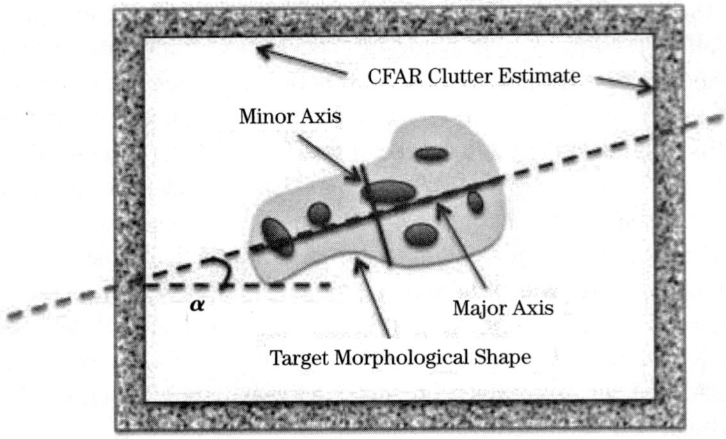

FIGURE 6–8

Estimating size and shape features for target characterization

The *second moment* of the image (i.e., its inertia) is now given in three dimensions, based on the distance between each detected image pixel and the center of mass of the image:

$$U_{xx} = \frac{1}{N}\sum_i (x_i - M_x)^2 \tag{6.34a}$$

$$U_{yy} = \frac{1}{N}\sum_i (y_i - M_y)^2 \tag{6.34b}$$

$$U_{xy} = \frac{1}{N}\sum_i (x_i - M_x)(y_i - M_y) \tag{6.34c}$$

The critical parameters for target discrimination are next determined by estimating the *pose* (i.e., orientation) of the shape with respect to the x- and y-axes. Based on these measurements, the pose of the object is determined as

$$\alpha = \frac{1}{2}\tan^{-1}\left(\frac{2U_{xy}}{U_{xx} - U_{yy}}\right) \tag{6.35}$$

The coordinate system can now be transformed from (x, y) by an α-rotation in the image plane to (x', y') along the pose axis. The length of the major and minor axes are next determined centroiding the locations of the N_C bright pixels:

$$U_{Major} = \sqrt{\frac{1}{N_C}\sum (x_i')^2} \tag{6.36a}$$

$$U_{Minor} = \sqrt{\frac{1}{N_C}\sum (y_i')^2} \tag{6.36b}$$

- *Contrast:* Obtained from the statistics in the CFAR processing. The ratio and distribution of the scattering center returns in the target shape in Figure 6–8 is compared with the background clutter statistics. The *maximum* statistic is the greatest intensity pixel in the target distribution. The *mean feature* is the average of the CFAR target returns within the morphological shape, and the *percentage bright* feature is the percentage of the pixels within the object that exceeds a CFAR threshold.
- *Polarization:* Has been found to be a very effective discrimination between man-made and natural scatterers, as summarized in Section 6.2. *Odd-bounce* statistics are representative of flat plate or trihedral objects, and *even-bounce* returns are associated with the RADAR scattering from a dihederal. Because few dihedral structures exist in natural clutter but are very prevalent for man-made targets, *even-bounce* is a strong discriminant. Hence, the polarimetric features are calculated from a transformation of the HH–VV data collection coordinates to even-bounce and odd-bounce images. This transformation is given by

$$E_{even} = \frac{|S_{HH} + S_{VV}|^2}{2} \tag{6.37}$$

$$E_{odd} = \frac{|S_{HH} - S_{VV}|^2}{2} - 2|S_{HV}|^2 \tag{6.38}$$

The *percentage-pure* feature is the fraction of the pixels within the target shape for which at least a threshold of the scattered energy falls within either the even-bounce or odd-bounce metric. The *percentage-bright-even* feature is the fraction of the pixels within the target shape that exceed a threshold in the CFAR image and which are predominantly even-bounce scatterers.

After image segmentation or designation of ROIs, evaluation of the pixels in a target chip provides additional quantitative scoring to the even- and odd-bounce features to support additional image characterization. First the S_{HH}/S_{HV} and S_{HH}/S_{VV} ratios are calculated for each pixel. Then the average and standard deviation of the measurements are calculated for the number of pixels in the ROI. The six additional polarimetric features that need to be calculated are

$$\begin{aligned} P_1 &= \text{Average}\left(\frac{|S_{HH}|}{|S_{HV}|}\right) & P_4 &= \text{Average}\left(\frac{|S_{HH}|}{|S_{VV}|}\right) \\ P_2 &= \text{Std.Dev.}\left(\frac{|S_{HH}|}{|S_{HV}|}\right) & P_5 &= \text{Std.Dev.}\left(\frac{|S_{HH}|}{|S_{VV}|}\right) \\ P_3 &= \left(\frac{\text{Average}|S_{HH}|}{\text{Average}|S_{HV}|}\right) & P_6 &= \left(\frac{\text{Average}|S_{HH}|}{\text{Average}|S_{VV}|}\right) \end{aligned} \tag{6.39}$$

The final step in the target discrimination process combines the appropriate features from the previously provided extensive list into a single discrimination statistic. These features can then be collected and examined in terms of known classes of objects for either classification or elimination as false alarms. The discrimination statistic is calculated as a quadratic distance metric:

$$d_T(X) = \frac{1}{N}(X - M)^T \Sigma^{-1} (X - M) \tag{6.40}$$

The parameters are

N	Number of features used for discrimination
M	Estimate of the mean target vector class
Σ	Estimate of the standard deviation target vector class
$d_T(X)$	Statistical distance of the unknown object from a target class

The estimates of M and Σ are often collected on targets in the open to obtain a fundamental understanding of the target features. The measured quadratic distance d_T(X) will be small for targets that are close to the correct class. For natural clutter, it is anticipated that the quadratic measure will be large for scattering that does not exhibit the dominant even or odd bounce characteristics of man-made objects.

The challenge in FOPEN ATD/C has always been the occlusion (i.e., shadowing) of the targets by dense foliage. If the targets statistics are collected for man-made objects in the open, there will be an inherent error in the mean and standard deviation of the target classes. These errors will be larger if the target SCR is low or if there are very large natural objects in the direct path. As a result, any collection of data to test ATD/C capabilities needs to be carried out in a variety of foliage conditions, including types of forests, terrain slope, and RADAR incidence angles.

6.4 RADCON Processing Development

To assess the feasibility of automatic target detection and cueing, the Air Force Research Laboratory started a program to evaluate ATD/C algorithms and to provide benchmarks of real-time processing on a highly parallel processor. Loral Corporation, in Goodyear, Arizona, conducted the Radar Detection of Concealed Targets (RADCON) program and demonstrated an end-to-end target detection and characterization tool employing a 104-node Mercury i860 processor [14].

The Defense Advanced Research Agency (DARPA) had flown the P-3 UWB FOPEN SAR system for four extensive data collection campaigns, as

Table 6–4 Summary of P-3 UWB FOPEN data campaigns for ATD/C [14]

Target	Length [m]	Width [m]	Maine	Camp Roberts	Grayling	Ft Indiantown Gap
HEMMT	10.0	2.4	4	2	6	5
5-ton	8.3	2.5	8			
M-60	6.9	3.6		6		
M-109	5.9	3.2		6	6	6
M-1	7.9	3.7			12	12
M-113	4.7	2.5			6	4
M-548	5.7	2.5			2	3
Total			12	14	32	30

summarized in Table 6–4. Over 80 vehicles and in excess of 1,000 square kilometers of clutter were collected over a 4-year span. The combination of these efforts was instrumental in bringing together alternative image formation and detection tools to examine the P-3 UWB FOPEN radar imagery. More importantly, the effort provided a qualitative assessment of the maturity of real-time ATD/C to guide future development efforts.

Figure 6–9 shows the processing flow and analytic elements of the RADCON development, which was tested using a Mercury massively parallel processor for near real time evaluation. Most of the techniques described in the previous sections were included. Significant innovations were made in the following:

- *Bayesian neural network (BNN):* The principal classifier in RADCON was a BNN developed over several years of SAR image processing. The BNN was used to train on know targets and features, and to determine the probability of correct association. For the RADCON program, representative prescreened image chips were collected for BNN screening. The final results were then applied to the segregated chips to obtain an unbiased estimate of the target feature correlation [16].
- *Local context:* Use of local context is an important technique in determining the likelihood of tactical targets. Tree lines and large terrain returns from trees and rocks can closely resemble a large vehicle. By using the image rotational inertia metric and a database of terrain features, the probability of correct association can be determined between anomalous terrain features and a tactical object.
- *Clutter segmentation:* This technique applies segmentation of target-to-clutter for prefiltering the discriminants. After the CFAR prescreening, the potential targets are masked and pattern matching is used to specify

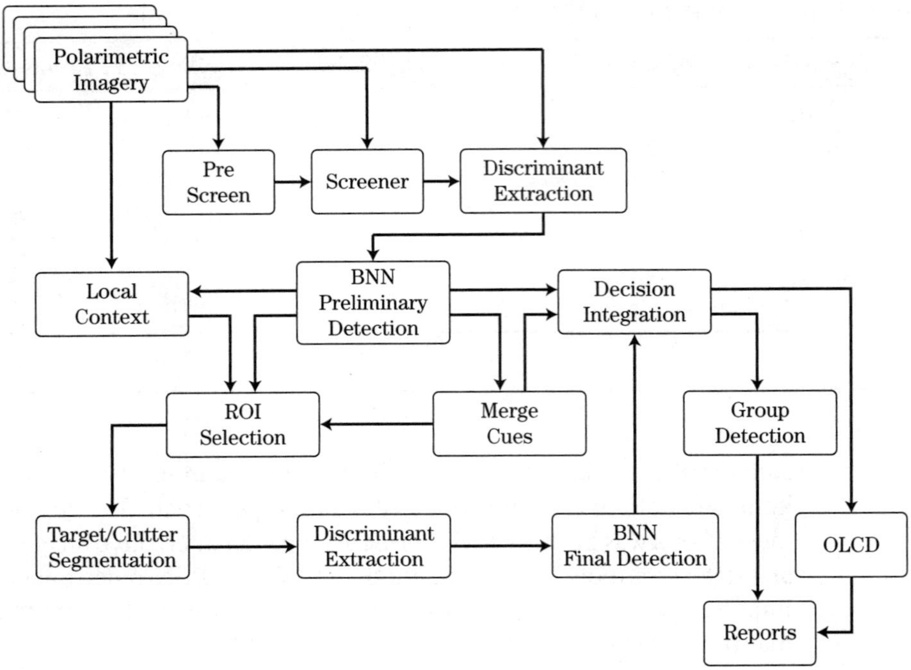

FIGURE 6–9

RADCON target detection and characterization functional flow [14]

the specific orientation. Because targets have enhanced returns at cardinal angles, knowing the pose of the targets allows the discriminator to apply statistics the returns to the image recognition.

- *Group Detection:* Tactical objects normally exist in groups. If there are several detections of similar shape and target features, they can be aggregated into a group. Metrics found to provide important target discrimination are *area per group* and *mass per group*. The use of these features and an association distance between groups is then used to verify other detections in the area.
- *Object level change detection:* A FOPEN SAR system will normally be tasked to revisit an area over a series of collections, often separated by days. When tactical objects move into an area, there will be a distinct difference in the return. By comparing SAR target images as being objects (rather than a single pixel), it is possible to improve the detection probability and to reduce false alarms. The change detection approach will be developed in more detail in Section 6.4.

The results of the various tests were assembled into an inferential assessment of the object classification process. When there was more than one possible explanation or answer, each could then be presented to the image analyst (IA) along with a probability of correct association.

Figure 6–10 presents the output of the RADCON processor for one of the Grayling, Michigan, images. The image on the left is the PWF image from the image formation processor, and the one on the right is the screen with target nominations superimposed. An experienced IA would be able to compare the two screens using a flicker feature and to improve their focus of attention. The technical objective that was successfully demonstrated included signal processing algorithms that provided a 90% detection probability with less than 0.1 false alarm per square kilometer against time-critical targets [17].

It is difficult to examine the receiver operating characteristic (ROC) based on a single image. The importance of the steps after the BNN final detection to meet the overall RADCON objectives of 0.1 FA/km^2 can be seen in Figure 6–11. The three UWB images are converted via the PWF filter and are then passed through the CFAR detection step, as previously described. A morphological filter then examines each pixel, and only pixels of appropriate size are passed on to the discrimination stage. The number of pixels sent from the CFAR to the size estimation can be qualitatively seen.

After size estimation, the remaining image chips are sent to the BNN classifier[16]. A scoring of the image is shown with the circles (valid targets)

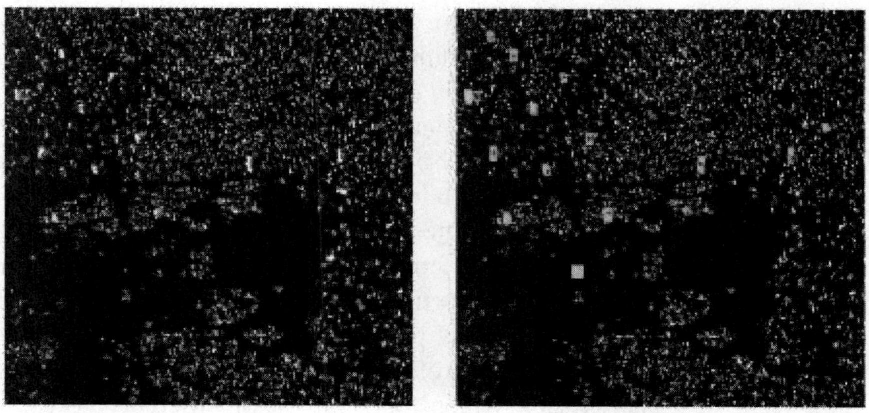

a. Input PWV Image

b. RADCON Target Nominations

FIGURE 6–10

RADCON nomination of tactical targets—Grayling, Michigan, collection [14]

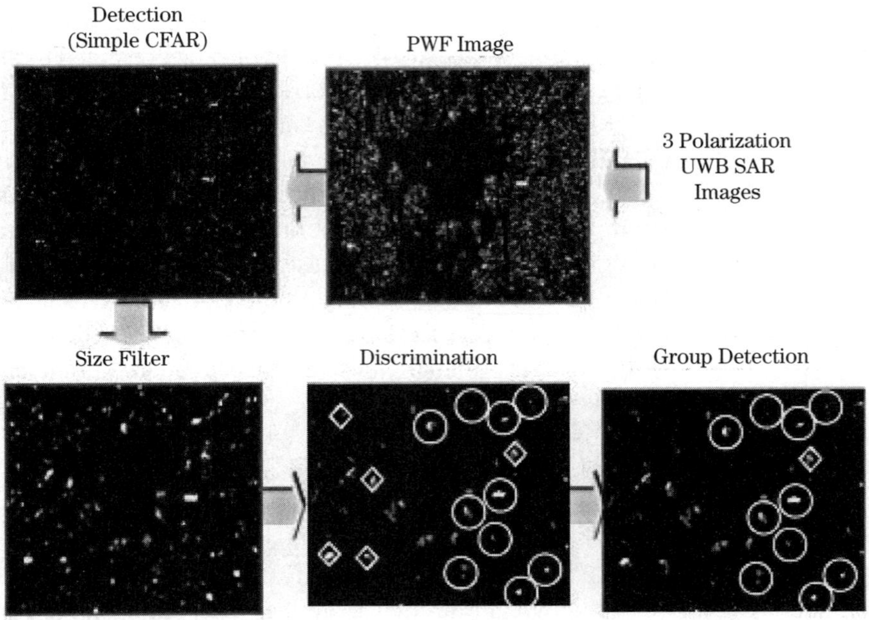

FIGURE 6–11

RADCON processing—Grayling, Michigan, FOPEN images [15]

and diamonds (false alarms). It should be noted that one of the "false alarms" was a known instrumented trihedral in the image.

The next step in the RADCON processing was to do group detection. All of the prior false alarms failed the group detection test; thus, the final image shows only one false alarm for of the approximately 1 km^2 image. The ROC performance of the final three processing steps is shown in Figure 6–12. The small diamond represents the overall goal of 0.1 FA/km^2. Based on these data, the ATD/C algorithms tested on RADCON achieved 1.4 FA/km^2 at a 90% probability of detection, but without using change detection.

The RADCON program demonstrated the use of image formation and ATD/C algorithms of targets concealed under foliage in real time. The program successfully demonstrated the use of a Mercury 9U commercial off-the-shelf (COTS) chassis populated with 104 I860 processors to process this real radar data. The image formation was accomplished in 4.7 seconds and the ATD/C in 6.1 seconds. To meet the real-time requirements, each stage was required to process several frames of data, with each frame of data not exceeding 7.4 seconds [18].

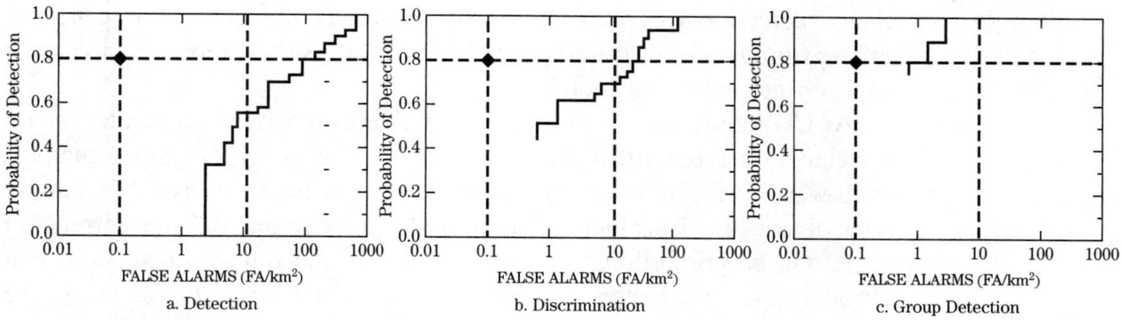

FIGURE 6–12

Receiver operating characteristics RADCON FOPEN processing [15]

6.5 Change Detection

The previous section suggests that change detection was an essential part of the RADCON program and was required to achieve a FOPEN SAR performance objective of less than 0.1 FA/km^2. Change detection has been used in many SAR systems to rapidly recognize differences in the SAR imagery due to changes in the terrain or the introduction or subtraction of objects between passes. With the improvements in navigation accuracy from global positioning systems (GPSs) coupled with laser gyro instrumentation systems, it is possible to obtain relative navigation accuracies between passes of well under 10 meters. However, the FOPEN SAR systems need absolute accuracy for many practical science and military applications. GPS will provide positional accuracy, but the navigation systems need to provide attitude, heading, and velocity accuracies commensurate with a 1 meter circular error probability at slant ranges over 10 kilometers to meet these objectives. With image resolutions below 1 meter, subpixel registration introduces the need for using very accurately known scene feature locations to provide fiducial points for achieving improved absolute accuracy.

Change detection is the process of comparing two images to determine fine changes due either to object movement, to terrain or target decorrelation, or to specific differences in viewing angles. These processes can be categorized as noncoherent, coherent, polarization, pixel, or group depending on the image characteristic being exploited. Adaptive change detection has been used to noncoherently compare FOPEN images from several passes [19]. Noncoherent change detection measures the amplitude of each pixel in the image and makes a comparison between scenes. The amplitude of a FOPEN SAR image is reasonably insensitive to small differences in the collection viewing geometry but provides little discrimination in low SCR conditions. Coherent change detection uses phase differences between two views of the scene, but suffers from loss of discrimination if the two views decorrelate due to aspect angle or internal clutter motion.

At UHF and VHF, the signatures of objects having sizes greater than a wavelength in size remain correlated over significantly wider angles than for microwave SAR. The correlation angle can be approximated as λ/D, where D is the dimension of an object's feature. At longer wavelength λ, the correlation angle will be proportionately larger. However, the foliage fluctuations can seriously affect the image correlation due to variations in path loss and phase perturbations as the signal passes through the foliage.

Pixel-level and group-level change detection are useful when the resolution of the image is comparable with the size of the objects being imaged. Pixel-level change detection requires fine image registration between the two images, often much finer than the resolution of the SAR image. If there are a large number of false alarms in the image, the use of group change detection can eliminate many of the isolated detections caused by discrete clutter returns. Group-level change detection is also less sensitive to image registration constraints when the image registration is finer than the size of the images.

6.5.1 Single-Pass Change Detection

One of the earliest experiments in FOPEN SAR pixel-level change detection was carried out with data from the 1993 Maine experiment, using the SRI FOLPEN II sensor [3]. The concept of adaptive change detection was introduced as applied to VHF FOPEN images, where consideration was for both repeat-pass change detection and change detection within a single pass. Repeat-pass change detection was used for objects in the scene that have moved on a time scale of hours or days. Single-pass change detection was used to discriminate objects that are moving during the SAR image collection timeframe, such as moving targets. This technique requires either two antennas or a scanning antenna to collect an image along the collection path. Chapter 8 will cover some details of along-track interferometry, a potentially effective ground moving target indicator (GMTI) technique.

For change detection within a single pass, two images are formed on an image plane but at different squint angles, as shown in Figure 6–13. The processing for change detection within a single pass greatly facilitates accurate georegistration accuracy, in that the incident angle and location of the image was common between subapertures. This approach leverages the assumption that natural clutter will have little geometric variance with respect to the look angle, as opposed to man-made targets, which are known to have significant variation, previously illustrated in Figure 3-28.

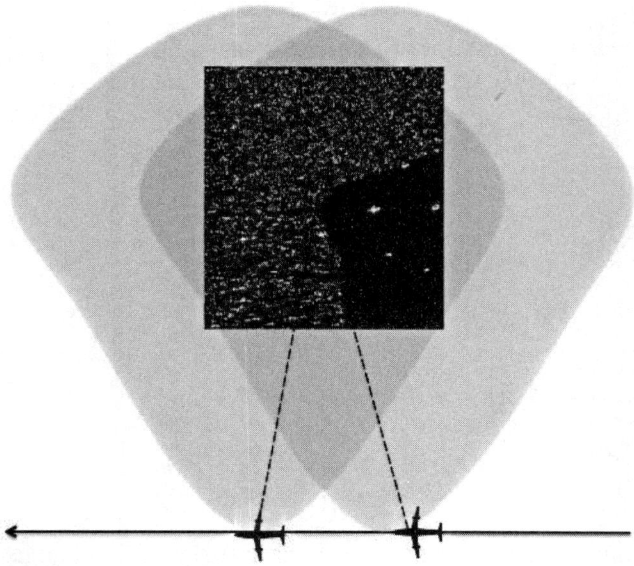

FIGURE 6–13

Change detection within single-pass FOPEN SAR collection

The change detection algorithm typically uses a minimum least squares adaptation of the returns from an image $D(m, n)$[20]. A reference image $X(i, j)$ is formed on one of the image planes and typically will have previously detected targets. The clutter in the reference image is convolved with a weight matrix W to predict the corresponding clutter in the current (i.e., mission) image. This predicted image is subtracted from the reference image to yield an error image ε_j [3,21]:

$$\varepsilon_j = D(m, n) - \sum_{l=0}^{M-1} \sum_{k=0}^{N-1} W_j(l, k) X(m - l, n - k) \tag{6.41}$$

The difference in (6.41) is used to adaptively calculate each element of the weight matrix $\mathbf{W}$, using a method of steepest descent:

$$\begin{aligned} W_{j+1} &= W_j - M\nabla_j \\ &= W_j - 2M\varepsilon_j X_j \end{aligned} \tag{6.42}$$

$\mathbf{M} = [\mu_{ij}]$ is iterated to converge the error ε_j to a minimum value and is itself updated with the estimate of the gradient of the alignment steps:

$$\nabla_j(l, k) = -2\varepsilon_j X(m - l, n - k) \tag{6.43}$$

The sign of the gradient ∇_j is measured after each iteration to assess whether the subsequent iterations are approaching a minimum in the error surface.

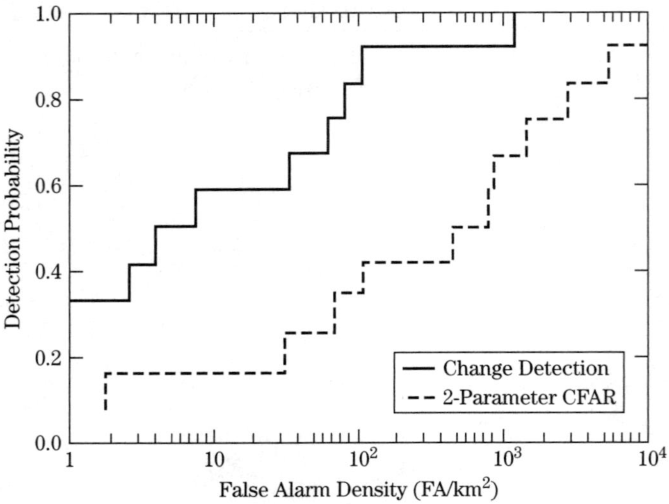

FIGURE 6–14

Improvement of detection probability with change detection [3]

To make the convergence more efficient, the following two control changes are made: (1) after m_0 consecutive sign changes $\nabla_j(l, k)$ is divided by an adjustment factor, α; and (2) after m_1 consecutive identical signs $\nabla_j(l, k)$ is multiplied by α. The constants m_0, m_1, and and the minimum and maximum values of μ were determined empirically. Upon image difference convergence and when the background clutter is highly correlated, the resultant change detection image should contain only targets [22].

Multiple aperture change detection was evaluated for the SAR image from the 1993 FOPEN Maine data collection. The principal advantage of multiple apertures on the same SAR collection path is that there will be a limited temporal decorrelation between the two images; thus, the two images will be well aligned. However, there are likely to be targets in both of the images. Hence, the two images were used alternatively as reference and mission images, and two adaptive change detection processes were performed. The final difference image is taken by using the maximum of the two difference images on a pixel-by-pixel basis. An ROC comparison of the multiple aperture detection is presented in Figure 6–14 [3]. It is important to note that this ROC is two orders of magnitude from the objective in RADCON. Single-pass change detection, using only navigation-based image alignment, was not effective in meeting the demands of FOPEN SAR ATD/C.

6.5.2 Repeat-Pass Change Detection

Repeat-pass FOPEN SAR change detection is typically carried out after a significant time interval, with the objective of determining a new or missing

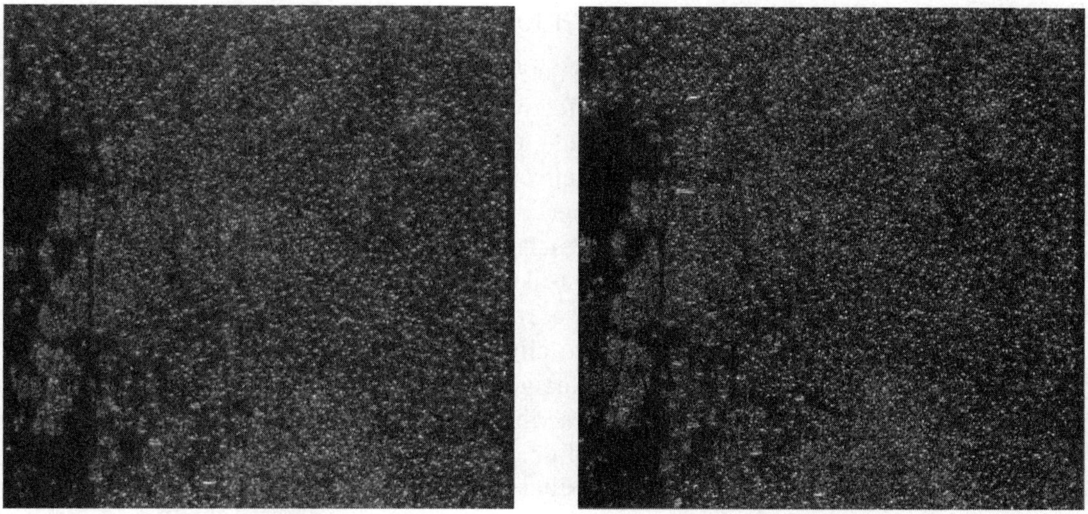

a. Sept 25, 1995 b. Sept 29, 1995

FIGURE 6–15

FOPEN SAR image collected on separate days—Grayling, Michigan
Source: MIT Lincoln Laboratory

tactical target. The changes in the collection geometry and any decorrelation of the clutter between images were at first considered to be a challenge to the effectiveness of this approach. Two enabling technology advances that facilitated this technique were the improved absolute geolocation accuracy between two collections and the alignment of the images to a fraction of a pixel. Both were needed to provide the minimum modulation of background clutter and improve the subclutter visibility of targets attenuated by the volumetric foliage clutter.

Extensive data were collected to evaluate change detection in the P-3 UWB data collections in 1995 at Grayling, Michigan. Over 30 tactical targets were placed in dense forests and were moved between collections separated by several days. Figure 6–15 shows the same area collected at horizontal polarization 4 days apart with horizontal polarization. Careful examination of the image shows that where the targets were placed broadside to the illumination there was a strong enhancement of the signal-to-clutter ratio. In these cases, the movement of vehicles can be observed. However, it is very difficult to detect all of the targets, especially those that are at an oblique angle. More importantly, the strong tree clutter provides many similar appearing image shapes that cause false alarms.

6.5.3 Image Registration

To obtain effective change detection performance, it is important to align the two images to provide effective cancellation of the stationary background clutter. These strong ground—tree trunk interactions create point like returns that resemble the top-hat reflectors. As was shown in Section 3.3.2, a top-hat return is insensitive to aspect angle look angles. Hence, the forest itself can provide the fiducials to align the background clutter from one pass to another, if sufficiently isolated point-like clutter can be identified.

To achieve sufficiently accurate image registration, the characteristics of the tree clutter are used to align individual subimages. Several techniques have been developed for the alignment of two images. The minimum least squares error, which is described in (6.41), is an effective technique, when there is a moderate level of image decorrelation between the two SAR image collections [3]. A Doppler filter approach has also been examined in which the spatial frequency of the image has a small amount of aliasing. This registration algorithm exploits the fact that the phase of the Fourier spectra of an image pair contains sufficient information to determine the translation offset difference of the images. The technique has successfully demonstrated subpixel registration accuracy between two microwave images.

A third technique demonstrated with FOPEN SAR for both autofocus and registration is the use of minimum entropy between two images. This approach exploits the localized scattering of the ground–trunk interaction to obtain both fine translational and rotation alignment [7,8]. Figure 6–16 depicts the process of comparison and subtraction of the two FOPEN SAR images. The weight matrix could be based on least mean square (LMS), minimum entropy, or Fourier-based correlation depending on registration accuracy and on computational complexity considerations.

Because there is generally some difference between the flight paths and aspect angles to the two scenes, it is important that the scene be segmented to understand the spatial orientation differences. One polarization is typically used to create a transformation mapping, which is then applied to the other polarization channels of SAR data.

In the new image, each block of clutter is convolved with a weight matrix obtained from the reference image. The difference between the two provides a difference image. If the targets have arrived in the scene from the past collection, the pixel will be bright or white. If the targets have left the scene, the pixel will be dull or black. The change detection has been most effectively used to focus the attention on those areas of the scene that indicate change in activity. Change detection is not a substitute for target characterization but will focus the attention on those areas that are most likely to contain tactical targets.

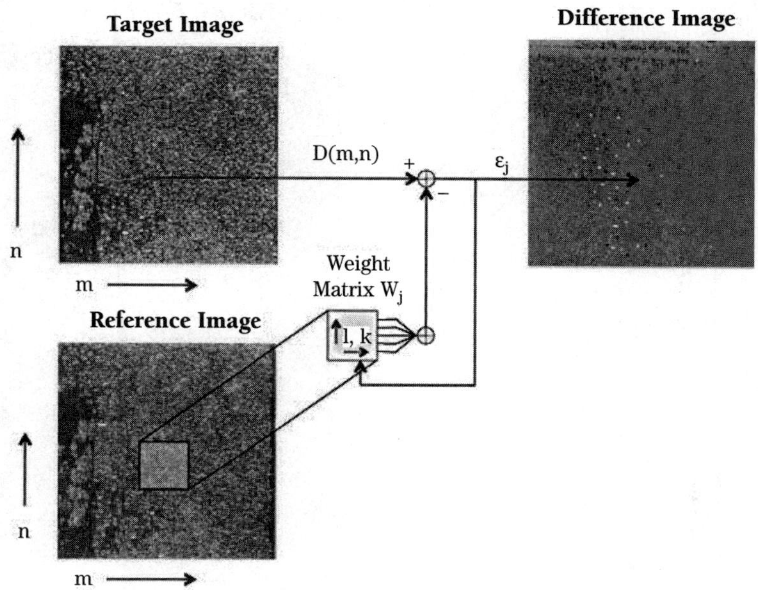

FIGURE 6–16

Adaptive change detection of FOPEN images—Grayling, Michigan
Source: MIT Lincoln Laboratory

FOPEN SAR images, by the nature of their varying terrain height and wide-angle collections, are difficult to align by a global transformation. This is the major impediment to minimizing the stationary clutter residue for detecting changes in target location and characteristics between passes. A straightforward approach to improve the alignment is to first conduct a multiple step correlation and subtraction of the two images [23]. The correlation coefficient of the image return or the minimum entropy is calculated for each sparse segment of the two SAR images as indicated in Figure 6–17. At this point the two images can be coarsely aligned by a global transformation. Next, each segment of the image is further divided into a dense set of subareas. Area correlation of the subareas is carried out with rotation and translation, and the two images are then subtracted. Figure 6–18 results from this refined subtraction of the images and shows the large areas of canceled clutter.

The next step in the adaptive change detection is to smooth the images using a boxcar filter, as shown in Figure 6–19. Within this stage, any small features that exhibit residual clutter due to internal clutter motion can be removed by averaging. An empirical assessment of the size of the boxcar filter, which was made for each of the three polarization channels, is shown in Figure 6–20. It is interesting to note, however, that each of the three polarizations is affected by the size of the boxcar smoothing filter.

The ROC curve illustrates the benefit of change detection for reducing false alarms in FOPEN SAR imagery. Figure 6–19 provides the false alarm density

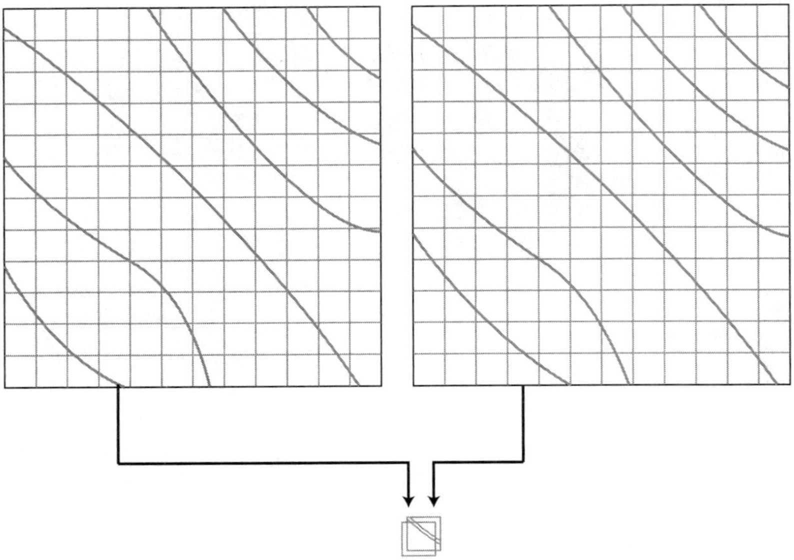

FIGURE 6–17

Alignment of two images based on subimage comparison [23]

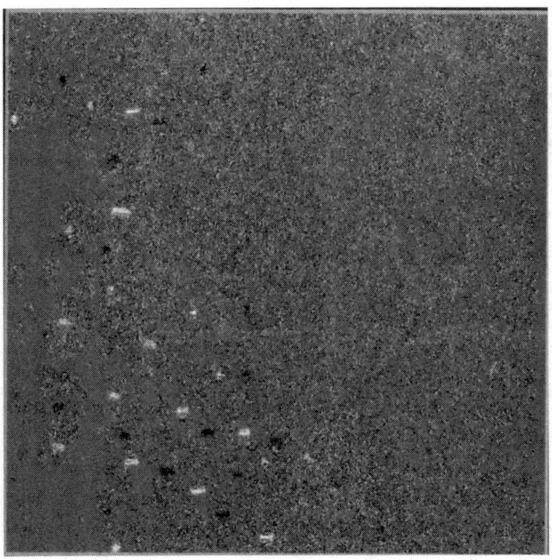

FIGURE 6–18

Change detection image from P-3 UWB after three-step image alignment [23]

at an 80% probability of detection. The three polarizations are independently shown versus the size of the boxcar filter. At the bottom of the figure is an indication of the median length and width of the vehicles in the images. It is clear that if the association filter is much smaller than the expected, or if the objects are much larger, the change detection performance is significantly affected.

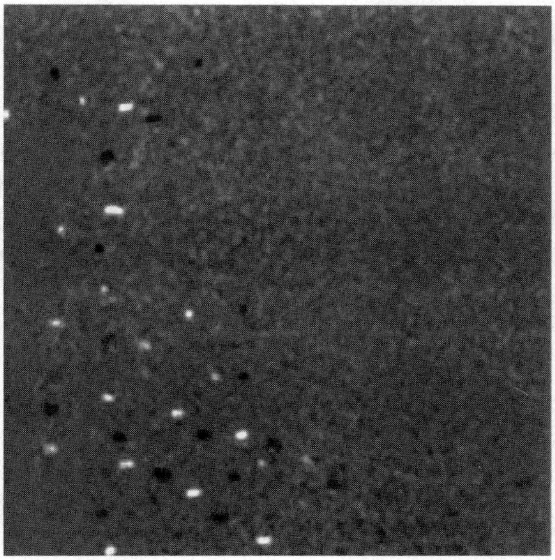

FIGURE 6–19

Change detection image from P-3 UWB after boxcar filter [23]

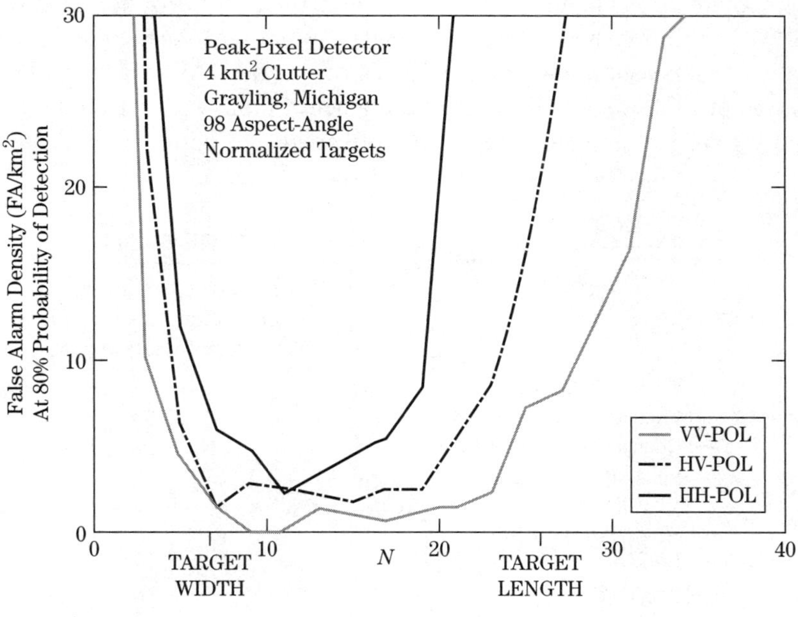

FIGURE 6–20

Impact of boxcar filter dimension on change detection false alarms [23]

The results do not include the effect of polarization whitening filters. In the data analyzed there is not statistical difference between polarizations. However, when these results are compared with the PWF results, there is an order of magnitude reduction in the false alarms. The major advantage is seen in the change detection results.

6.6 FOPEN ATD/C Summary

In summary, polarization diversity and PWF have been demonstrated in many data collections. These technologies are instrumental in reducing false alarms in FOPEN SAR images when compared with a single polarization system. Processing of the added degrees of freedom in the RADAR data can reduce speckle and provide strong indicators of the types of scattering in the resolution cell. Moreover, the difference in vertical and horizontal polarization scattering can be used to differentiate between trees and man-made objects.

For civilian applications polarization has been used for characterizing terrain. Scattering from various crops, sloped terrain, and cultural features can be used for land use determination. There has also been a significant amount of research conducted in use of long wavelength polarimetric scattering to examine ice flows, geologic faults, and health of crops and forests.

However, it is important that the system be specified in terms of the antenna and waveform characteristics. If the spatial and spectral characteristics of the two polarization channels are not matched, the benefits of polarization diversity will be diminished. These characteristics are necessary to maintain the independent measurement characteristics of the RADAR channels. The specification of FOPEN antennas and the impact of polarization channel effects on image processing are covered in more detail in Chapter 7.

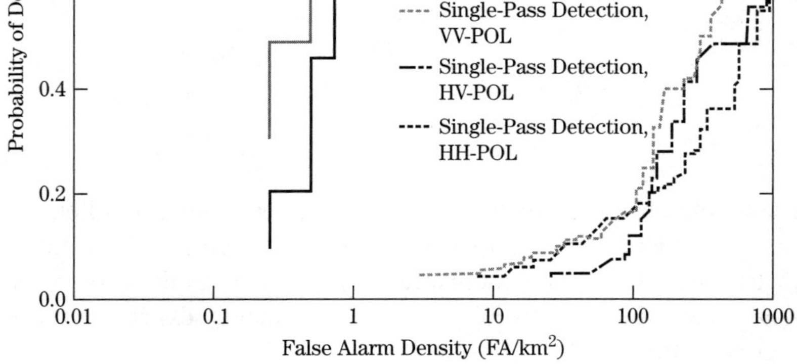

FIGURE 6–21

Receiver operating characteristics for FOPEN change detection [23]

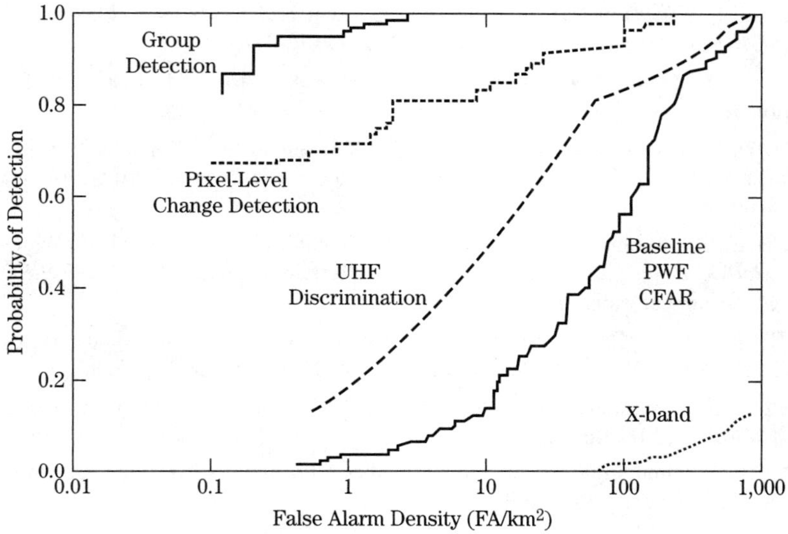

FIGURE 6–22 Summary of FOPEN detection and characterization algorithms on ROC curve [23]

To obtain the overall objective for detection probability at an acceptable false alarm rate, the ROC curve has been developed for several stages of the image conditioning and processing. It has been shown that no single technique will provide the desired capabilities. This is due in large part to the non-Gaussian statistics of foliage clutter. It is also due to the variation in the signatures of man-made targets under foliage.

Figure 6–22 shows the ROC characteristics for a typical FOPEN SAR, based on data from a P-3 UWB SAR collection. Both the PWF and UHF feature discrimination provide improvement in the false alarm density at a given Pd. However, it requires a combination of pass-to-pass change detection and group detection to eliminate the majority of foliage-induced false alarms.

There has been very encouraging progress made in image processing for UWB FOPEN SAR to reduce the number of false alarms. The performance of ATD/C is continuing to improve with research in algorithms for target feature discrimination and image registration. To date the capabilities have not matched those of microwave SAR for targets in the open. However, it can be clearly seen by the ROC performance of an X-band SAR taken during the same collection as the UHF SAR, that microwave SAR is inadequate for detecting any targets under trees.

6.7 References

[1] Novak, L. M., Burl, M. C., and Irving, W. W., "Optimal Polarimetric Processing for Enhanced Target Detection," *IEEE Trans. On Aerospace and Electronic Systems* Vol. 29, No. 1, January 1993, pp. 234–43.

[2] Novak, L. M, "Effects of Various Image Enhancement Techniques on FOPEN Data," *Proc. 2001 IEEE Radar Conference,* Atlanta, GA, May 2001, p. 87.

[3] Nanis, J. G., Halversen, S. D., Owirka, G. J., and Novak, L. M., "Adaptive Filters for Detection of Targets in Foliage," *IEEE AESS Magazine,* August 1995, p. 34.

[4] Schuler, D. L., Lee, J. -S., and De Grandi, G., "Measurement of Topography using Polarimetric SAR Images," *IEEE Trans. On Geoscience and Remote Sensing* Vol. 34, No. 5, September 1996.

[5] Davis, M. E., Tomlinson, P. G., and Malone, R. P., "Technical Challenges in Ultra Wide-band Radar Development for Target Detection and Terrain Mapping," *Proc. 1999 IEEE Radar Conference,* Boston, MA, April 1999, p. 1.

[6] Ertin, E. and Potter, L. C., "Polarimetric Classification of Scattering Centers using M-ary Bayesian Decision Rules," *IEEE Trans. AESS* Vol. 36, No. 3, July 2000, p. 738.

[7] Yegulalp, A. F., "Minimum Entropy SAR Autofocus," *Adaptive Sensors Array Processing (ASAP) Workshop,* MIT Lincoln Laboratory, Lexington, MA, March 1999, pp. 25–36.

[8] Kragh, T. J., "Minimum Entropy autofocus for Three-Dimensional SAR Imaging," *Proc. SPIE* Vol. 7337, Orlando, FL, April 2009, pp. B1–11.

[9] Hermes, L. and Buhmann J. M., "A Minimum Entropy Approach to Adaptive Image Polygonization." *IEEE Trans. On Image Processing* Vol. 12, No. 10, October 2003, pp. 1243–58

[10] Cloude, S. R. and Pottier E., "An Entropy Based Classification Scheme for Land Applications of Polarimetric SAR" *IEEE Trans. on Geoscience and Remote Sensing* Vol. 35, No. 1, January 1997.

[11] Cloude, S. R. and Pottier, E., "A Review of Target Decomposition Theorems in Radar Polarimetry," *IEEE Trans. Geoscience and Remote Sensing* Vol. 34, No. 2, March 1996, pp. 498–518.

[12] Lee, J. -S. and Pottier, E., *Polarimetric RADAR Imaging—From Basics to Applications*, CRC Press, Boca Raton, FL, 2009, Chapt. 7.

[13] Novak, L. M., Halversen, S. D., and Owirka, G. J., "Effects of Polarization and Resolution on the Performance of a SAR Automatic Target Recognition System," *MIT Lincoln Laboratory Journal* Vol. 8, No. 1, January 1995.

[14] MacDonald,D., Chang,C. F., Roman,J., and Koesel,R., "Automatic Target Detection and Cueing for Foliage-Concealed Targets," *Proc. SPIE* Vol. 2757, Orlando, FL April 1996, pp. 152–62.

[15] Gorman, J. D. and Marble, J. A., "A Low Complexity Multi-Discriminant FOPEN Target Screener," *Proc. SPIE* Vol. 3370, Orlando, FL, April 1998, pp. 307–317.

[16] Hilbert, E. E. and Chang, C. -F., "Bayesian Neural Network ATR for Multi-feature SAR, *Proc. SPIE,* Orlando, FL, April 1994, pp. 344–355.

[17] Department of Defense, "Defense Science and Technology RELIANCE—Defense Technology Objectives Success Stories," Report ADM001269, Publicly released January 2001, p. 41.

[18] Air Force Research Laboratory, "FY98 Avionics Technology Area Plan," Publicly released report, Wright Patterson Air Force Base, OH, May 1997, p. 9, http://www.fas.org/man/dod-101/usaf/docs/taps98/avionics98.pdf

[19] James, R. R. and Hendrickson, C. R., "Efficacy of Frequency on Detecting Targets in Foliage Using Incoherent Change Detection," *Proc. SPIE* Vol. 2230, Orlando, FL, April 1994, pp. 220–231.

[20] Hadmoud, M. M. and Thomas, D., "The Two-Dimensional Adaptive LMS (TDLMS) Algorithm," *IEEE Trans. Circuits and Systems* Vol. 35, No. 5, May 1988, pp. 485–494.

[21] Widrow, B., McCool, J., and Ball, M., "The Complex LMS Algorithm," *Proceedings of the IEEE* Vol. 63, No. 4, April 1975, pp. 719–720.

[22] Stone, H. S., Orchard, M. T., Chang, E. -C., and Martucci, S. A., "A Fast Direct Fourier Based Algorithm for Subaperture Registration of Images," *IEEE Trans. On Geoscience And Remote Sensing* Vol. 39, No. 10, October 2001, pp. 2235–2244.

[23] Crooks, S. M., "FOPEN Change Detection," unpublished presentation, FOPEN Seminar, MIT Lincoln Laboratory, Lexington, MA, November 1996.

CHAPTER 7

FOPEN SAR Design

This chapter will present the major system design of a tactical foliage penetration (FOPEN) synthetic aperture RADAR (SAR) that could be built with emerging technology. It will go through the steps of the concept of operations (CONOPS) of FOPEN SAR with high area coverage rates (ACRs) and with focused attention for more accurate target detection and characterization. The RADAR power-aperture product can be chosen, given a standoff range and characteristics of the foliage and targets, to meet the requisite resolution and signal-to-noise ratio (SNR) requirements. Next, guidelines will summarized for the subsystem specifications, the grazing angle and receiver dynamic range for minimizing the foliage loss and RFI effects, and several modes to meet the CONOPs requirements.

The design of a tactical FOPEN SAR system, as illustrated in Figure 7–1, was postulated in the early 1990s for operation on either a manned or unmanned vehicle. The platform needs to be flown at a standoff range and altitude that provides both the proper grazing angle for efficient operation in the foliage environment and with sufficient swath width to meet the user's need for ACR. Two classes of FOPEN SAR platforms are shown with their respective altitude and standoff ranges: a high altitude long endurance (HALE), and a lower altitude tactical vehicle. The later is shown as a piloted vehicle that would be appropriate for geoscience measurements, but not be survivable in a military operation. As such, the designs will consider a tactical unmanned air vehicle (TUAV).

The scenario shows a representative data link (i.e., the lightning bolt) for tasking and dissemination of FOPEN SAR images from either a mission control element (MCE) or a ground communications and distribution system (GCDS). The former would be a fixed site with significant power and data link bandwidth (e.g., over 50 Mbps) to provide tasking and exploitation of the

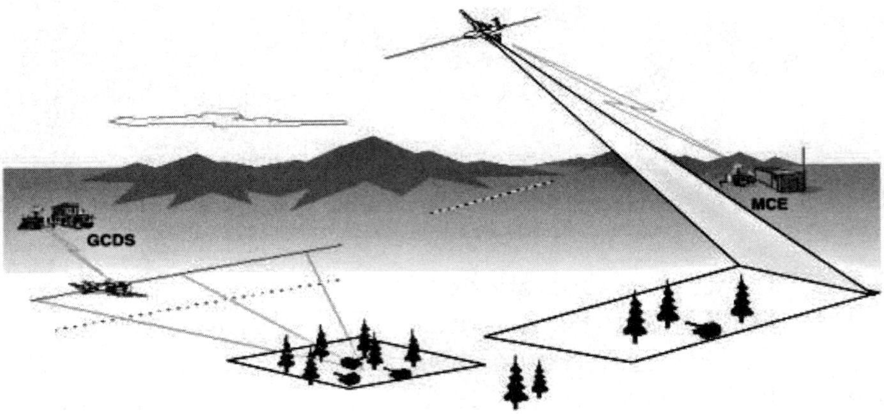

FIGURE 7–1

FOPEN SAR concept of operations

FOPEN SAR. The latter was envisioned as a mobile, tactical capability with lower data link bandwidth (e.g., less than 1 Mbps). If high-performance computing could provide adequate real-time onboard processing, the more tactical GCDS would be able to operate with a greatly reduced data link bandwidth. Hence, image processing operations count is an important measure of performance (MOP) to meet the bandwidth constraints of desired tactical data links.

Based on the FOPEN SAR parameters of this notional RADAR, the design for critical subsystems are presented. The ultra wideband (UWB) antenna design is most critical for providing the desired image illumination and for providing calibrated polarimetric returns for effective target detection and characterization. Examples of prior FOPEN SAR antennas will be provided, along with a consideration of the channel match over wide frequency ranges. In addition, the performance requirements will be evaluated for orthogonal polarization and sidelobe effects on the SAR images. Next the transmitter and receiver subsystems will be covered in terms of supporting the UWB waveform and operation in the dense RFI environment. Finally the mode management will be presented in terms of the synchronization of transmit and receive timing and the critical RADAR measures of performance.

7.1 Concept of Operations

Two types of unmanned aerial vehicles (UAVs) are depicted in Figure 7–2 with different operational altitudes and ground ranges, and considered in previous FOPEN SAR system studies , as shown in Figure 7–2 with the altitude versus ground range [1]. A HALE UAV would enable the longest standoff range

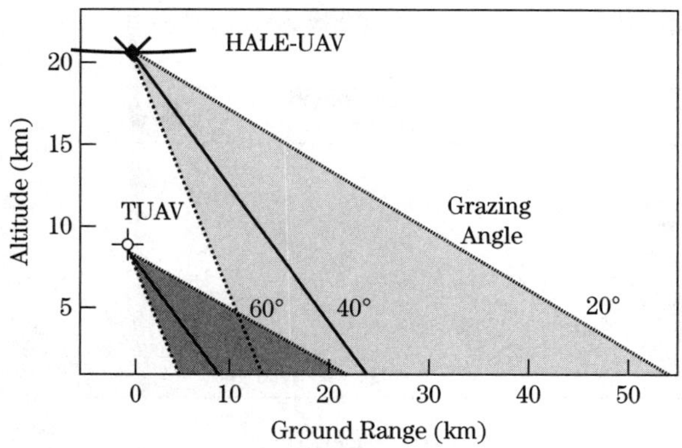

FIGURE 7–2

Ground range versus grazing angle for two platform altitudes © 1999 IEEE [2]

and swath width. However, the long range to the scene makes the greatest demands on the RADAR power-aperture product and motion compensation. Conversely, a TUAV could operate at modest ranges and provide close-in support for troops. The shorter range would enable a smaller, less expensive payload; if the demands for onboard processing, and low weight and volume can be achieved.

The ground range R_G is measure from the nadir point under the SAR platform to the SAR image point. When a platform is flying at altitude H_T, R_G is related to the slant range R_S and the grazing angle γ_g. For moderate ranges ($H_T \ll$ Earth radius), the relation can be approximated by a triangle with the relation

$$R_G = R_S \cos \gamma_g \tag{7.1}$$

Ground range is an important system parameter in that it establishes the separation from any threats or countermeasures near the SAR image point. Slant range is important to sizing the FOPEN SAR, since it determines the range and integration angle for obtaining system sensitivity and cross-range resolution. The two platforms examined for FOPEN SAR design sizing are summarized in Table 7–1. A TUAV candidate platform is examined at an altitude of 7.5 Km and ground velocity of 60 m/sec. For comparison, the HALE UAV is characterized at an altitude of 20 Km and ground velocity of 150 m/sec [2].

The foliage loss models as a function of grazing angle and frequency were given in (3.33). These models are shown in Figure 7–3 applied to each of the candidate FOPEN SAR platforms, and operating at either VHF (60 MHz) or UHF (350 MHz). The two-way slant range losses are greatly increased for grazing angles below 20 degrees, with a practical limit of a 15 dB two-way loss at UHF band. Above 60 degrees grazing angles, the SAR image collection is inefficient, due to the reduced Doppler cone angle and ground range resolution. Using these grazing angles as practical limits, the limits of ground range

Table 7–1 Generic unmanned air vehicles for FOPEN SAR tradeoff © 1999 IEEE [2]

Platform	Units	TUAV	HALE
Altitude	Km	7.5	20
Ground Speed	m/sec	60	150
Max. Ground Range (@ 20° grazing)	Km	20.5	54.3
Min. Ground Range (@ 60° grazing)	Km	4.3	11.5

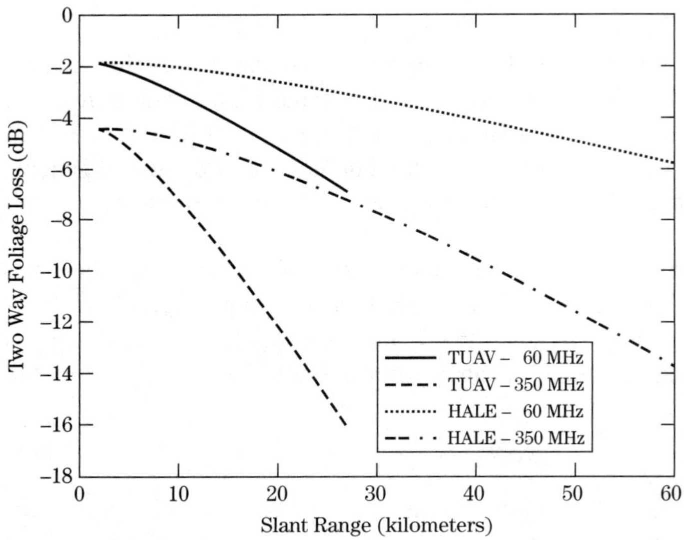

FIGURE 7–3

Two way losses for UAV operation at VHF and UHF bands

coverage for the two platforms are summarized in Table 7–1. From Figure 7–2, it is seen that image swath widths are practical for more than 20 km at 20 km altitude and for approximately 10 km at 7.5 km altitude. However, the loss and SNR characteristics of the target features (e.g., polarization) under the foliage need to be evaluated more thoroughly in terms of the probability of detection.

The SAR integration times from (4.3) are shown in Figure 7–4 for the alternate platforms and two integration angles −30 and 60 degrees. It is important to note that the effective UWB SAR length L_{eff} from (4.2) is used in calculating the integration time. Moreover, these integration times are shown versus ground range to illustrate the effect of standoff range on the performance of SAR collection.

The first integration angle analyzed is 30 degrees, representative of a UHF FOPEN SAR with 1 meter cross-range resolution. The second integration angle of 60 degrees is representative of VHF with approximately 3 meters cross-range resolution. The VHF example exhibits both longer integration times and

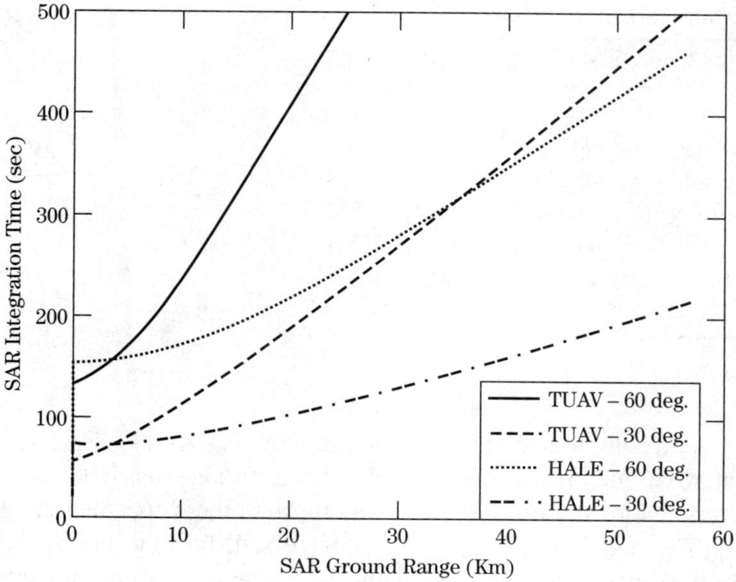

FIGURE 7–4

SAR integration times for TAUV and HALE FOPEN SAR platforms

a lower cross-range resolution, whereas the UHF example has modestly lower integration times and cross-range resolution below a meter. So the choice of frequency comes down to weighing the need for high detection probability versus high target probability of characterization.

Power-aperture trade-off is also a major factor when considering both the standoff range and the foliage loss. The SAR image SNR for a target size σ_T was given by (3.15) in terms of the area of the transmit and receive antennas and the SAR integration length L form (3.10) for narrow integration angle SAR. By substituting for antenna gain from $G = 4\pi A/\lambda^2$ and the UWB SAR length L_{eff} from (4.2), the UWB required SNR versus P_{av} can be expressed by

$$SNR = \frac{P_{av} G^2 \lambda^3 \sigma_T}{2(4\pi)^2 R_S^3 k T_0 F_n L_{tot} v_P k_{CR} \delta_{CR}} \left(1 - \left(\frac{k_{CR}\lambda_c}{4\delta_{CR}}\right)^2\right)^{-\frac{1}{2}} \tag{7.2}$$

where

P_{av}	Average transmit power (watts/polarization)
G	Antenna gain (dBi)
λ	Wavelength (meters)
R_S	Slant range from platform to the image pixel (meters)
kT_0F_n	Noise power spectral density (watts/Hz)
L_{tot}	Total loss, including foliage loss (dB)
v_P	Platform velocity (meters/second)
k_{CR}	Weighting factor for cross-range resolution
δ_{CR}	Cross-range resolution (meters)

Table 7–2 Parameters for candidate SAR tradeoff

Factor	Symbol	VHF	UHF
Wavelength	λ_0 [m]	5.0	0.83
Noise Figure	F_n [dB]	28	6
Cross Range Resolution	$\delta_{CR}[m]$	3.0	0.5
Platform Velocity	v_P [m/s]	150	150
Antenna Gain	G [dBi]	0	5.0
System Loss	L_R [dB]	4.0	4.0
Polarization		HH	HH, VV

Table 7–2 provides a comparison of two candidates HALE UAV FOPEN SAR systems: one at VHF, and one at UHF. VHF has lower foliage loss but higher noise temperature due to sky noise propagating through the troposphere and ionosphere [3]. The antenna aperture gain at VHF will be low due to the need for integration over wider angles to achieve a moderate resolution. UHF has significantly lower background noise, assuming the RFI is adequately removed. UHF operation also provides finer range resolution, due to the potential of wide absolute bandwidth. However, the foliage loss contribution to L_{tot} is expected to be significantly higher at UHF than at VHF, as demonstrated in past experiments summarized in Chapter 3.

From (7.1), the average power required, given in watts per polarization, can be determined:

$$P_{av} = R_S^3 \left[\frac{2(4\pi)^3 k T_0 F_n L_{tot} v_P k_{CR} \delta_{CR}}{G^2 \lambda^3 \sigma_T} \right] \left(1 - \left(\frac{k_{CR} \lambda_c}{4\delta_{CR}} \right)^2 \right)^{-\frac{1}{2}} \tag{7.3}$$

To evaluate (7.3) for alternative system characteristics, the total loss L_{tot} is divided into two parts:

$$L_{tot} = L_R + L_{fol} \tag{7.4}$$

where L_R is the system loss between the transmitter and receiver, and L_{fol} is the foliage loss at a given polarization, frequency, and grazing angle, which is given by (3.33).

Figure 7–5 evaluates the required average power required for a FOPEN SAR for the parameters in Table 7–2 . There is not a significant difference in the average power for the two frequencies, despite the increased UHF loss and choice of multiple polarizations. However, double-canopy loss greatly reduces the operational range of the system. Additionally, vertical polarization has a

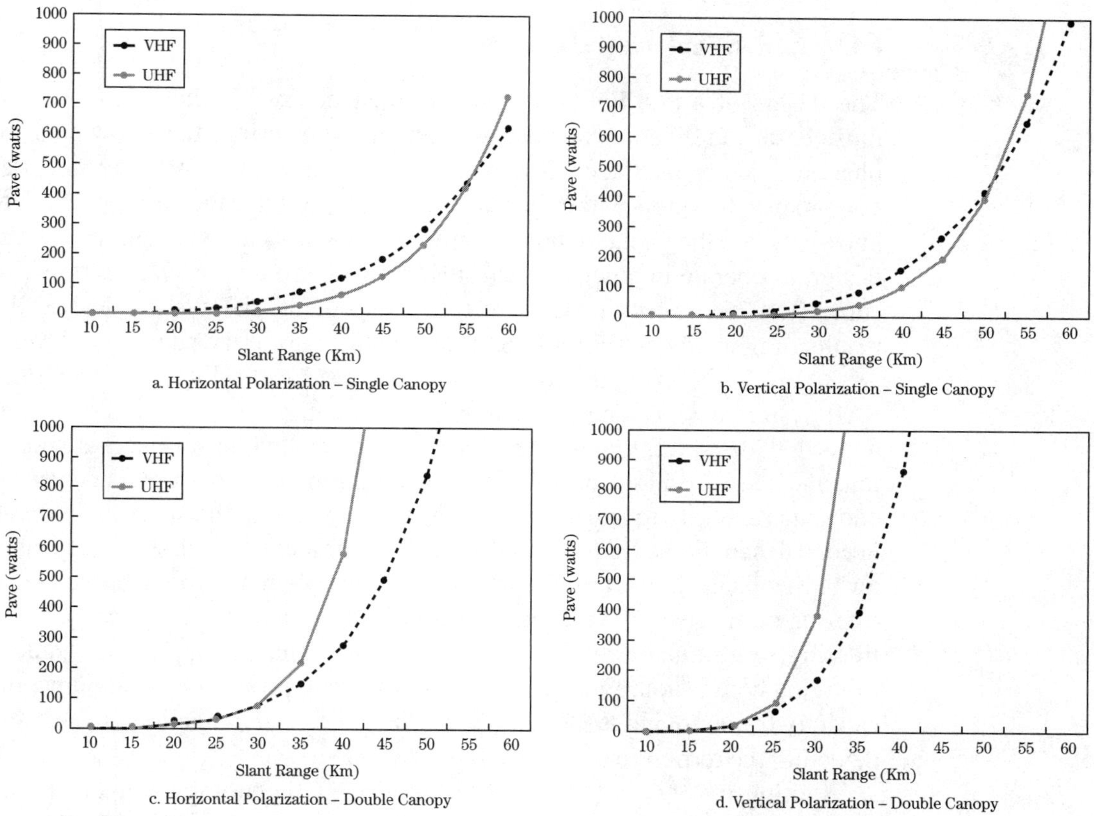

FIGURE 7–5

Required FOPEN average power per polarization versus slant range

reduced operational range at UHF due to the higher foliage loss. It should be noted that the VHF vertical polarization is included for completeness. Unfortunately, a practical V-pol antenna has not yet been developed for an airborne platform.

A fully polarimetric FOPEN SAR requires transmitting each polarization and receiving on two channels (i.e., copolarization and cross-polarization). One practical approach is transmit orthogonal polarizations on alternative pulses, and to obtain the full complement of transmit and receive polarization senses. In effect the pulse repetition frequency (transmitter PRF) is doubled for full polarization coverage, and the usable imaging range is halved. Thus, the total UHF average power requirement (and effective PRF) will be doubled for full polarization operation.

7.2 FOPEN SAR Hardware

The design of a FOPEN SAR system, intended for installation on a UAV, must be a combination of detailed system engineering, the knowledge of phenomenology and the environment, and allowance for future algorithm developments. Chapters 3 through Chapter 6 have provided the current technical knowledge on the phenomenology and the necessary hardware and waveform design to operate in a dense electromagnetic environment. This section will summarize, for the purpose of selecting an operational system design, the results of many recent FOPEN SAR system designs at the subsystem level.

Figure 7–6 depicts the major subsystems to be considered in a FOPEN SAR system design. The antenna must provide the illumination signal over an ultra wideband frequency expanse, a wide beamwidth for the SAR imaging, and a low backlobe return. The transmitter must have the responsiveness and spectral purity to support the waveform, along with the spectral notching needed to satisfy the frequency allocation constraints. As was shown in Chapter 6, the exciter must digitize the signal with coherency to enable matched filtering on receive. The receiver must have the dynamic range, linearity, and filtering to eliminate out-of- band interference and to support the analog-to-digital conversion. Finally, the on-board signal processor should have the capacity for removing the RFI, creating the multiple polarization images, and detecting the terrain features or targets.

Starting in 1997, a FOPEN SAR Advanced Technology Demonstration (ATD) system was developed to be flown on an army RC-12 aircraft. Figure 7–7 shows the weight and power of each of the subsystems. The combined UHF and VHF system produces real-time image formation of both frequency SAR images on board the aircraft.

The relatively high-resolution images coupled with multiple polarization processing provide the necessary image fidelity to separate targets of interest from tree trunks and other discrete scatterers. The onboard processor has the capacity to process three polarization high-resolution images. The FOPEN

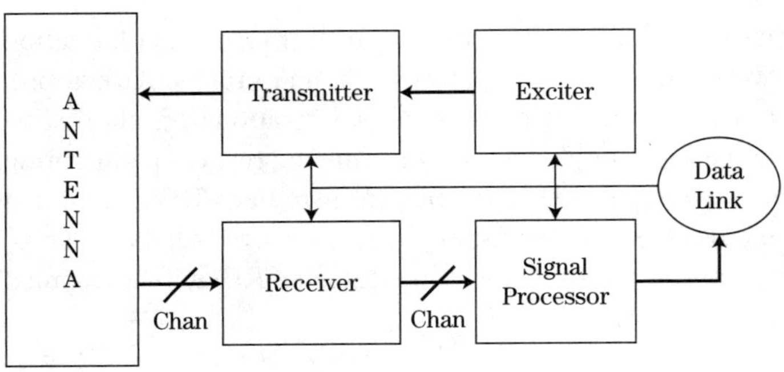

FIGURE 7–6

FOPEN SAR generic block diagram

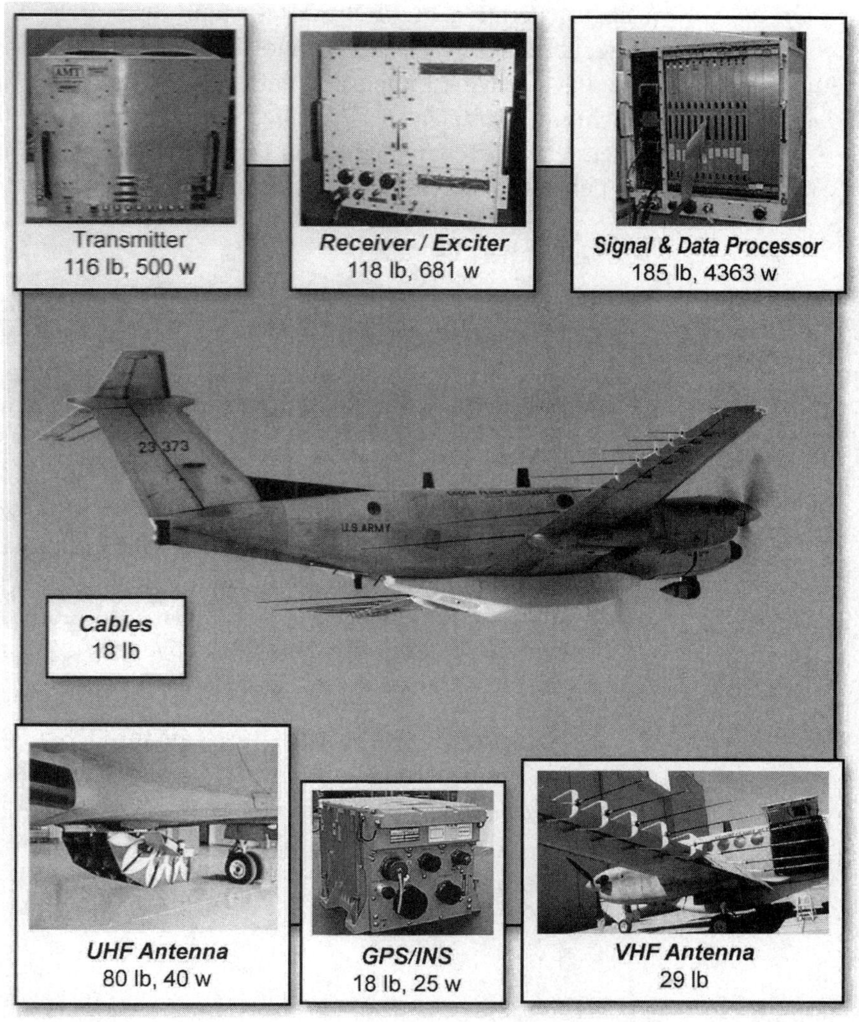

FIGURE 7–7

FOPEN advanced technology demonstration system
Source: Lockheed Martin, Goodyear AZ [4]

ATD system has been deployed with over 600 flights. Image formation and subsequent target detection processing are performed in real time onboard the aircraft [4]. The physical characteristics of the subsystems are summarized for comparison with future technology, and the capability to integrate into smaller platforms.

The following sections provide the historical development of many critical FOPEN subsystems along with a generic development process to produce a next generation FOPEN SAR system. Several of these subsystems have been subsequently integrated into FOPEN SAR systems. All of the subsystems have provided important details for system integration and calibration to achieve significant performance in real-time image formation, automatic target detection/characterization (ATD/C), and, most importantly, operation in a dense and sensitive RF environment.

7.2.1 Antenna

The design of the FOPEN SAR antenna is an important factor in the design of a RADAR so that it will be able to detect and characterize terrain and man-made objects. The system's design provides high-resolution imaging through a combination of wide-bandwidth and wide-angle target illumination. The specification and selection of antenna components for these SAR systems cannot rely on the same methods developed for narrowband SAR in the microwave regime. A UWB antenna presents clear mapping between antenna pattern trade-offs and UWB SAR image quality. When the bandwidth approaches an octave, the antenna beamwidth, gain, efficiency, and phase-center all vary significantly with frequency.

Antennas to be used in UWB SAR have the following, often conflicting, requirements, as illustrated in Figure 7–8:

- An adequate level of real-aperture gain throughout the coherent aperture, which should ensure sufficient signal energy from individual pixels to provide adequate signal- or clutter-to-noise ratios
- Adequate beamwidth or scene angle θ_I at each frequency to give the required resolution

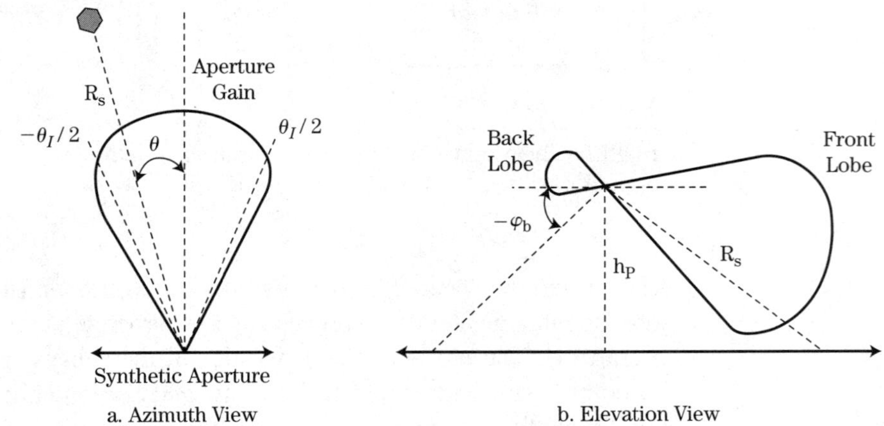

FIGURE 7–8

UWB SAR aperture characteristics

- Controlled or minimized response in extraneous directions, which is needed to minimize the slow-time Doppler spectrum, wrong-side ambiguities, sky noise, and RFI to the extent practicable.
- A "well-behaved" impulse response in angle and frequency; excessive variation in gain and phase versus angle and frequency can lead to excessive sidelobes or require equalization that will degrade the SNR
- Minimum backlobe at elevation angle $-\phi_b$, to reduce SAR artifacts from the opposite side of the platform.
- Adequate polarimetric purity over the beamwidth to provide the required isolation in the images

The most important feature is the beamwidth of the antenna, which determines the angle over which SAR data can be collected. Wide integration angles are needed to maintain fine cross-range resolution, shown in Figure 7–9. When the antenna gain varies as a function of frequency, the support for cross-range resolution can suffer. The cross-range resolution δ_{CR} can be expressed in terms of the frequency dependent azimuth integration angle θ_I as [5]

$$\delta_{CR}(f) = \frac{c/f}{4\sin(\theta_I(f)/2)} \tag{7.5}$$

Two prototype FOPEN SAR antennas were developed to provide fully polarimetric operation. The first was the GeoSAR antenna developed by the Jet Propulsion Laboratory, which is still flying on the GeoSAR platform. The second was the Ball Aerospace circles array, a technology project to reduce the weight of fully polarimetric antennas, compatible with a HALE UAV platform.

A notional objective for antenna gain was derived to provide for integration angles of 30 to 45 degrees and operation over frequencies between 200 to 600 MHz. This is the solid line termed "specification" in Figure 7–9. The measured peak gain for the two experimental antennas is shown as dashed lines in the figure. Details of these two antennas will be provided in following sections. However, it is important to note the variation of gain over frequency and the deviation from the optimum antenna gain for full cross-range resolution. Antenna design MOPs are derived next to aide in characterizing the antenna performance and in the attendant impact on UWB SAR.

The difference in gains between the two developmental antennas can be characterized by the constraints on installation in their respective aircraft. The GeoSAR antennas are housed in wingtip pods and are inclined to point at the nominal depression angle to the SAR image. As a result, the gain is larger due to the full projection of the aperture area. As a comparison, the Ball FOPEN or *Circles Array* antenna was designed to fit into a centerline pod, beneath either a commercial aircraft (e.g., a King Air) or into a HALE

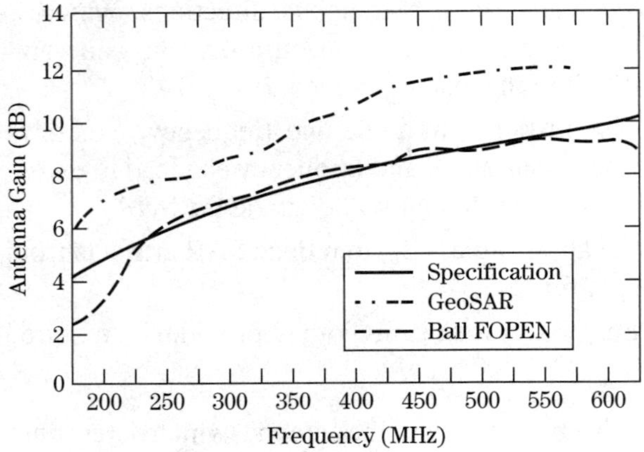

FIGURE 7–9

FOPEN SAR antenna performance versus gain specification © 1998 IEEE [6]

UAV. The height of the enclosure was not adequate for full elevation gain; thus, the antenna was designed to be conformal to the volume. The antenna fully satisfies the bandwidth for FOPEN SAR, but has some falloff in gain at the band edges. These characteristics are not unexpected for UWB systems. Their effects can be compensated in signal processing, as will be covered in the following discussion [6].

To evaluate the impact of antenna design on UWB SAR image collection, it is important to consider the illumination power spectral density on the image surface. During a coherent dwell, the total energy E received from an image resolution cell is the integral of the received power density over both the aperture time and the frequency spectrum of the waveform as [7]

$$E = \frac{P_{av}\sigma_T}{(4\pi)^3 R_s^3 v_P}\Psi(\varphi, \theta_1, \theta_2) \tag{7.6}$$

where

$$\Psi(\varphi, \theta_1, \theta_2) = \int_{f_L}^{f_H}\int_{\theta_1}^{\theta_2} |W_t(f)|^2 |G_\varphi(f, \theta)|^2 \left(\frac{c}{f}\right)^2 \cos^2(\theta)\, d\theta\, df \tag{7.7}$$

In evaluating the energy at the image resolution cell, the critical parameters are

$[\theta_1, \theta_2]$	Azimuthal integration angle for the SAR image
φ_b	Depression angle from the RADAR to the image pixel
$W_t(f)$	Normalized spectral power density for the antenna pattern
$G_\varphi(f, \theta)$	Antenna power gain at depression angle φ_b , over frequency and azimuth angle
$(c/f) = \lambda(f)$	Wavelength of the waveform, over the design frequency band

The function $\Psi(\varphi, \theta_1, \theta_2)$ is the geometric characterization of the antenna pattern that affects the SAR image quality and has dimensions of square meters. The Ψ term is obtained by integrating the antenna gain over great circle contours of frequency and angle. As such it is the effective gain-aperture of the antenna, in terms of the real antenna aperture and the synthetic aperture. The first MOP of a UWB antenna is the gain-aperture GA_d on the desired side of the platform evaluated over $\pm\pi/2$ angular interval

$$GA_d = \Psi(\varphi_b, -\pi/2, \pi/2) \tag{7.8}$$

While the RADAR platform is flying a synthetic aperture of length L, there is the potential for forming a synthetic aperture on both sides simultaneously. An undesirable, ambiguous image can result due to energy received through the backlobe of the antenna, as shown in Figure 7–8b. Hence, it is an important measure of the installed antenna pattern that the antenna gain in the undesired direction GA_ube significantly lower than the front lobe. Otherwise, the ambiguous SAR image will fold into the desired image. The ratio of left-to-right ambiguity, which determines the amount of unwanted contribution to the UWB SAR, is given by

$$U_{LR} = \frac{GA_u}{GA_d} \quad \text{where} \quad GA_u = \Psi(-\varphi_b, \pi/2, 3\pi/2) \tag{7.9}$$

There is also a Doppler ambiguity due to the PRF f_{PRF} of the RADAR, which is similar to left–right ambiguity of the aperture. The angle where the Doppler ambiguity folding begins is given by

$$\theta_{dop}(f) = \sin^{-1}\left(\frac{cf_{PRF}}{4v_P f}\right) \tag{7.10}$$

The Doppler ambiguity ratio U_{dop} is determined by the ratio of the gain-aperture at the Doppler ambiguity angles θ_{dop} to the desired gain-aperture as

$$U_{dop}(f) = \frac{GA_{dop}}{\Psi(\varphi_b, -\theta_L(f)/2, \theta_L(f)/2)} \tag{7.11}$$

It should be noticed that U_{dop} is evaluated between the azimuth angles $[-\theta_L/2, \theta_L/2]$ for the SAR integration length L.

Digital processing (inverse filtering) can relieve some of the burden of UWB SAR antenna design. However, such processing induces a mismatch loss that must be considered, since angles or frequencies with low gain must be boosted at the expense of increased noise. The SAR metrics need to be expressed in terms of frequency and angle to correctly evaluate the antenna's impact upon image quality, SNR, and Doppler ambiguities.

The following sections will provide the characteristics of the antennas used for the early FOPEN SAR systems, described in Chapter 2. In addition, details of the prototype Ball Aerospace circles array will be presented.

BoomSAR Antenna

The Army Research Laboratory (ARL) boom synthetic aperture RADAR (BoomSAR) antenna, shown in Figure 7–10, represents a major development to match the characteristics of the antenna with the impulse transmitter. A transverse electromagnetic (TEM) horn was developed, such that the wave propagation impedance was matched to the output of the transmitter. The wave is launched with a balanced line into the apex of the horn, and propagates from the open end of the horn. As indicated in Figure 7–11 the E-plane and H-plane are orthogonal. By matching the fields within the two plates, the impedance remains well controlled. The electromagnetic pulse is launched at the point in the flare that matches the wave impedance of free space. This was especially important in an impulse waveform to obtain monocyclic characteristics over the 40 to 1400 MHz bandwidth [8].

The gain characteristics versus frequency for the individual horns are shown in Figure 7–12 for the H-plane and Figure 7–13 for the E-plane. The plots, which show the angular gain pattern over a frequency band of 300 MHz

FIGURE 7–10

Army research laboratory BoomSAR antenna © 1996 IEEE [8]

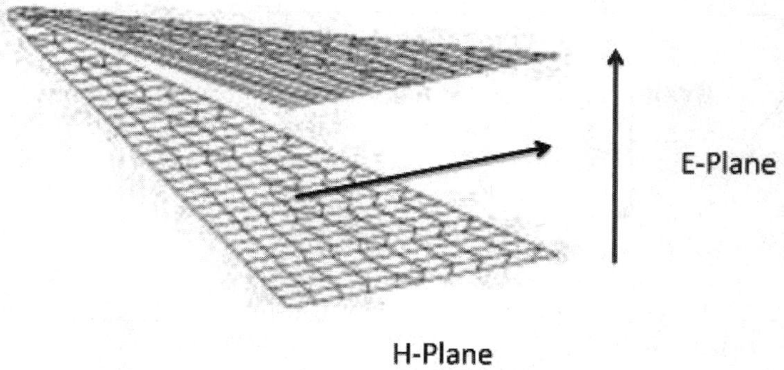

FIGURE 7–11

Flared TEM horn and field orientation © 2002 IEEE [9]

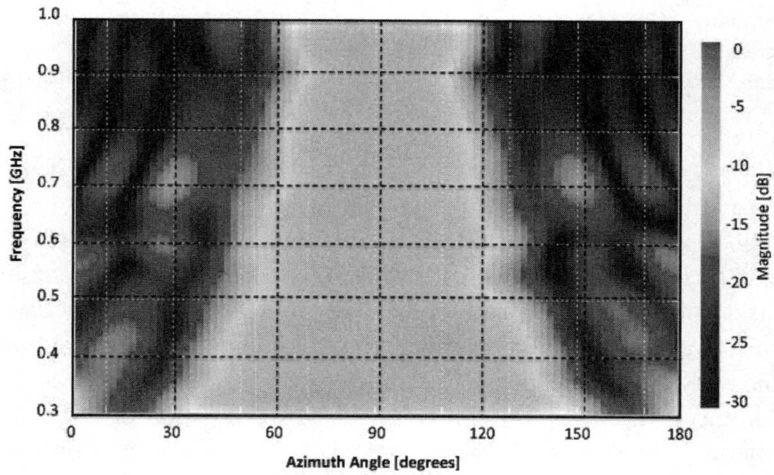

FIGURE 7–12

ARL BoomSAR H-Plane antenna gain characteristics © 2002 IEEE [9]

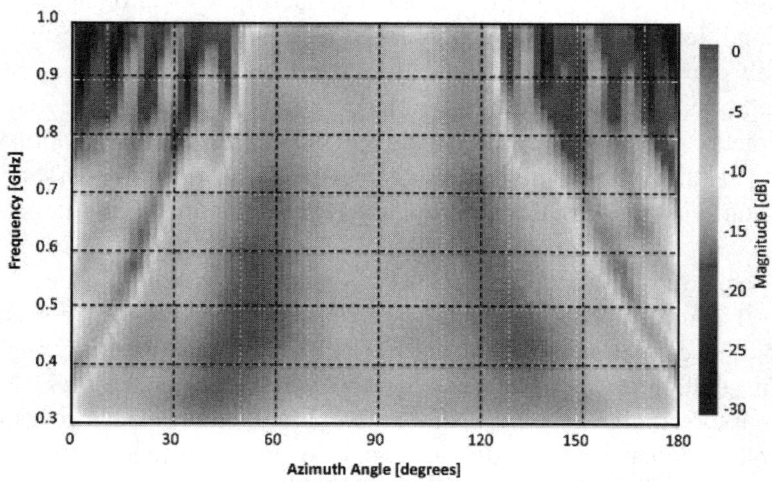

FIGURE 7–13

ARL BoomSAR E-Plane antenna gain characteristics © 2002 IEEE [9]

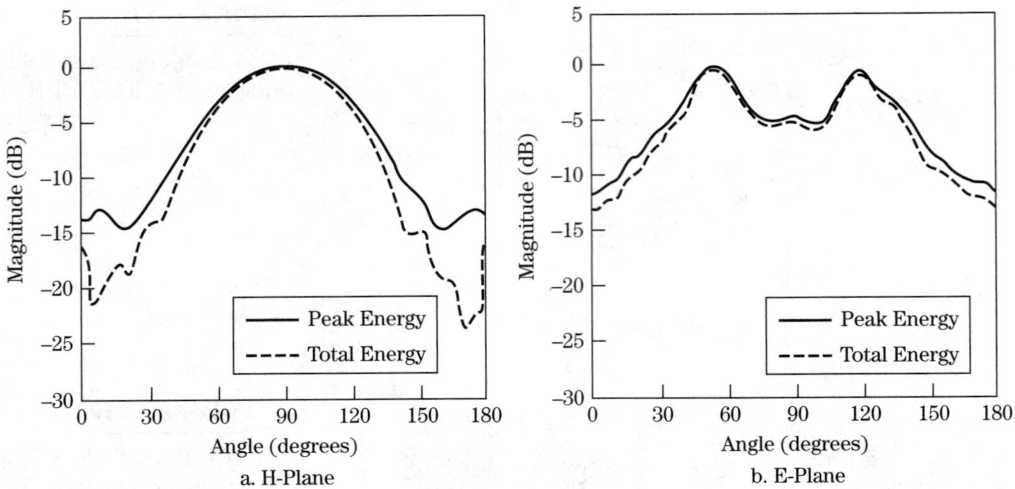

FIGURE 7–14

Comparison of peak and total energy in ARL UWB antenna © 2002 IEEE [9]

to 1,000 MHz, represent well-controlled transmit characteristics over the ultra wideband and illumination angles. However, there is a ±45 degree phase dispersion as a function of angle and frequency within the pattern. This dispersion required compensation in the image formation process, such that the effects on the SAR impulse response were minimized.

The peak and total energy in the UWB pulse as a function of angle is shown in Figure 7–14. For the H-plane, the energy is uniform in both frequency and angle. However, for the E-plane there is an enhancement in the energy density at the edge of the flared antenna due to the electromagnetic boundary conditions. This might have been better controlled in the design, but was easily compensated in the signal processing.

The four horns shown in Figure 7–10 are combined to synthesize the four fully polarimetric components. The spacing and isolation between the horns provided outstanding cross-polarization characteristics, which is necessary for conducting research into the polarimetric effects on foliage and target scattering.

The principal drawback of the antenna was a large size of 1 m wide by 1 m high by 1.5 m long. This was not an issue, however, as the antenna was designed for data collection on the ground and not for installation on a flying platform. In addition, the antenna's mass was over 100 kilograms, which would make it too cumbersome for operation on virtually all tactical aircraft. The data collection carriage could operate only over a path that had been prepared and smoothed sufficiently for safe passage of the 50 m high boom.

FIGURE 7–15

CARABAS II antenna installed on a sabreliner aircraft [11]

CARABAS II Antenna

The Swedish coherent all radio band sensing (CARABAS) II SAR required a significant development effort to provide a physically stable antenna at low VHF (i.e., 20 to 90 MHz) that would not cause flight envelope problems on the Sabreliner aircraft. The CARABAS II antennas are 5.5 meters in length and cantilevered from the nose section of the airplane, as shown in Figure 7–15. The boom antennas were constructed to minimize any vibration modes that could interfere with flight operation. To achieve an acceptable vibration characteristic, the Swedish Defence Research Agency (FOA) designed a unique composite fiber structure, in which the UWB radiators were integrated into the material [10].

A single horizontal antenna element has a symmetrical radiation pattern around the boom, as shown in Figure 7–16. Without compensation, the backlobe of the antenna would be identical to the front lobe. FOA used a pair of parallel antennas, combined with an active matching network, to cancel the backlobe. An active matching circuit, shown in Figure 7–17, was used to interface between the transmitter and the antenna. The switching was done dynamically so that SAR maps could be obtained from both sides of the flight line on a pulse-interval-to-pulse-interval basis. Because the two antennas were parallel in the horizontal plane, it was possible to create a backlobe null over a significant region of the SAR integration [12].

The effectiveness of this backlobe suppression can be appreciated by considering the two images in Figure 7–18. The image on the left is a SAR map obtained by using the signal received on only one of the antennas. The bright streaks in the image are artifacts caused by strong features on the opposite side of the flight path. As a consequence of cross-track platform motions, the

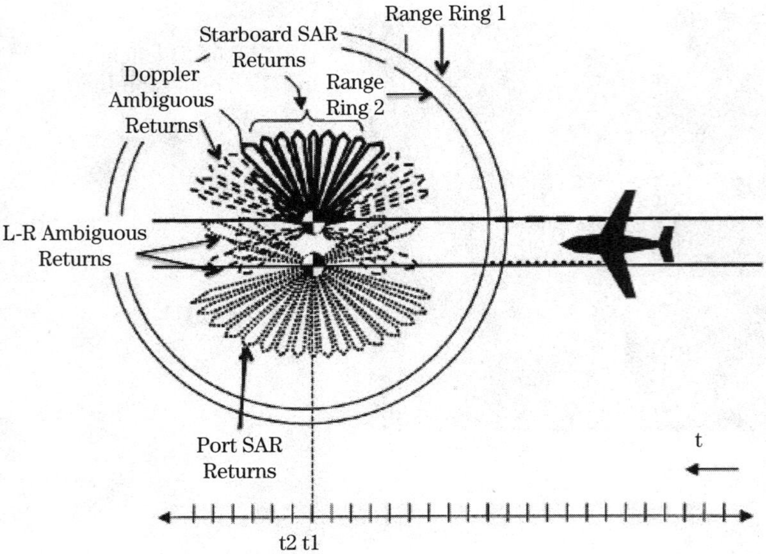

FIGURE 7–16

CARABAS II backlobe effects and dual boom compensation approach [12]

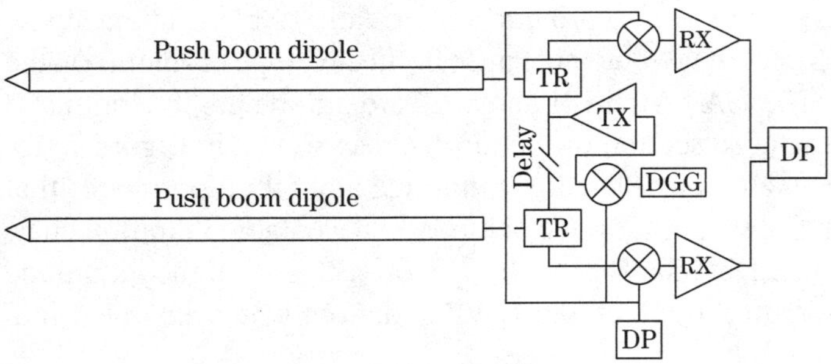

FIGURE 7–17

Compensation circuit for left right ambiguity cancellation [12]

Doppler frequency at each point in the backlobe image is not appropriately summed in the back projection algorithm (BPA). Therefore, tree lines and extended structures in the backlobes produce unfocused artifacts in the image. The right image is from the same collection and was formed by applying the appropriate cancellation coefficients to the second antenna channel. It is clear that the artifacts have been suppressed and that improved fidelity of the SAR image was achieved [13].

P-3 UWB Antenna

The UWB antenna for the P-3 FOPEN SAR was designed using the quad-ridged horn shown in Figure 7–19 [14]. The antenna employed a flared-notch radiator integrated into the brass structure to maintain the polarimetric characteristics over a wide bandwidth. The beamwidth varied from 40 degrees at the high end of the transmit spectrum to 100 degrees at the low end, which is

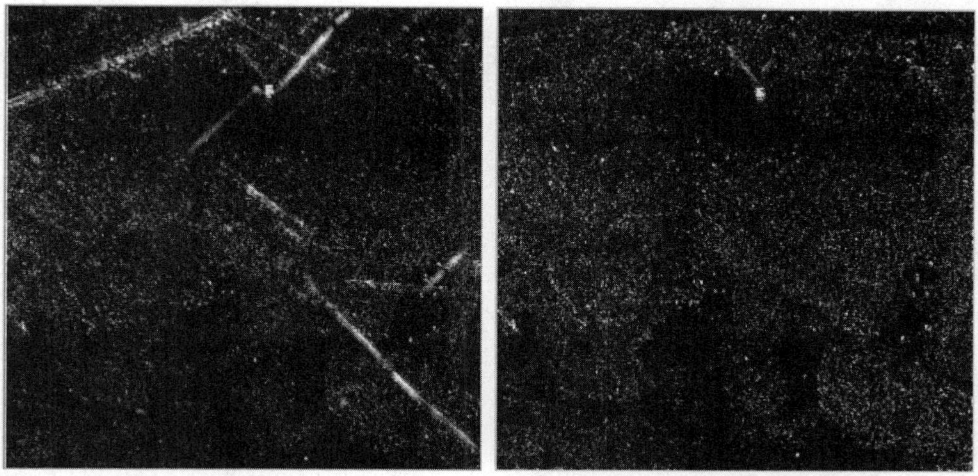

a. Uncompensated Image with Ambiguities b. FOPEN SAR image without Ambiguity

FIGURE 7–18

CARABAS II Left/Right ambiguity cancellation—Ft Indiantown gap PA (1996)
Source: MIT Lincoln Laboratory [13]

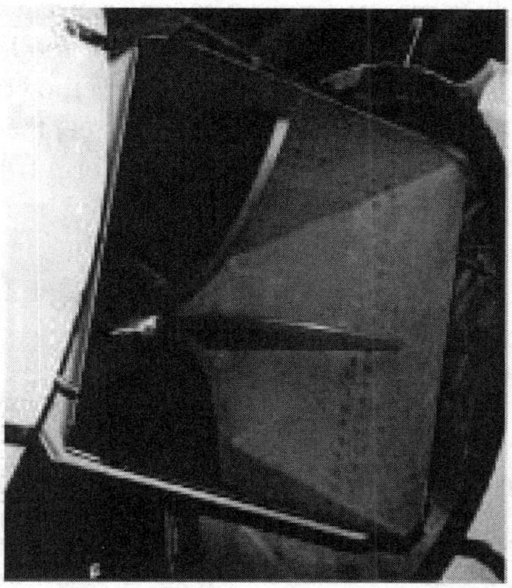

FIGURE 7–19

Polarimetric UWB antenna designed for Navy/ERIM P-3 data collection platform [14]

consistent with the desired 0.66 meter cross-range resolution. Moreover, the use of the quad-ridged waveguide insured the ability to maintain a minimum of 20 dB cross-polarization isolation over the band.

However, the gain of the device was not able to achieve the $1/\lambda$ objective over the full bandwidth. Figure 7–20 shows the measured H-pol and V-pol gain response versus the desired gain, over a 200 to 900 MHz band. The gain

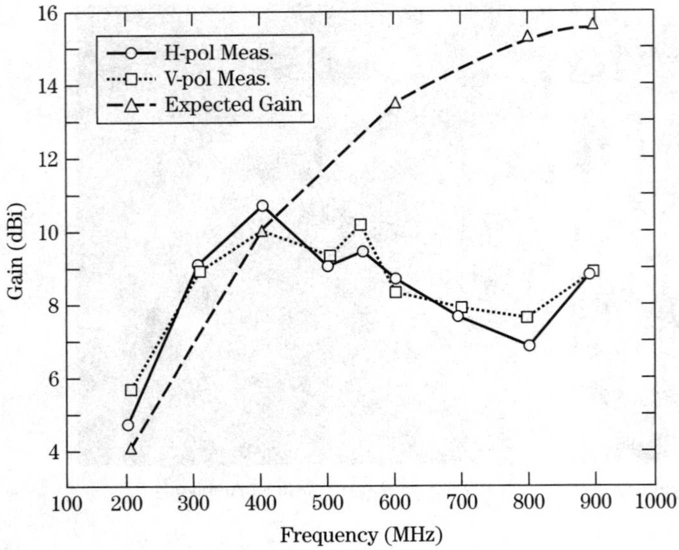

FIGURE 7–20

Predicted and measured gain P-3 UWB antenna [15]

exceeded the requirement at the low end and was significantly lower at the upper end of the band. This was known to have a significant impact on the impulse response of the system, since the beamwidth was too small at the low end. At the higher frequencies, the beamwidth was too large, increasing the ambiguity ratio and intercepted RFI. However, a method of compensating for the antenna pattern was developed to partially correct the impulse response.

The frequency response of the transmit waveform from a point scatterer is proportional to the product of the transmit power, the transmit gain, and the receive effective aperture area [15]:

$$\Upsilon(f) = \overline{P}(f)G(f)A_e(f) \tag{7.12}$$

The average power $\overline{P}(f)$ will depend on the design of the specific transmitter over the band. For the initial assessment, the average power will be assumed to be uniform over the band. The effective aperture A_e, which is a function of the measured antenna gain, is given by

$$A_e(f) = \frac{G(f)\lambda^2}{4\pi} \tag{7.13}$$

Thus the impulse response function will be proportional to

$$\Upsilon(f) \propto G^2(f)\lambda^2 = \frac{G^2(f)c^2}{f^2} \tag{7.14}$$

The impulse response function, assuming Hamming weighting, is given by

$$I^2(t) = \frac{f^2(2R/c)}{f^2(0)} \tag{7.15}$$

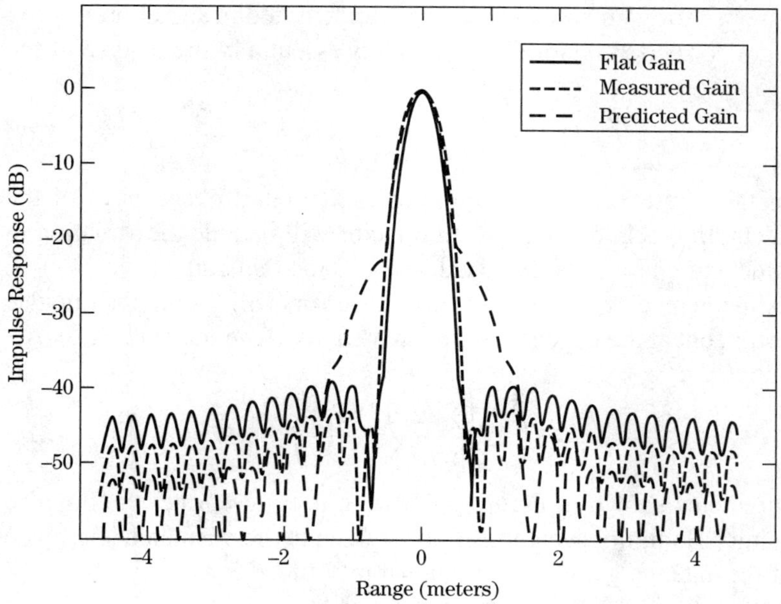

FIGURE 7–21
Waveform impulse response with antenna frequency characteristics [15]

where $f(t)$ is the inverse Fourier transform of the waveform

$$f(t) = \int_{f_1}^{f_2} \exp(-j2\pi ft)\sqrt{\Upsilon(f)}H(f)\,df \qquad (7.16)$$

and $H(f)$ is the Hamming window function.

The impulse response for the 200 to 715 MHz band is shown in Figure 7–21. The solid line is the desired response for uniform gain and antenna response $\Upsilon(f) = 1/f$. The short dashed curve represents the case that uses the predicted antenna gain to compute $\Upsilon(f)$, and the long dashed curve is the case using the measured antenna gain. The important factors to note are as follows:

- There is little difference in the 3 dB width of the IPR mainlobe among the three curves.
- There is some broadening of the mainlobe pattern at the −20 dB points for the measured pattern.
- The analysis considers only the amplitude of the antenna pattern function and not the phase. When phase errors are included, the sidelobe levels will be adversely affected.

The response for gain variation can be compensated in signal processing by applying a correction factor $C(f)$, which is essentially the inverse of the errors:

$$C(f) = \frac{1}{\sqrt{\Upsilon(f)}\exp(j\phi(f))} \tag{7.17}$$

This factor takes into account the amplitude $\Upsilon(f)$ and phase $\phi(f)$ of the pattern effects. In practice, the correction factor will include the variation in the amplitude and phase of both the antenna and the transmitter.

Processing on receive for the gain variation errors will restore the impulse response errors but at the expense of a mismatch loss. The mismatch loss Λ_P can be computed as

$$\Lambda_P = \left(\frac{|\sum_i s_i w_i^*|^2}{\sum_i |w_i|^2 \sum_i |s_i|^2} \right) \tag{7.18}$$

The measurements shown in Figure 7–20 were used to correct the impulse function using a Hamming weighting. An estimate of the mismatch loss made for each of the antenna gain values is shown in Table 7–3.

In summary, the P-3 UWB antenna was sized to provide the desired gain at the low end of the spectrum. A quad ridged horn structure was chosen to provide for full polarization with sufficient isolation between copolarization and cross-polarization for image processing development. For experimental operation, the antenna was installed in the *tail cone* of the P-3 aircraft, with an installation configuration designed to minimize the backlobe of the radiation pattern. The 100 kilogram high mass of the antenna made for a stable instrument over the altitude and temperature of flight operation. However, it was clear early in the development cycle of tactical FOPEN SAR systems that a much lighter system would need to be developed for UAV applications.

GeoSAR Antenna

The GeoSAR interferometric mapping system operates over a 160 MHz bandwidth centered at 350 MHz. GeoSAR can be used for either IFSAR mapping or polarimetric terrain characterization. As such, significant design requirements were placed on the antenna for cross-polarization matching. A significant

Table 7–3 Mismatch loss for compensation of gain in the P-3 UWB antenna [15]

Gain	None	Hamming	$1/\sqrt{\Upsilon(f)}$	Hamming$/\sqrt{\Upsilon(f)}$
Uniform	0.60	2.7	2.2	3.3
Measured	0.53	1.0	2.6	1.8
Predicted	0.52	1.4	2.8	2.0

FIGURE 7–22

GeoSAR P-Band interferometric SAR antenna
Source: NASA Jet Propulsion Laboratory © 1998 IEEE [17]

GeoSAR challenge was to integrate two antennas into each wingtip pod to provide simultaneous port-starboard operation from the Gulfstream aircraft [16].

The GeoSAR antenna design uses four cavity-backed radiators, as shown in Figure 7–22. Four symmetric feeds on each radiator provide the fully polarimetric excitation of the antenna over the 46% fractional bandwidth. This cavity-backed radiator was found to be much wider bandwidth than required, providing for better-than-anticipated operation at the band edges. Moreover, since these antennas were installed in the wingtip pods, their gain could be optimized in elevation toward the swath center. The GeoSAR antenna is 173 cm long by 43 cm high by 10 cm deep, and its weight is less than the 20 kilogram objective [17].

Figure 7–9 showed the GeoSAR peak gain over the 160 MHz design bandwidth met the 10 dBi specification at band center. Figure 7–23 shows the performance of the four polarimetric components as a function of the azimuth and elevation cuts at a 350 MHz center frequency. The elevation cuts indicate that the one-way backlobe was better than −15 dB for both polarizations. More importantly, the cross-polarization (dotted lines in each figure) is below than −30 dB at center frequency. Measurements verified that the cross-polarization component was better than −20 dB over the full bandwidth.

The azimuth patterns were equally well behaved. The azimuth sidelobes for horizontal polarization were well below −20 dB. For vertical polarization, the first azimuth sidelobes were at −13 dB due to the uniform illumination but then fell below −20 dB. Both of these characteristics fully achieved the image Doppler ambiguity requirements.

Circles Array Antenna

The Circles Array FOPEN SAR antenna was developed by Ball Aerospace to be installed in a centerline pod and to operate from either side of the aircraft. As a result the antenna, shown in Figure 7–24, was conformal in design. Since the objective was to image from both sides of the aircraft, a switching circuitry was incorporated to use three rows of circles on each side. The two rows are at 45 degrees inclination, and the third row is nadir looking. Because of this

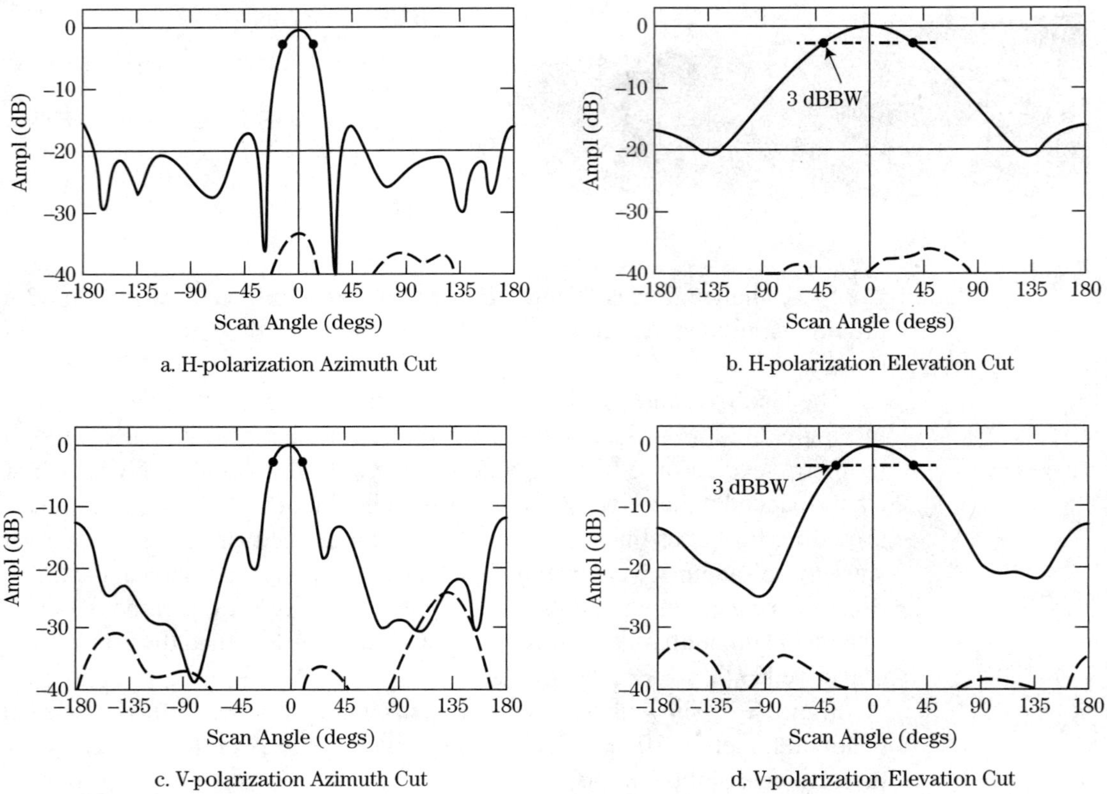

FIGURE 7–23

GeoSAR antenna characteristics at 350 MHz
Source: NASA Jet Propulsion Laboratory ©1998 IEEE [17]

nonplanar arrangement, the gain at the 35 degree depression angle could not be optimized.

The *circles array* design, as shown in Figure 7–24, was unique in meeting requirements for a small conformal installation as well as providing UWB performance with low depth [18]. It approximates a flared notch by conformal mapping to a flat surface, with a ground plane behind the array. A balanced feed to each circle is located at the tangent between pairs of circles. The polarization diversity is obtained by the excitation modes in the circles, with the pairs of circles in the vertical direction forming the V-pol radiators and pairs in the horizontal direction forming the H-pol radiators.

The patterns for the array of five elements in azimuth and two elements in elevation are given for H-polarization in Figure 7–25. The elevation cuts, which are shown at 200, 400, and 600 MHz, exhibit a strong lobe at

FIGURE 7–24

Circles Array, Ball aerospace UWB FOPEN SAR antenna [19]

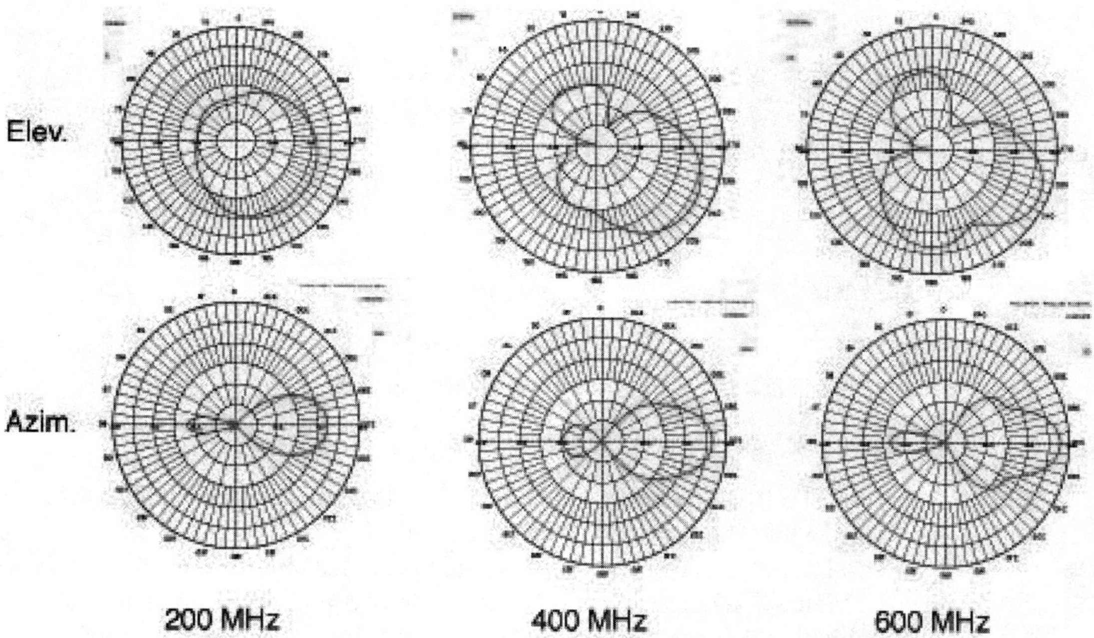

FIGURE 7–25

Circles ArrayBall FOPEN antenna—horizontal polarization [19]

−35 degrees that is based on correct phasing between the two pairs of circles. It should be noted that the backlobe at the same −35 degree cone angle is down over 10 dB (one-way gain), thus yielding the 20 dB backlobe suppression.

The patterns for the array of five elements in azimuth and two elements in elevation are shown for V-polarization in Figure 7–26. Similar to the H-polarization, the elevation cuts at 200, 400, and 600 MHz show a lobe at –35 degrees. However, the lobing is not as well behaved due to the vertical polarization being conformal around the antenna. The backlobes are marginally

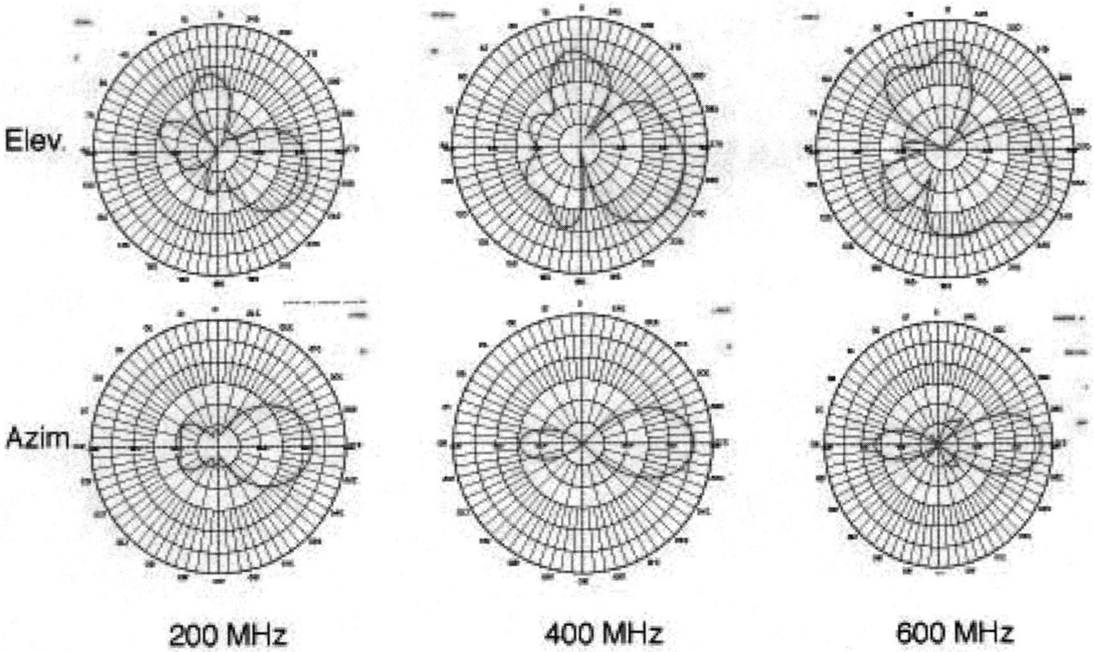

FIGURE 7–26

Circles ArrayBall aerospace FOPEN antenna—vertical polarization [19]

less than 10 dB (one-way gain). The azimuth beam is well controlled at broadside, with a gain variation shown in Figure 7–8.

The patterns from Figure 7–25 are assembled into the contour plot shown in Figure 7–27. The variation of the mainlobe gain versus frequency and the azimuth sidelobes were quantitatively examined. The contour plots of both the H-pol and V-pol antenna structure over frequency and angle were analyzed to obtain the antenna MOPs covered in section 7.2.1. Table 7–4 summarizes the key parameters.

The left–right ambiguities were well behaved for both polarizations, as was the Doppler ambiguity. However, the projected area in the H-pol was low due to the fact that one of the rows was nadir pointing. This could be compensated with the V-pol elements by steering the radiation circle pairs. As a result of the same issue on element area projection, the cross-polarization characteristics did not meet the −20 dB specification.

The circles array was a success in that it demonstrated that a conformal array with over 60% fractional bandwidth could be fabricated for installation in a relatively small volume. The entire assembly weighed about 20 kilograms

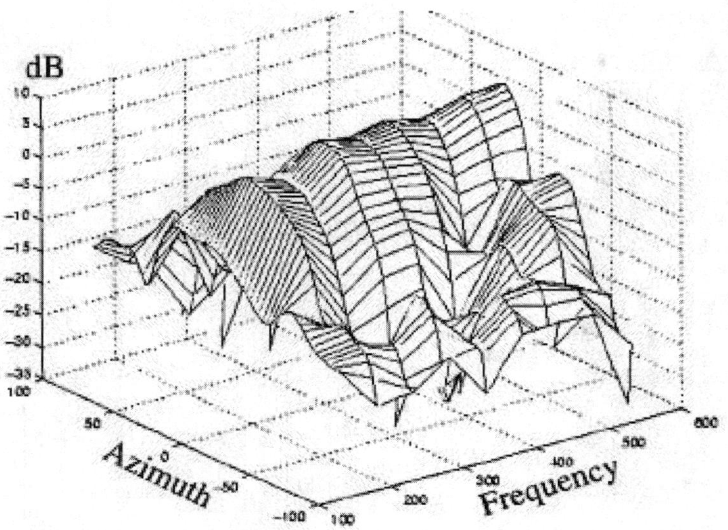

FIGURE 7–27

3-Dimensional contours of circle array © 1998 IEEE [7]

Table 7–4 Circles array measures of parformance for UWB SAR © 1998 IEEE [7]

Measure	Symbol	Ideal	Circle HH	Circles VV
Gain Aperture [m^2]	GA_d	15.5	3.7	8.6
Cross Range Resolution [m]	δ_{CR}	0.8	1.07	0.95
Left/Right Ambiguity Ratio [dB]	U_{LR}		−20.0	−25.5
Doppler Ambiguity Ratio [dB]	U_{dop}	−14.1	−12.4	−13.7
Polarimetric Coupling Ratio [dB]	U_{HV}, U_{VH}		−9.1	−13.4

for two-sided operation, a major improvement over the P-3 UWB antenna. The development of UWB array measures of performance was significant in characterizing these critical components for FOPEN SAR applications [7].

7.2.2 FOPEN Transmitter

The FOPEN transmitter is typically the second most critical subsystem after the antenna. The requirement for high-efficiency linear amplification of the UWB waveforms is essential in operating in the complex RF environment. It is not desirable to design a transmitter that uses saturated amplification,

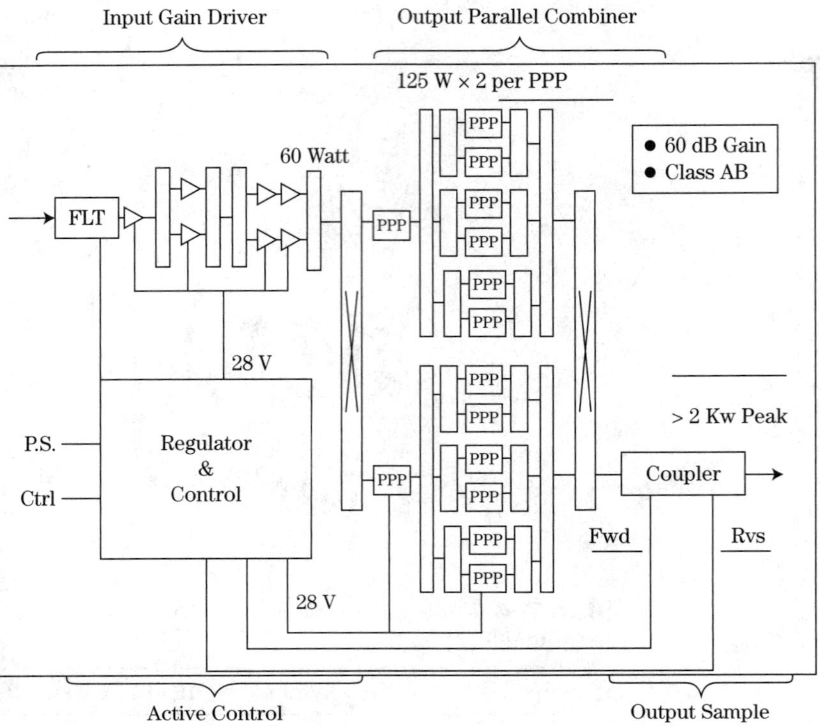

FIGURE 7–28

Solid state power amplifier suitable for FOPEN SAR

as with Class C solid-state devices. The rapid modulation of the pulse in either amplitude or phase creates harmonics and spectral spreading that are unacceptable, given the spectral constraints of the NTIA regulations.

Significant progress has been made over the past decade in high-efficiency solid-state transmitter modules. More importantly, these modules have been designed for Class AB operation with modest dynamic range (15–25 dB) and low intermodulation products [20]. Figure 7–28 illustrates a power amplification chain for conditioning the FOPEN SAR signal from the waveform generator to interface to the antenna. Figure 7–5 indicates that a useful FOPEN SAR system needs between 100 and 200 watts of average power. With a typical 10% duty factor, the transmitter chain must generate greater than 1,000 watts of peak signal power and must be capable of operating over at least a 200 MHz bandwidth [21].

The FOPEN transmit amplifier has four important attributes that need careful specification, depending on the application and environment. First, the input signal from the exciter needs to be filtered to eliminate any out of band noise or harmonics. The input gain driver should be a highly linear (i.e.,

FIGURE 7–29

Solid-state power amplifier [22]

Class A) design, with pairs of amplifiers to handle any impedance mismatch over the band and to provide for load pulling into the power stage.

A parallel combiner can be built with very wideband stripline circuits to provide the high output power. Again, it is important to specify and fabricate these combiners with the known characteristics of the high-power Class AB amplifiers to achieve both amplitude and phase match and for maximizing the efficiency of the total transmitter chain. A specification of −60 dB in band intermodulation products along with −60 dB out-of-band harmonics suppression is typical, and may require more for operation into the antenna impedance or in environments with more stringent control of radiation outside of the FOPEN specified band.

The basic building block of the solid-state transmitter will be a high peak power solid-state amplification gain block as shown in Figure 7–29 [22]. This device must operate over the full band of the RADAR, have the time-bandwidth characteristics that it can amplify the LFM pulse without introducing amplitude and phase ripple, and accommodate the need for spectral notching within the pulse. Depending on the notching approach, a Class AB amplifier is highly recommended. Class AB has the characteristics that transistors are biased just below saturation for linearity but do not incur the large penalty in efficiency of Class A amplifiers. Fortunately, the digital television and mobile communications industries have invested in reliable devices that provide 200 to 300 watt building blocks with gains approaching 10 dB.

Solid-state amplifiers are preferred for many airborne applications because of their low-voltage power supply requirements and their reliability. One of the most important characteristics of a solid-state gain stage is the power-added efficiency, defined by [23]

$$\eta_{PA} = \frac{P_{out} - P_{in}}{P_{DC}} \tag{7.19}$$

where P_{out} is the average output power, P_{in} is the average RF input power, and P_{DC} is the total direct current (DC) power input to the device. When multiple amplifiers are combined into a hybrid combiner, the combining loss will directly affect the overall power added efficiency, as illustrated in Figure 7-30. In addition, the match of the amplifier into the combiner and output

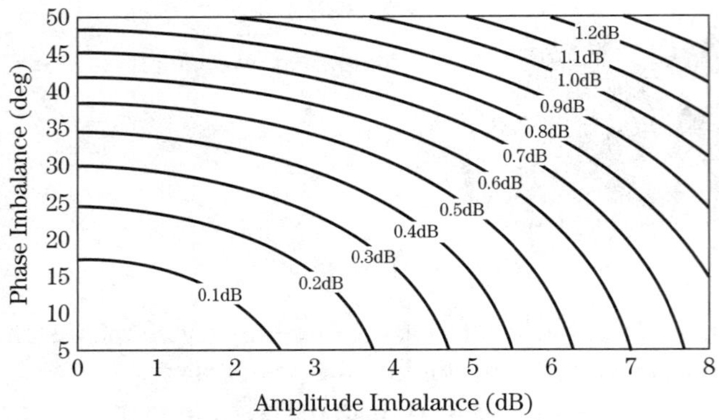

FIGURE 7–30

Contours of power loss by combining two solid state power amplifiers [23]

circuitry will cause reflections and reduce the power delivered to the antenna. The loss of power for two amplifiers, P_1 and P_2, is given by the amplitude and phase match relationship

$$P_{lost} = 20 \log \left(\frac{\sqrt{P_1^2 + P_2^2 + 2P_1 P_2 \cos(\theta_P)}}{P_1 + P_2} \right) \tag{7.20}$$

In general, the power combiner should have the following characteristics [16]:

- Low insertion loss to maximize transmitter efficiency
- High isolation between ports, such that a failed module does not affect load impedance or combining efficiency
- Well-matched impedance across the tunable band, such that the gain and efficiency are not affected
- Good thermal match and power dissipation to limit the temperature rise in the integrated combination of several amplifiers

The regulation and control circuitry should be designed to sample the output signal and to provide long-term stability in power and phase compensation in the drive circuitry. The design needs fast control algorithms to accommodate the frequency notching for minimum phase perturbation. In addition to the sampling of the input signal for control, it is important to sample the reflected signal from the antenna interface to compensate for antenna voltage standing wave ratio (VSWR) effects on the waveform. If the control cannot adequately provide the amplitude and phase compensation within the pulse, the transmitter needs to be instrumented to measure these parameters over frequency for subsequent waveform reconstruction.

Figure 7–31 shows a commercially available 1,500 watt integrated amplifier from the digital television industry that can be easily adapted for FOPEN SAR applications. The output power is rated from 470 MHz to 960 MHz with

FIGURE 7–31

Integrated high-power amplifier from mobile communications [20]

an overall conversion efficiency of 25%. Spurious levels and intermodulation products are below −60 dBc. The eight solid-state output stages are matched to better than 2 degrees phase and ±0.5 dB amplitude.

7.2.3 Receiver Exciter

The receiver and exciter designs have to be closely coordinated with those of the waveform generator and the transmitter. When the FOPEN SAR system is operating in the fully polarimetric mode in a dense RFI environment, the time must be allocated for transmitting the two orthogonal polarizations and for opening the receivers for reception of the copolarization and cross-polarization signals. In addition, the receiver must be opened for the reception of the RFI environment to estimate and cancel the interference.

The interfaces for a UHF fully polarimetric FOPEN SAR system are depicted in Figure 7–32, showing the two separate antennas for port and starboard. Each antenna will have two orthogonal polarization channels—one for horizontal and one for vertical. A common exciter signal is interfaced to the transmit chain to insure coherency and waveform fidelity between the separate polarization channels. It is also very important to have fine timing on the transmit and receive diplexers so that any timing errors are compensated for accurate geolocation and correlation of the SAR image channels.

Figure 7–33 illustrates the timing for the transmitter starting at the PRI strobe, shown as the vertical time lines. Transmit pulses alternate between horizontal and vertical polarizations, transmitted through the appropriate antenna. There will be a delay from the end of the pulse until the near range SAR returns are received. The two orthogonal polarization receivers are kept closed to protect the receiver from the ground bounce. For the alternate pulses, the vertical channel can be used to sample the RFI environment, as indicated, if

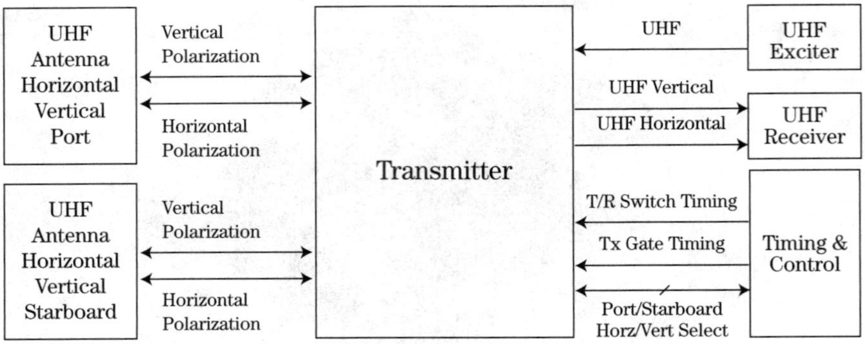

FIGURE 7–32

Interface between antenna, transmitter and receivers

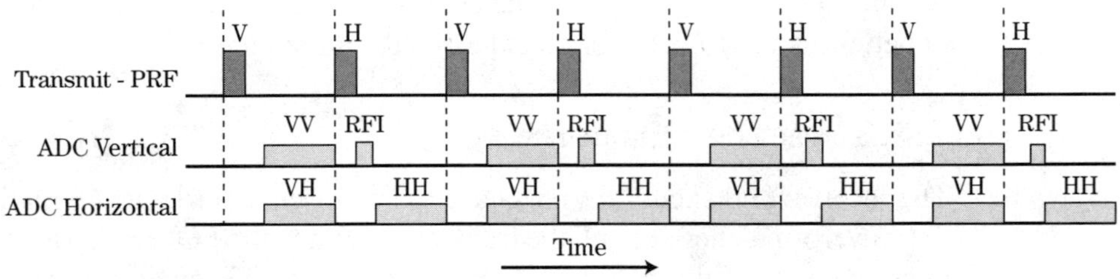

FIGURE 7–33

Timing of transmit pulse and receiver windows for collecting SAR and RFI

the high-voltage pulse return is not processed. It is important that the returns do not have the clutter return corrupting the RFI *sniff*.

A generic receiver is shown in Figure 7–34, with the input into and out of the antennas. The inclusion of a transmit and receive switch and limiter is essential to protect the input to a wideband low-noise amplifier (LNA). There will typically be 1.5 dB of RF loss before the LNA. However, if there is any mistiming of the transmit pulse, the nadir bounce return or strong RFI will desensitize the receiver or, worse, destroy it. At UHF it is common to obtain receivers with LNA noise figures approaching 3 dB and at least 20 dB of gain. This is an important step to establish the SNR for the processing and to allow downstream filtering and automatic gain control to achieve spurious-free dynamic range operation.

Figure 7–34 shows a direct down conversion from the UHF signal to baseband. Currently, analog-to-digital converters (ADCs) with sampling frequencies over 500 MHz and of 8 or 10 bits (i.e., 59 to 71 dB of dynamic

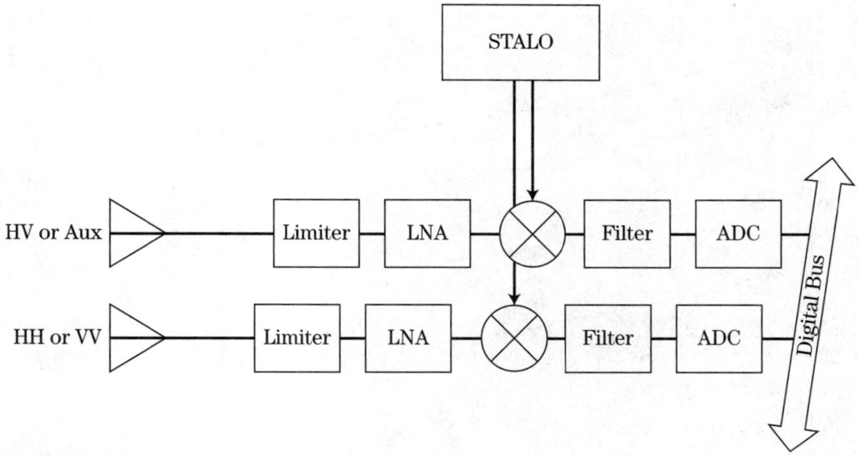

FIGURE 7–34

Two channel polarimetric SAR receiver

range) are available from many sources. With direct digital downconversion, there is no need for multiple IF conversion stages and the attendant impact on dynamic range through component intermodulation products.

The most important part of selecting the ADC is establishing the dynamic range requirement based on the RFI environment. SAR systems normally do not require high dynamic range receivers when the environment is composed of small numbers of moderate power emitters. But, as has been shown in Chapter 5, the UHF environment having a large number of commercial TV and radios is severe. Figure 7–35 shows the impact of various classes of emitters as a function of peak power and range to the platform. The interference-to-noise ratio at the front end of the receiver can reach 60 dB and must be contained by the analog filtering to meet the receiver's dynamic range and intermodulation product specification. But if the RFI removal is to be done digitally, the front end must remain linear.

A trade-off between the ADC sampling rate and dynamic range continues to be a system engineering challenge. If the maximum signal bandwidth for range resolution is expected to be on the order of 500 MHz and direct sampling of the SAR returns is desired for wide swath width, the ADC must be capable of sampling at a rate of over 1 Gsps. The current state of the art is challenged to provide 10 bit, 1 Gsps ADCs in the size factor required for a UAV. The FOPEN SAR design examples in the next section will use deramp-on-receive (i.e., stretch) processing to reduce the IF bandwidths to the 180 MHz regime. This is well within the current state-of-the art ADC capability.

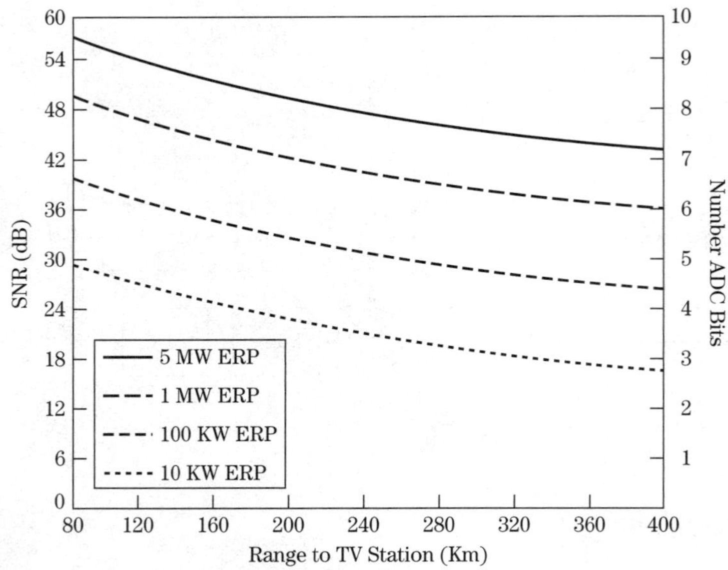

FIGURE 7–35

Impact of strong RFI sources on receiver dynamic range

7.3 FOPEN SAR System Design

This section will show the details of a system design for UHF fully polarimetric SAR, based on a qualitative comparison of two operating frequencies. VHF SAR will be a subset of the UHF design but will use only a single polarization. However, a separate assessment of UHF versus VHF performance will be made due to the significant differences in resolution and SAR integration times. The platform altitude was chosen as a midpoint between the TUAV and HALE UAV platforms. This was done to maintain a modest slant range for SAR operation. A variety of swath width requirements will be considered to illustrate the impact on signal processing and data link requirements. The key performance parameters are summarized in Table 7–5; where there are three FOPEN SAR modes to be considered in the UHF design trades: strip map, area coverage, and point modes.

7.3.1 Multiplicative Noise Ratio

A major design consideration in the selection of the FOPEN SAR modes is the effect of distributed clutter and waveform design on the target detection characteristics. Many of these factors have been presented in the sections concerning the waveform and antenna design. A primary performance factor is total multiplicative noise ratio (MNR) for the system. MNR is the summation

Table 7–5 Key FOPEN SAR performance parameters

Parameter	Value	Units
Platform Velocity	150	m/sec
Platform Altitude	15	Km
Peak Power	1,500	watts
Pulse Length	190	μsec
Waveform Broadening (R, Cross-R)	1.2	
Wavelength (Center)	0.7	m
Target Cross Section	10	dBsm
Clutter Cross Section	−8.0	dBsm

Table 7–6 Contribution to FOPEN SAR multiplicative noise ratio

Multiplicative Noise Ratio—Factors	Value
Range ambiguity level	−18.0 dB
Doppler ambiguity level	−23.0 dB
Left/right ambiguity	−21.0 dB
Impulse response ISLR	−16.5 dB
ISLR from Notching	−18.0 dB
ISLR from motion sensing residuals	−20.0 dB
ILSR from amplitude ripple	−28.0 dB
ISLR from phase ripple	−23.0 dB
ADC Quantization noise	−28.0 dB
Net MNR	**−10.7 dB**

of the background clutter or noise due to waveform (ISLR), ADC quantization noise ratio (QNR), and the antenna/waveform ambiguity ratio (AMBR) [24]:

$$MNR = ISLR + QNR + AMBR \tag{7.21}$$

A summary of the primary contributions for FOPEN SAR is given in Table 7–6; where it is assumed that all are common for each mode of the RADAR. This is not generally true, and the particular waveform needs to be examined based on the local RFI environment and the range from various emitters to the platform as well as the reflectivity and variation of the clutter in the area surrounding the target.

The additive noise factor σ_n is commonly represented as the effective backscatter contribution of the surrounding clutter and interference environment. It is equivalent to the terrain backscatter σ_0 that produces a signal in the receiver equal to the noise level:

$$\sigma_n = \frac{\sigma_0}{CNR} \tag{7.22}$$

CNR is the average intensity of a diffuse target return divided by the average noise intensity:

$$CNR = \frac{P_{ave}G^2}{2kT_0F_nL_{rec}L_{fol}}\left(\frac{\lambda_c}{4\pi R_c}\right)^3\left[\frac{\sigma_0\rho_R}{k_Rv_P\sin(\alpha_{dc})\cos(\gamma_g)}\right] \tag{7.23}$$

It should be noted that the receiver loss has been separated into two factors:

L_{rec} All the losses following the LNA

L_{fol} Explicit loss due to propagation through the foliage

The foliage loss, which was given in (3.33), depends on the foliage type, grazing angle, and center frequency.

The signal-to-clutter ratio (SCR) is given by the ratio of the SNR to CNR:

$$SCR = \frac{\sigma_T\cos\gamma_g}{\sigma_0\delta_R\delta_{CR}} \tag{7.24}$$

The system factors that directly affect the image quality for FOPEN are the left–right ambiguity and ISLR from notching. These factors currently (and most probably will continue to) represent areas needing technology development to improve the MNR.

The most important contribution to MNR, as seen from Table 7–6, is the waveform ISLR by combining the notching on transmit and the receive impulse response after RFI is removed. It is important to synthesize a waveform that has minimum interference to any sensitive receiver in the environment, as covered in Chapter 5. However, this is governed by the local NTIA control over frequency allocation. The next most important contribution to the MNR is the range ambiguity. Because of the limitation on antenna pattern elevation directivity on most platforms, the range ambiguity is often obtained solely by selection of the PRF. A very low PRF will reduce the range ambiguity to those scatterers at very far ranges, ideally over the local terrain horizon. But this may also affect the achievable Doppler ambiguities, depending on the design for polarization (requiring multiple transmit pulses at orthogonal polarization) and the speed of the platform.

7.3.2 FOPEN SAR Modes

The strip map is the highest area coverage rate at a 10 km swath width, as illustrated in Figure 7–36. A 1 meter range and cross-range resolution was

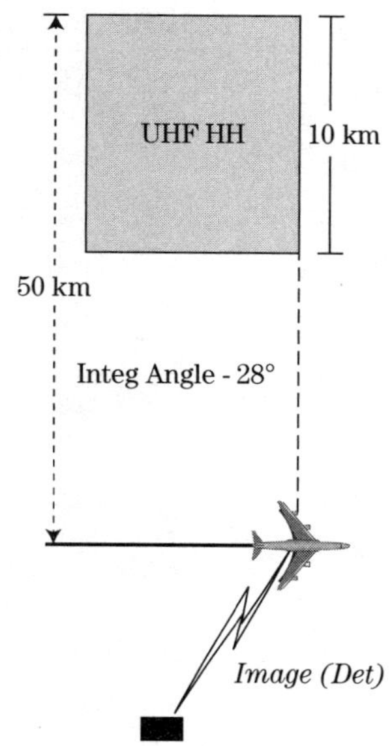

FIGURE 7–36

FOPEN SAR strip-map mode scenario

selected to determine the design approach for modest resolution SAR. In addition, this mode uses a single polarization to enable real-time throughput onboard with the very wide swath width.

The rationale and summary characteristics for the strip-map mode are given as follows:

- Principal advantage is the higher area coverage rate as the background for future mission planning
- Moderate resolution is weighed against the need for high detection probability for tactical targets.
- Single horizontal polarization provides adequate return from cultural features and enables doubling the average power for this long-range mode
- Principal disadvantage of long-range, low-resolution operation is the loss of target characteristics of many tactical objects under foliage

The area mode, shown in Figure 7–37, will be designed to provide improved target detection by increasing the resolution and using polarimetric processing. This mode can be used as the primary wide area surveillance technique for tactical targets under trees. By combining polarimetric imaging with strip-map collection, the discovery of buildings and vehicles under dense foliage is made possible. The features of area mode are as follows:

- Operates at a shorter slant range to increase the grazing angle for lower foliage loss and signal sensitivity

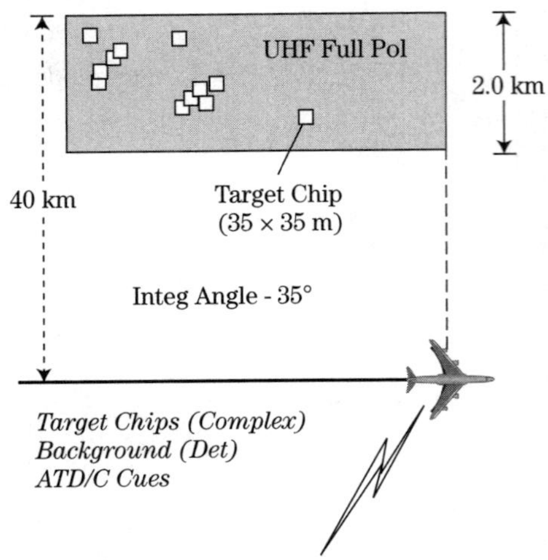

FIGURE 7–37

FOPEN SAR area mode scenario

- Increased bandwidth and collection angle to provide 0.7 meter resolution
- Improved target-to-clutter ratio (TCR) for higher probability of detection
- Fully polarimetric processing is provided to reduce the false alarm rate and to enhance the target discrimination
- Disadvantage of increased clutter cell area, due to a steeper grazing angle, which offsets some of the advantage of wider bandwidth
- Disadvantages of the reduction of the average power in individual polarizations and increased onboard signal processing complexity

Finally, the point mode, shown in Figure 7–38, has the finest resolution by collecting over the widest possible angle. The point mode characteristics are summarized as follows:

- Significant enhancements in both the target detection and characterization potential by operating with a 60 degree integration angle
- Shorter range, steeper grazing angle, and higher resolution improves the SCR by reducing foliage loss and reducing the clutter cell area
- Disadvantage is the very long integration time to capture 60 degrees of operation
- Typically, this type of mode is scheduled only when the user has knowledge of a particular area of interest, where more detailed image processing is desired

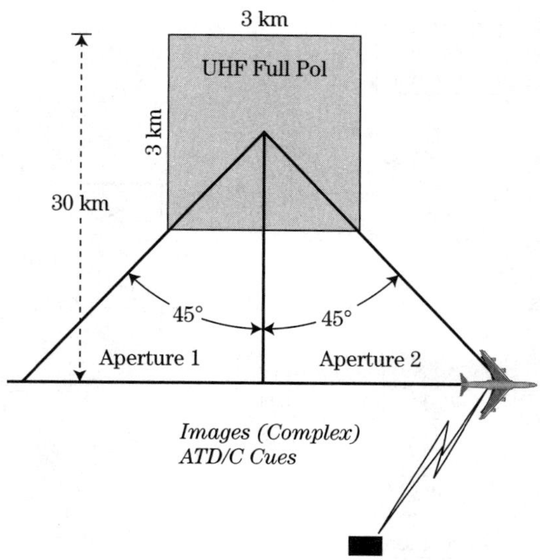

FIGURE 7–38

FOPEN SAR point mode scenario

Table 7–7 System performance parameters for UHF SAR design trades

RADAR Parameter	Strip	Area	Point	Units
Azumuth Integration Angle	28.0	35.0	60.0	deg
Grazing Angle	19.5	22.6	31.8	deg
Range Resolution	1.0	0.6	0.5	m
Azimuth Resolution	1.0	0.8	0.5	m
Foliage Loss (H-pol)	13.1	11.4	9.6	dB
PRF each Polarization	1000.0	500.0	500.0	pps
Average Power/Polarization	285.0	142.5	142.5	w
Range	45.0	39.0	28.5	Km
Bandwidth	180.0	300.0	400.0	MHz
Integration Time	149.6	163.9	219.4	sec
Performance Metric	**Strip**	**Area**	**Point**	**Units**
Additive Noise Backscatter	−41.2	−35.8	−36.0	dB
SNR on Target under foliage	38.1	37.4	42.1	dB
SCR under foliage	4.8	9.5	14.0	dB

Details of the individual modes are provided in Table 7–7, along with the FOPEN SAR measures of performance. The azimuth integration angles are derived from (4.8) and push the capabilities for antenna gain and ambiguity ratio. The foliage losses are obtained from (3.33) and are for single-canopy forest and horizontal polarization. Slightly higher losses (e.g., 2 dB) are expected

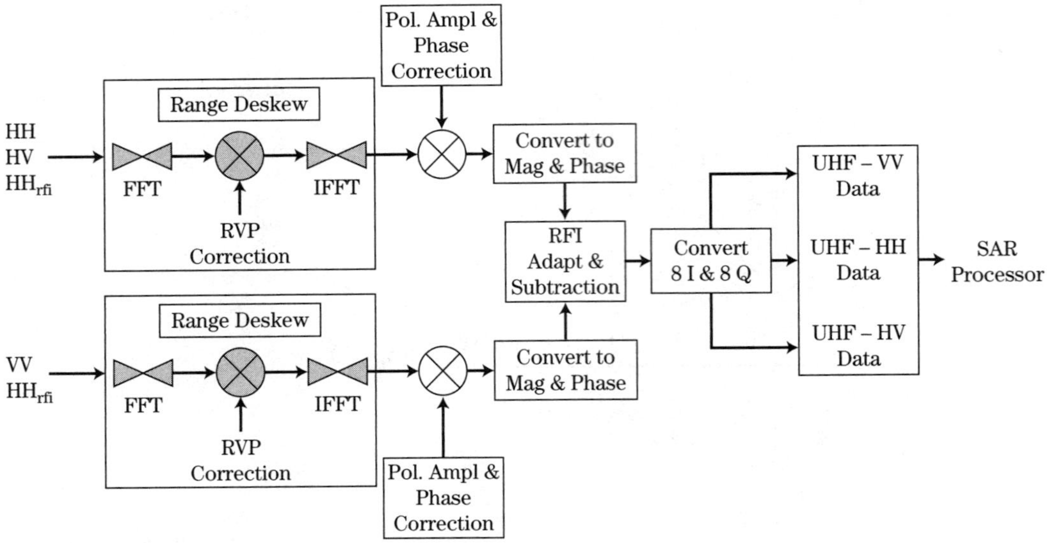

FIGURE 7–39

FOPEN SAR preprocessing steps

from the vertical polarization component along with an increase in clutter backscatter.

The major impact on signal return is the effect of volumetric clutter around the target. The noise equivalent clutter reflectivity σ_{ne}^{0} is very low. However, the CNR is established by the strong backscatter from the tops of the forest foliage. This coupled with the added foliage loss L_{fol}, provides a reduced SCR on the targets in all but the point mode.

7.3.3 Signal Processing Throughput

The block diagram in Figure 7–39 illustrates the preprocessing operations prior to image formation. The first step is to remove the residual video pulse term that remains after stretch processing, and to motion compensate establishing a constant range to the scene center. A series of processing steps were outlined in Chapter 5 for estimating of the computational load.

The correction terms are applied for known polarization, amplitude, and phase errors or artifacts. These terms are normally known from calibration runs and are low-complexity processing. After the stretch deskew process, the RFI is identified and subtracted from the phase history. As indicated earlier this can be accomplished by several algorithmic approaches. The objective is to have all three channels (i.e., HH, VV, HV) be correctly motion compensated

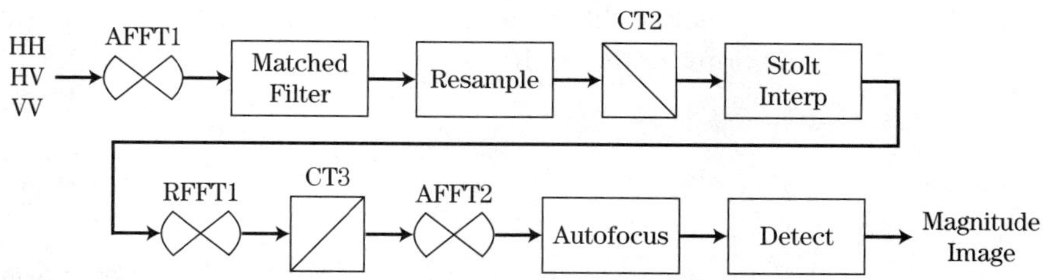

FIGURE 7–40

Image formation processing using RMA

Table 7–8 FOPEN SAR processing summary—three illustrative modes

	CELL $[m]^2$	Area $[Km^2]$	No Chnls	Range [Km]	Integration Time [s]	Proc. Thruput [Gops]	Memory [GB]
STRIP	1	70	1	45	150	12.41	8.42
AREA	0.49	14	3	39	164	11.27	10.31
POINT	0.25	9	3	28.5	157	13.43	12.98

with as much RFI removed as possible. After the RFI has been removed, the complex (I and Q) pulse samples can be scaled to 8 bits for image formation.

The range migration algorithm (RMA) is illustrated again in Figure 7–40 providing the key processing sequential steps. The number of operations can be calculated for the fast Fourier transform (FFT), Stolt interpolation, autofocus, and target detection processing based on the operations in Figure 4–18.

Table 7–8 quantifies the important image metrics that determine the amount of processing required. These include the number of additional data points (including the range curvature and along-track samples) needed to support the cross-range resolution. Both the processing throughput and the memory requirements need to be estimated, along with any inefficiencies in the parallel processing of the data. It is typical to size a process at twice the estimated throughput and memory to allow for requirements growth in the real-time processing.

7.3.4 Data Link Bandwidth

The area coverage rate and SAR resolution drives the ability to transmit a SAR image to the ground in near real time. For a strip-map SAR system, with onboard image formation processing, the ACR is determined by the swath width W_S and the aircraft velocity v_P:

$$ACR_{strip} = W_S v_P \tag{7.25}$$

For a spotlight mode, the area of the spot and the total integration time t_I determines the ACR:

$$ACR_{spot} = \frac{W_L W_S}{t_I} \tag{7.26}$$

For real-time transmission of the imagery, the conventional definition of *real-time* is that the data must be sent to the ground in the same time or less than the collection time. If this is not accomplished, then the imagery must be stored and transmitted during the time that the RADAR is not operating. Examples of nonoperating include aircraft turnaround time or transition to a new collection waypoint. In this case the onboard processing is commonly called *near real time*.

The SAR image data volume depends on several factors:

- *Pixel area:* The area of the SAR image pixel in either the ground or slant plane. For UHF SAR the pixels are typically spaced by 0.5 to 2 meters in dimension, depending on the range and cross-range resolution, and the grazing angle. For VHF SAR the pixels are typically spaced by 3 to 5 meters.
- *Image format:* The images can be either *magnitude* or *complex*. A magnitude image might be sent as a background context for positioning detected targets on fine resolution spotlight images. The complex image is needed if further image processing is desired. A magnitude-only image will nominally have 8 bits of information per pixel, whereas the complex image will have twice that amount of data.
- *Polarization:* If the exploitation of the data is to be done in a ground station instead of onboard the UAV, provisions for sending the three components of complex polarimetric data must be made. Depending on the dynamic range of the signal and clutter, it is common to have between 40 and 60 bits of fully polarimetric data per pixel.
- *Image compression:* Bandwidth compression is normally provided onboard for any high-volume data to be transmitted to the ground. If secondary products are needed, it is important that the compression be nearly lossless. The most common secondary products are coherent or polarimetric change detection, interferometric image processing, or detailed analysis of nonstationary artifacts. Lossy image compression can provide over 20:1 reduction in the data bandwidth but at the cost of reduced features. Lossless image compression is normally restricted to less than a 10:1 compression.

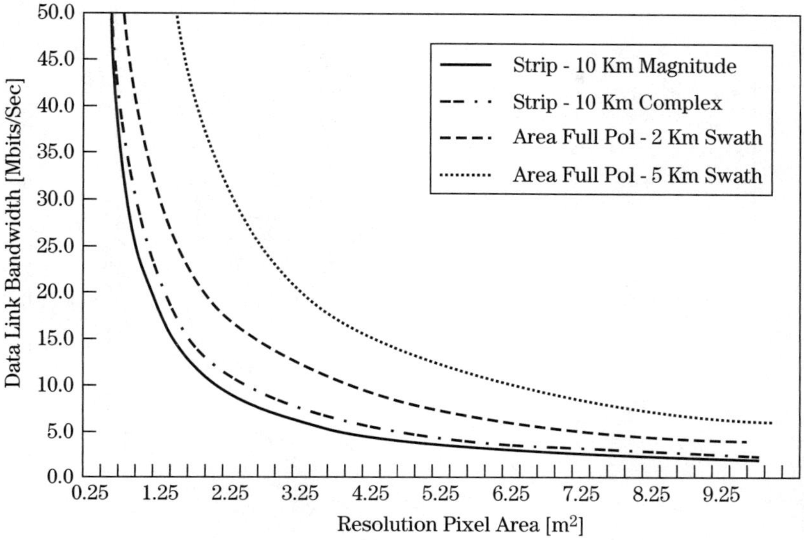

FIGURE 7–41 Data link bandwidth versus pixel area and SAR mode

The data link bandwidth can be approximated by

$$BW_{DL} = \frac{ACR}{\delta_R \delta_{CR}} \tag{7.27}$$

Figure 7–41 illustrated the data link bandwidth required for several image products versus the pixel area. The platform is assumed to be moving at 150 meters per second in a strip-map mode for each example. Common tactical data links have 10 million bits per second (Mbps) data rates. However, recent developments project a maximum of 50 Mbps, indicated as the maximum on the y-axis of the figure. The magnitude-only strip map can be accommodated within 50 Mbps, especially at the 1 m^2 pixel area for the FOPEN SAR mode illustrated in Figure 7–36. In fact, the complex strip map can be transmitted with little or no bandwidth compression. However, at the lower bandwidth of 10 Mbps, developments in lower loss image compression are required.

The area coverage mode for both 2 km and 5 km swath width are assumed to have 0.5 m^2 pixel area and be fully polarimetric and complex. With these parameters, the real-time image cannot be accommodated in the given data rate without very lossy compression. As a comparison, a VHF SAR system with 9 m^2 pixel area and 10 Km swath width is seen to fit within the 10 Mbps data rate.

The alternative to sending a complex polarimetric map over a data link is to only send target chips. The extraction of these chips depends on an efficient constant false alarm rate (CFAR) processor and moderate false alarm rates. Depending on the characteristics of the clutter and the size of the target, chips have been considered between 32 and 64 pixels on a side. Figure 7–42

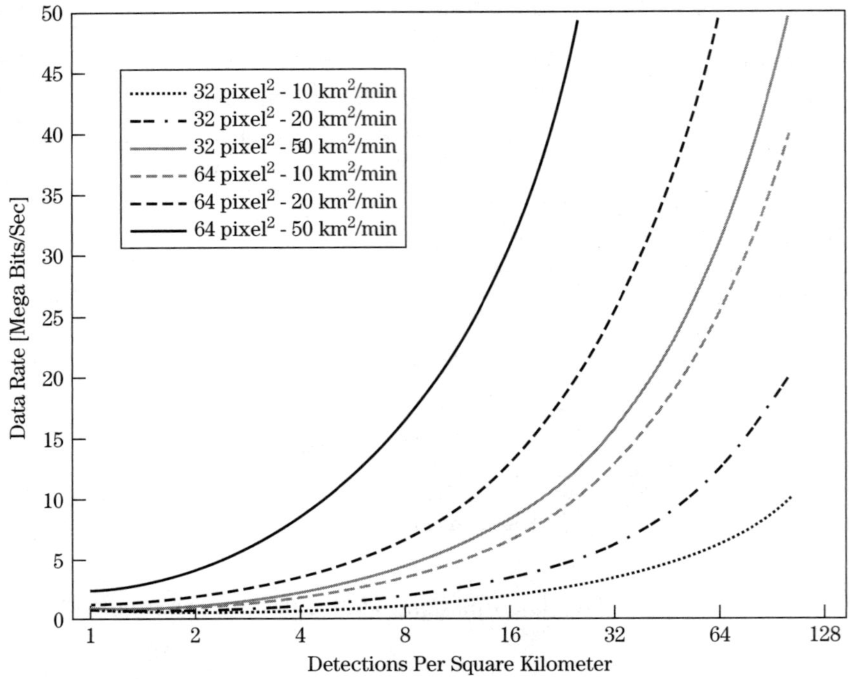

FIGURE 7–42

Capacity for transmitting detected chips versus detections per Km^2

illustrates the data link data rate for two chip sizes (32 and 64 pixels per side) and SAR area coverage rate (10, 20, and 50 Km^2/min). When these chips are sent down, complex and fully polarimetric data are needed. The number of detections per square kilometer drives the data rate. As such, it is important to have low false alarm rate processing prior to selection of chips.

7.4 References

[1] Sullivan, R. J., *Microwave RADAR—Imaging and Advanced Concepts*, Artech House, Boston, MA, 2000, Chapt.), 7.

[2] Davis, M. E., Tomlinson, P. G., and Maloney, R. P., "Technical Challenges in Ultra-Wideband RADAR Development for Target Detection and Terrain Mapping," *Proc 1999 IEEE RADAR Conference,* Boston, MA, April 1999, p. 1.

[3] Dobricic, D., "VHF Antenna Noise Temperature," *AntenneX,* No. 132, April 2008, http://www.yu1aw.ba-karlsruhe.de/vhfnoisetemp.pdf.

[4] Lasswell, S., "History of SAR (Synthetic Aperture RADAR) at Lockheed Martin (formerly Goodyear Aerospace)," Brochure from Lockheed Martin, Litchfield Park, AZ, 2009.

[5] Carrara, W. G., Goodman, R. S., and Majewski, R. M., *Spotlight Synthetic Aperture RADAR,* Artech House, Norwood, MA, 1995.

[6] Davis, M. E., "Technical Challenges in Ultra-Wideband RADAR Development for Terrain Mapping," , presented at *IGARSS'98,* Seattle, WA, July 1998.

[7] Ayers, E. L., Ralston, J. M., Maloney, R. P., Tomlinson, P. G., and McCorkle, J., "Antenna Measures of Merit for Ultra-Wide Synthetic Aperture RADAR," *Proc. IEEE 1998 National RADAR Conference,* Dallas, TX, May 1998, pp. 331–36.

[8] Ressler, M.,., "The Army Research Laboratory Ultra Wideband BoomSAR,"", Proc. IGARSS ,Vol. 3, May 1996, pp. 1886–1888.

[9] McCorkle, J., "So What's so Special about UWB Propagation," Presented at 2002 IEEE Communications Theory Workshop, Atlanta, GA, May 2002.

[10] Hellsten, H. and Ulander, L. M. H., "VHF/UHF Synthetic Aperture RADAR—Principles and Motivation,"", Proc. 2003 *Internl. RADAR* Conference, Adelaide, Australia, September 2003.

[11] Hellsten, H., "CARABAS II in Flight," photo courtesy of Swedish Defence Research Establishment (FOA), Linkoping, Sweden, 1998.

[12] Hellsten, H., "Antenna Backlobe Rejection," U.S. Patent Application Publication, US2009/0051584A1, February 26, 2009.

[13] Yegulalp, A. F., image courtesy of MIT Lincoln Laboratory, Lexington, MA, 1998.

[14] Ausherman, D., former manager P-3 SAR Development, Environmental Research Institute Michigan (ERIM), Ann Arbor, 1997.

[15] Tomlinson, P. G., "Implications of the Frequency Dependence of the P-3 UWB Antenna Gain," Unpublished memorandum, DSA Inc., Arlington, VA, July 21, 1994.

[16] Wheeler, K. and Hensley, S., "The GeoSAR Airborne Mapping System," *Proc. 2000 IEEE International RADAR Conference,* Washington, DC, May 2000, pp. 831–35.

[17] Thomas, R. E. and Huang, J., "Ultra Wideband Microstrip Array Antenna for GeoSAR Applications," *Proc. 1998 IEEE AP-S Symposium,* June 1998, p. 2096.

[18] Kalbasi, K., Plumb, R., and Pope, R., "An Analysis and Design Tool for a Broadband Dual Feed Circles Array Antenna,"*Proc. International Symposium on Antennas and Propagation* Vol. 4, Chicago, IL, July 1992, pp. 2085–2088.

[19] Lalazari, F., "FOPEN RADAR UWB Antenna Program Review," Unpublished", unpublished report, Ball Aerospace, Westminster, CO, May 1997.

[20] Sokal, N. O., ., "RF Power Amplifiers-Classes A through F,"", *Proc. ELECTRO '96. Professional Program,* Somerset, NJ, April 1996, pp. 317–321.

[21] ABE Electronics, "5KW UHF Solid State TV Amplifier," brochure, http://www.abe.it.

[22] Scarpa, F., Facco, A., Zviagintsev, V., and Lipeng, Z., "A 2.5 KW, Low Cost 352 MHz Solid State Amplifier for CW and Pulsed Operation," *Proceedings of EPAC 2002,* Paris, France, pp. 2314–2316.

[23] Borkowski, M. F., *RADAR Handbook,* Chapt. 11, McGraw Hill, New York, 2008.

[24] Carrara, W. G., Goodman, R. S., and Majewski, R. M., *Spotlight Synthetic Aperture RADAR—Signal Processing Algorithms*, Artech House, Boston, MA, 1995, pp. 331–336.

CHAPTER 8

FOPEN Ground Moving Target Indication

The detection of fixed objects has been covered in significant detail. However, moving objects are of great interest to tactical users. Other than the early foliage penetration (FOPEN) ground moving target indication (GMTI) RADARs of the 1970s, there has been little development for these applications until approximately 2005 [1]. This chapter will detail the approach for detecting moving targets in ultra high frequency (UHF) FOPEN systems. It is well known that a moving platform provides competing clutter spread in Doppler that masks all but the very fast ground targets. Moving target detection RADARs have been developed at UHF for detecting airborne targets. For ground moving target detection, the competing ground clutter can be canceled by adaptive processing and multiple antenna phase centers. However, a very large antenna is required at UHF to provide both accurate target location and to minimize the clutter masking of targets. These requirements will be explored in terms of the ability to achieve an acceptable minimum discernable velocity (MDV) and their impact on the RADAR design.

Because the targets under foliage experience significant attenuation and the clutter is between the RADAR and the object, FOPEN GMTI RADAR systems need to address low signal-to-clutter ratio (SCR) conditions. One technique that has been used in earth resource mapping is the application of an along-track interferometry (ATI) mode. By using high-resolution synthetic aperture RADAR (SAR) processing, the SCR can be improved. However, the target motion must be compensated. The ATI approach will be summarized, and then developed in terms of capabilities and limitations through detailed modeling and simulation.

8.1 FOPEN GMTI RADAR Design

FOPEN GMTI RADAR presents significantly different challenges from those experienced by GMTI RADAR operating against targets in the open. The foliage return consists of volumetric clutter, which will both attenuate the signal and overlap the target Doppler frequency. This competition for target detection is due to clutter spread from a moving platform, and the Doppler modulated clutter from internal clutter motion (ICM). It will be demonstrated that resolution improvement alone cannot mitigated the ICM effect. Therefore, any modulating effects of clutter on the measured signal are important in both detecting a target and estimating its position and velocity.

The problem of detecting a target via Doppler frequency discrimination in a real beam RADAR is well known and is illustrated in Figure 8–1. The antenna of length dimension D with weighting k_{AZ} will have a finite beamwidth ϕ_D [2]:

$$\phi_D = \frac{k_{AZ}\lambda_0}{D} \tag{8.1}$$

The ground clutter will have a Doppler frequency around the platform given by

$$f_{dop} = -\frac{2v_P}{\lambda_0} \sin\phi_P \cos\gamma_P \tag{8.2}$$

where

v_P Platform velocity
λ_0 Wavelength of signal
ϕ_t Azimuthal angle measured from the velocity vector
γ_g Grazing angle from the platform to the ground patch

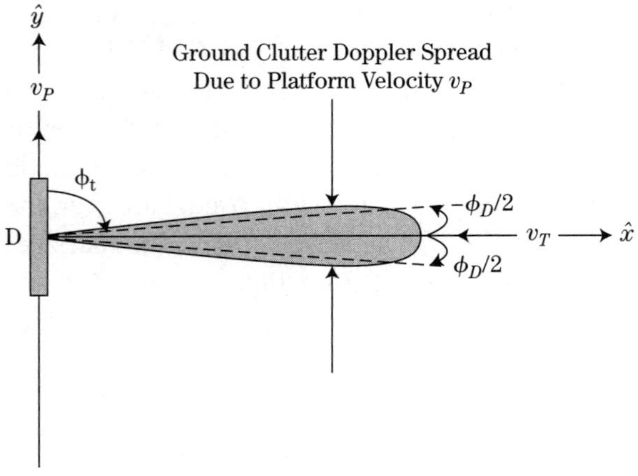

FIGURE 8–1

Target Doppler within RADAR main-beam

The convention is that the azimuth angle is measured off the velocity vector, so clutter spread is maximum at a broadside look angle of $\phi_t = \pi/2$. When the beamwidth is small, the clutter Doppler spread across the beam width can be approximated as

$$f_D \approx -\frac{2k_{AZ}v_P}{D} \tag{8.3}$$

The Doppler spread across the beamwidth is independent of frequency and depends only on the platform velocity and the antenna width.

Targets need to have a positive SCR (typically in excess of 10 dB) in order to be detected; which in turn implies a radial velocity greater than the clutter spread within the full, null-to-null azimuth beam [3]:

$$v_T \geq v_P \sin\left(\frac{k_{AZ}\lambda_0}{D}\right)\cos\gamma_g \tag{8.4}$$

Many techniques have been developed to reduce the effective main-beam spread through GMTI RADAR pulse-Doppler and space-time adaptive processing (STAP). GMTI RADARs will ideally cancel the clutter to the half beamwidth and STAP down to approximately 0.1 beamwidth, provided there is no significant internal clutter motion. These factors will be developed in the following sections.

Even with significant Doppler processing, there is going to be a residual Doppler width competing with low MDV targets. More importantly, both GMTI and STAP processing will attenuate slow-moving targets making it more difficult to detect the signal. A system engineering approach to FOPEN GMTI RADAR needs to quantify target signal-to-clutter level including both the foliage loss and the signal-to-interference-plus-noise-ratio (SINR) loss.

The most effective technique to reduce main-beam clutter spread in GMTI RADAR systems is to increase the aperture. At microwave frequencies, it is possible to achieve a main-beam spread below 0.5 m/sec, as illustrated in Figure 8–2. This ability has been demonstrated on several wide area surveillance systems, where the vehicles are not moving radially toward the RADAR [4]. However, at UHF frequencies the antenna must be physically quite long to minimize the main-beam clutter spread. The effect of UHF antenna size on the target detection MDV from (8.4) is illustrated in Figure 8–3, illustrating that the MDV can be excessively high and not useful for GMTI.

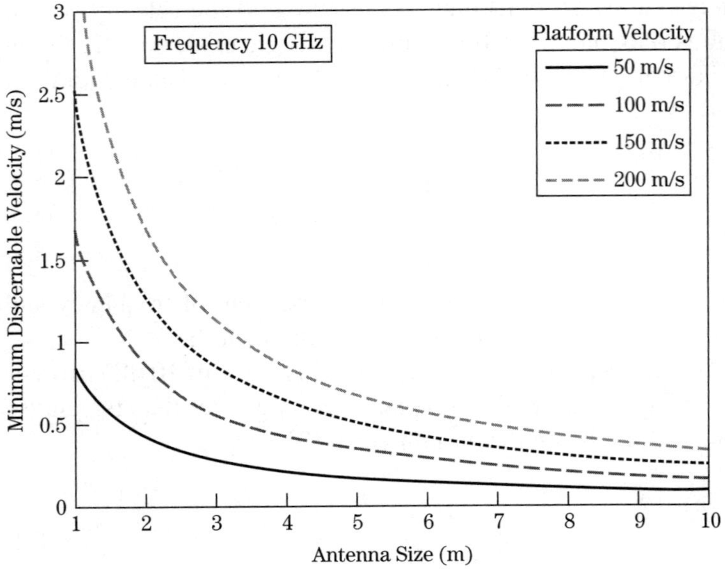

FIGURE 8–2

X-band minimum discernable velocity versus antenna size

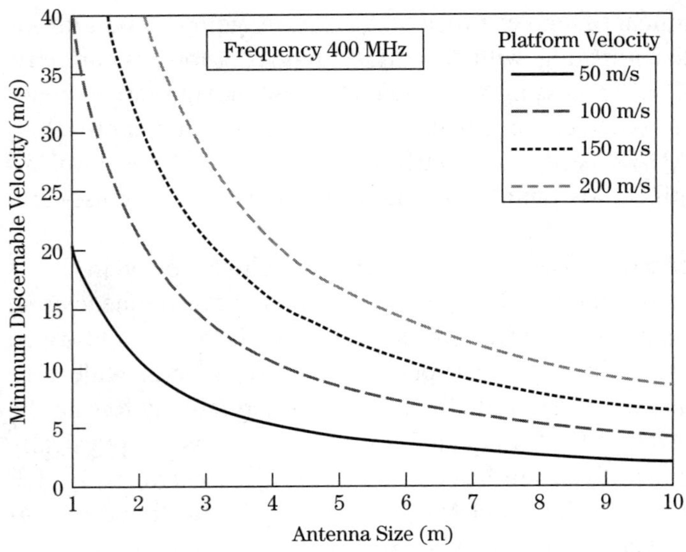

FIGURE 8–3

UHF minimum discernable velocity versus antenna size

Clutter rejection options are limited in a GMTI RADAR that has only a single-channel, real-beam antenna. Two approaches can be considered:

- Tether the antenna on an aerostat or helicopter to minimize the platform motion during the coherent integration period.
- Form a SAR to provide narrow azimuth resolution cells, and cancel the ground clutter with a multiple phase center antenna.

The first approach has been demonstrated in the early Camp Sentinel FOPEN system from a GMTI RADAR on a tall tower, but with limited range and no mobility. The very large antenna SOTAS RADAR, albeit at microwave frequency, was demonstrated on a helicopter at longer ranges. However, as has been established in Chapter 2, microwave frequencies do not achieve foliage penetration.

A novel system approach has developed a FOPEN GMTI system on an unmanned air vehicle. The FOPEN reconnaissance surveillance, tracking, and engagement RADAR (FORESTER) system operates at UHF with a 6 meter long antenna. When installed on a hovering A-160T unmanned rotorcraft, it can detect dismounts moving under foliage to ranges of 25 km. The developmental FORESTER system shown in Figure 8–4 entered testing in 2007, although few details have been published [5].

The system trade-off for FOPEN GMTI RADAR is complicated, as the choices of frequency, bandwidth, aperture size, and signal processing are all interrelated. Figure 8–5 qualitatively illustrates those relationships [7].

- Clearly, if the antenna size needs to be multiple wavelengths, with the frequency in the mid-to-high UHF band to minimize antenna length.

FIGURE 8–4

FORESTER GMTI RADAR on A-160T unmanned rotorcraft [6]

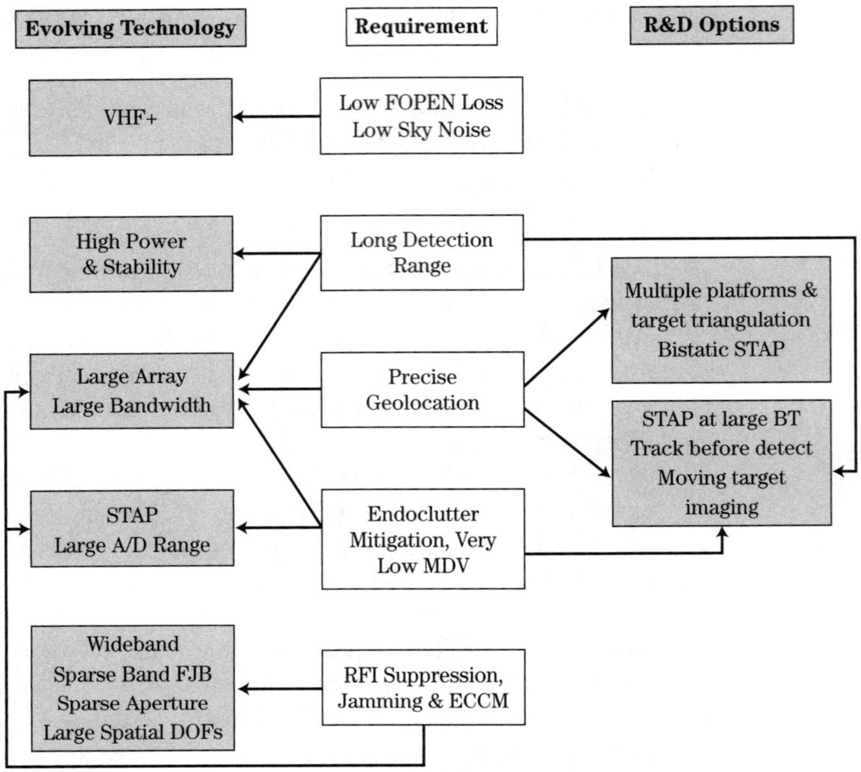

FIGURE 8–5

Trade-off for low MDV FOPEN ground moving target indication [7]

- The bandwidth of the system will be an issue for frequency allocation as we have seen in the FOPEN SAR design. It is possible to synthesize a wide bandwidth using frequency jump bursts (FJBs). But for a wide area GMTI RADAR, the range and Doppler ambiguities will be critical design factors.
- Processing to achieve a low MDV is critical, where a combination of STAP and wideband operation is needed. However, the development of efficient algorithms for wideband STAP processing continues to be an important research topic.
- Discrimination of moving vehicles and personnel from the variation in clutter, especially at tree lines, continues to be a challenge to signal processing.

The impact of system and technology choices needs careful consideration of maturity and system risk before going into a development phase. The following sections of this chapter provide the signal processing considerations for two approaches. First the application of STAP will be presented, along with first-order estimates of the benefits for improving MDV. Next, an introduction to GMTI using an ATI will be developed. ATI is a high-spatial resolution technique that can increase the target SCR, by using a sparse aperture technique to estimate both the target location and velocity from the phase of SAR pixels.

8.2 Space-Time Adaptive Processing

It has been stressed several times that FOPEN RADARs must operate at lower frequencies than microwave RADARS to enable detection of manmade targets and structures below the canopy. Moreover, when these systems are integrated on airborne platforms, the clutter return will be determined by the antenna directivity, and the range and Doppler characteristics of the waveform. Both the main beam and the sidelobes of the antenna affect the returns, as illustrated in Figure 8–6. The diagonal pattern in this figure is referred to as the *clutter ridge* [8].

Adaptive array processing was developed to eliminate strong interference from a specific angle, such as sidelobe jamming. However, these techniques were not able to cancel the distributed clutter that exists at all angles and exhibits a Doppler spread due to the platform motion. The effects of jamming and clutter masking of targets can be mitigated by the sidelobes of the antenna or by Doppler filters; if those interference sources are stationary and widely separated from the targets. However, the returns from the slow-moving target are often masked by clutter in the main beam of the antenna. With only one

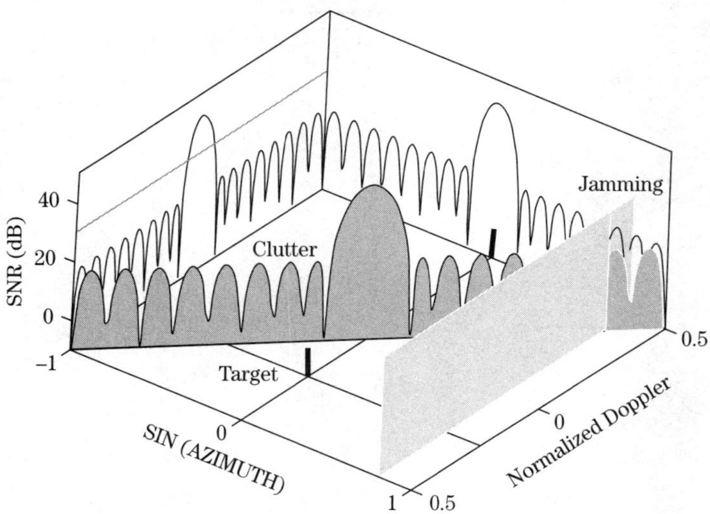

FIGURE 8–6

Angle and Doppler dimensions of signal visibility
Source: MIT Lincoln Laboratory [8]

adaptive processing domain, e.g. spatial or temporal, the target will likely be obscured over a wide region of surveillance space. STAP was developed to enable the cancelation of both directional interference and Doppler spread of clutter [8].

8.2.1 STAP Theory

STAP refers to the extension of adaptive antenna techniques to processors that simultaneously combine the signals received on multiple elements of an antenna array (the spatial dimension) and from multiple pulse repetition periods (the temporal dimension) of a coherent processing interval (CPI). The RADAR antenna is typically a uniform linear array of N elements. The pulse-Doppler waveform is a sequence of M pulses, which are transmitted at a constant PRI of T_r. When the transmit platform moves at a velocity v_P, the received signal will be given by the return from target signal and clutter (and possibly intentional jamming or RFI) [8].

Figure 8–7 shows the coordinate system for STAP analysis. A unit vector $\vec{k}(\phi_t, \theta_t)$ points in the slant range R_S to a point on the ground surface. This unit vector is expressed in terms of the azimuth and elevation coordinate angles (ϕ_t, θ_t) as

$$\hat{k}(\phi_t, \theta_t) = \sin\phi_t \sin\theta_t\,\hat{x} + \cos\phi_t \sin\theta_t\,\hat{y} + \cos\,\theta_t\hat{z} \tag{8.5}$$

The concept of spatial and temporal degrees of freedom (DOF) is illustrated in Figure 8–8. The spatial DOF can be generalized for the array having a uniform separation between elements of d. When the array is oriented such

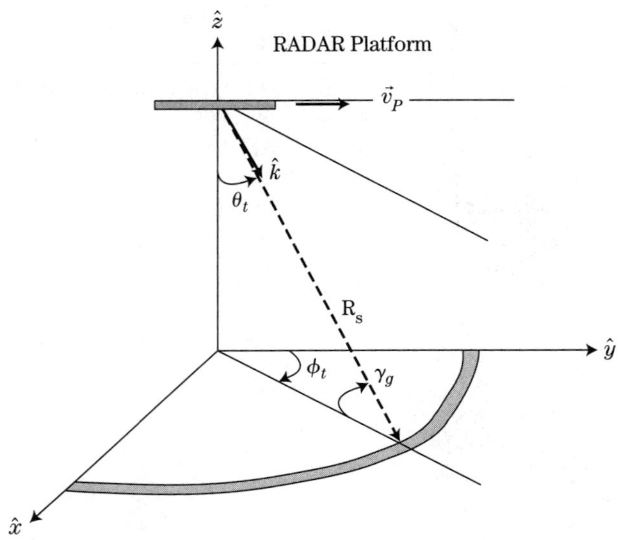

FIGURE 8–7

STAP platform coordinate system [8]

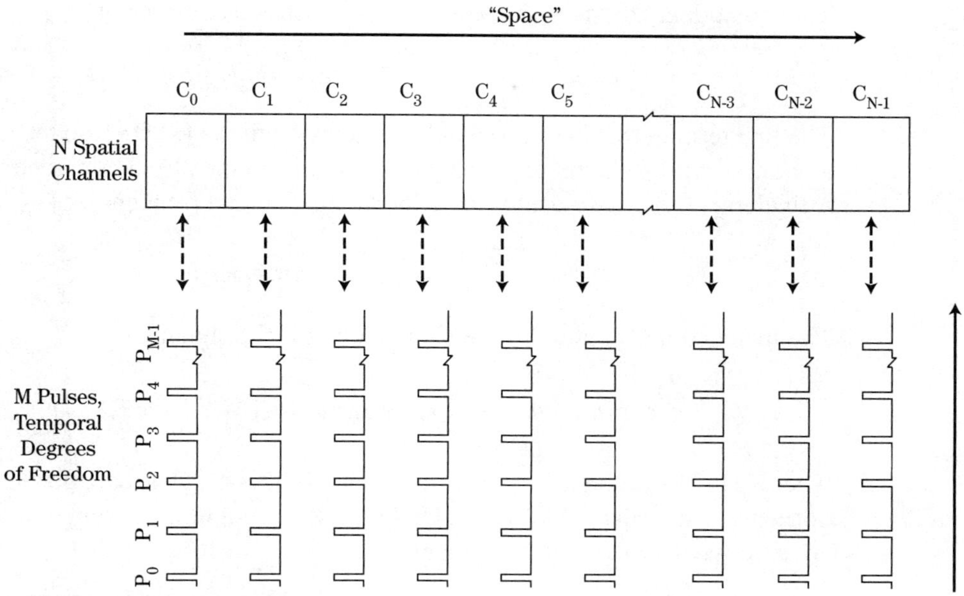

FIGURE 8–8

Spatial and temporal degrees of freedom for STAP processing

that the n-th element has a position at a reference time along the y-axis of r_n, the location of each spatial DOF is given by

$$\vec{r}_n = nd\hat{y} \tag{8.6}$$

The target echo received on each of the N receive channels is given as

$$s_n(t) = a_r u(t - \tau_n) \exp[j2\pi(f_0 + f_t)(t - \tau_n) + j\psi] \tag{8.7}$$

where a_r is the echo amplitude. The incremental times τ_n represent the temporal DOF, and f_t is the Doppler frequency imparted by the target motion.

The Doppler frequency f_t due to target motion v_t in (8.7) is expressed by:

$$f_t = \frac{2v_T}{\lambda_0} \tag{8.8}$$

Measured at the instantaneous frequency λ_0, the signal function is defined as

$$u(t) = \sum_{m=0}^{M-1} u_p(t - mT_r) \tag{8.9}$$

It is useful to define a normalized target Doppler frequency by the PRF (i.e. $1/T_r$):

$$\varpi = f_t T_r = \frac{f_t}{f_r} \tag{8.10}$$

The target delay to the n-th element consists of two components:

$$\tau_N = \tau_I + \tau_n = \frac{2R_T}{c} - n\frac{d}{c}\cos\phi_t \sin\theta_t \tag{8.11}$$

The first term in (8.11) is the round-trip delay from the RADAR to the target, and the second is the relative delay measured from the phase reference to the n-th element. It is also useful to define a target spatial frequency by

$$\vartheta_t = \frac{\hat{k}(\phi_t, \theta_t) \cdot \hat{d}}{\lambda_0} = \frac{d}{\lambda_0}\cos\phi_t \sin\theta_t \tag{8.12}$$

The matched filter output of the n-th receiver is the signal:

$$x_n(t) = a_t \exp[jn2\pi\vartheta_t] \sum_{m=0}^{M-1} \exp[jm2\pi\overline{\omega}_t]\chi(t - \tau_t - mT_r, \overline{\omega}_t) \tag{8.13}$$

where $\chi(r, f)$ is the waveform ambiguity function. This equation can be normalized such that $\chi(0, 0) = 1$. Under these assumptions, the target samples are expressed by the matrix elements:

$$x_{nm} = \alpha_t \exp(jn2\pi\vartheta_t)\exp(jm2\pi\overline{\omega}_t) \qquad \begin{matrix} n = 0, \ldots, N-1 \\ M = 0, \ldots, M-1 \end{matrix} \tag{8.14}$$

The target data can now be assembled into a space-time matrix snapshot as [8]

$$\chi_t = \alpha_t v_t(\vartheta, \overline{\omega}) \triangleq \alpha_t b(\overline{\omega}_t) \otimes a(\vartheta_t) \tag{8.15}$$

where

$$\vec{a}(\vartheta) = [1; \exp(j2\pi\vartheta); \ldots; \exp(j(N-1)2\pi\vartheta)] \tag{8.16}$$

$$\vec{b}(\overline{\omega}) = [1; \exp(j2\pi\overline{\omega}); \ldots; \exp(j(M-1)2\pi\omega)] \tag{8.17}$$

are the space-time steering vectors to the target. The term $\boldsymbol{v}_t\ (\theta, \overline{\omega})$ in (8.15) is the target steering vector.

The clutter return is characterized by dependency of the target's Doppler frequency in angle from the velocity vector. When the clutter return is received at a slant range R_S and azimuth angle θ_c, the Doppler frequency from the clutter patch is be given by

$$f_c(\phi_c, \theta_c) = -\frac{2\hat{k}(\phi_c, \theta_c) \cdot \vec{v}_P}{\lambda_0} \tag{8.18}$$

In this case, the angles θ_c and φ_c are continuous variables due to the distributed nature of the clutter. When the aircraft velocity vector is aligned with the antenna's longitudinal axis, there is a direct relationship between the normalized Doppler frequency and the spatial frequency

$$\overline{\omega}_c = f_c T_r = \left(\frac{2v_P T_r}{d}\right)\vartheta_c \tag{8.19}$$

With these normalized coordinates, the slope of the clutter ridge is given by β_c:

$$\beta_c = \frac{2v_P T_r}{d} \tag{8.20}$$

This term represents the number of interelement spacing traversed by the platform during one PRI. For half wavelength spacing, β_c is equivalent to the number of times the clutter Doppler spectrum aliases into the unambiguous Doppler space.

The clutter component of the space-time snapshot is given by

$$\chi_c = \sum_{N_R}\sum_{N_C} \alpha_{ik} v(\vartheta_{ik}, \varpi_{ik}) \tag{8.21}$$

where N_R is the number of range bins, N_c is the number of channels, and α_{ik} is the random amplitude of the ik-th clutter patch. The power of each clutter contribution for area clutter is the resolution area times the clutter scattering coefficient:

$$\sigma_{ik} = \sigma_0(\phi_k, \theta_t) R_i \Delta\phi \Delta R \sec \psi_i \tag{8.22}$$

The clutter space-time covariance matrix is related to the space-time steering vector quantities by [8]

$$R_c = E\{\chi_c \chi_c^H\} = \sigma^2 \sum_{N_R}\sum_{N_C} \xi_{ik} (b_{ik} b_{ik}^H) \otimes (a_{ik} a_{ik}^H) \tag{8.23}$$

where $b_{ik} = \mathbf{b}(\varpi_{ik})$, and $a_{ik} = \mathbf{a}(\theta_{ik})$. The term ξ_{ik} is obtained from the solution of the RADAR equation for the clutter power at each angle:

$$\xi_{ik} = \frac{P_T T_p G_T(\phi_k, \theta_i) G_r(\phi_k, \theta_i) \lambda_0^2 \sigma_{ik}}{(4\pi)^3 N_0 L_s R_i^4} \tag{8.24}$$

where

P_T	Transmitter peak power
T_p	Transmitted uncompressed pulse width
G_T	Gain of transmitter antenna
G_r	Gain of element or subarray on receive
σ_{ik}	Cross section of clutter in range ring and PRI
N_0	Receiver noise power spectral density
L_s	System losses
R_i	Range to i-th range ring

It should be noted that the transmitter and receiver gain would be different, depending on the architecture of the adaptive array. Each scatterer contributes a term that is the Kronecker product of a temporal covariance matrix with a spatial covariance matrix. These two components are coupled because the clutter Doppler frequency is a function of angle.

Table 8–1 FOPEN GMTI system parameters

Frequency	360	MHz	Antenna	7.1	meters
Bandwidth	15	MHz	N_{az}	16	
PRF	1000	Hz	N_{el}	4	
Power	2500	Watts	d_{az}	0.53	Wavelength
Pulse Width	25	usec	d_{el}	0.74	Wavelength
Mp	64		Altitude	7.5	Km
Duty Factor	2.56	percent	Velocity	[1, 50]	m/s
Target	−3	dBsm	R_S	25	Km
Clutter	−8	dBsm/m^2			

The space-time snapshot of target and interference can be decomposed as

$$\chi = \alpha_t v_t + \chi_u \tag{8.25}$$

where χ_u is the undesired component, which includes clutter, noise, RFI, and any jamming. Because the components are mutually uncorrelated, the interference-plus-noise covariance matrix is given by

$$R_u = E\{\chi_u \chi_u^H\} = R_c + R_n + R_j \tag{8.26}$$

8.2.2 UHF STAP for GMTI RADAR Operation

To illustrate these relationships between a target and antenna at UHF FOPEN RADAR frequencies, a scenario with parameters outlined in Table 8–1 is simulated. The 7.1 meter wide antenna has 16 azimuth elements and 4 elevation elements. On transmit the channels are beamformed with uniform illumination, providing the pattern shown in Figure 8–9.

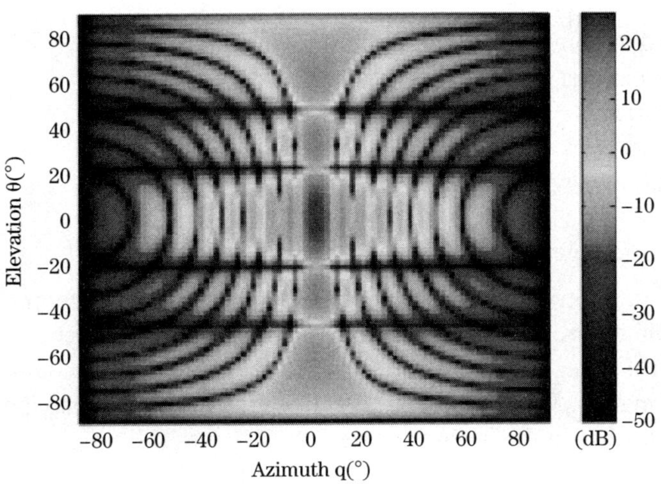

FIGURE 8–9

Transmit antenna pattern

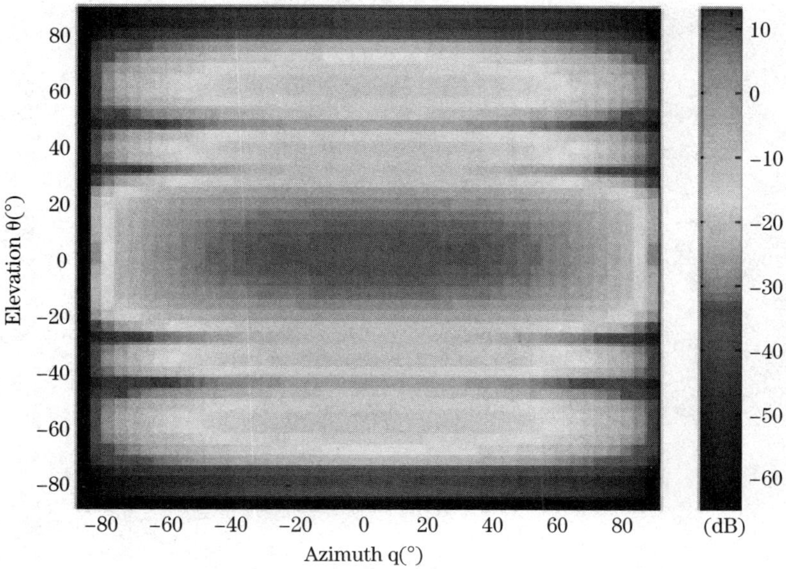

FIGURE 8–10

Receive channel antenna pattern (Single array element)

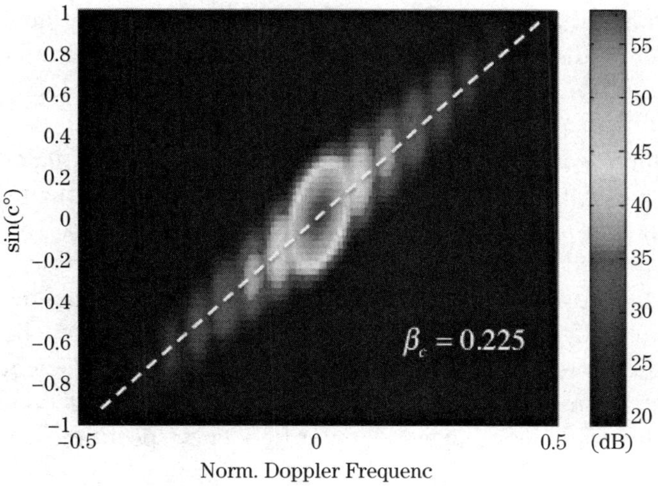

FIGURE 8–11

Clutter ridge for a 50 m/sec platform

On receive, the columns are beamformed into an elevation pattern having 20 dB Taylor weighting. Each column is used as an individual spatial degree of freedom for STAP processing, assuming the antenna pattern shown in Figure 8–10. The temporal degrees of freedom are provided by 64 pulses and used in a fully adaptive STAP algorithm. Two platform velocities are examined: 50 m/sec representative of a tactical unmanned aerial vehicle (TUAV); and 1 m/sec representative of a tethered or hovering platform.

Figure 8–11 illustrates the clutter ridge for the case of a 50 m/sec platform. The main beam of the clutter is represented by the oval at zero normalized

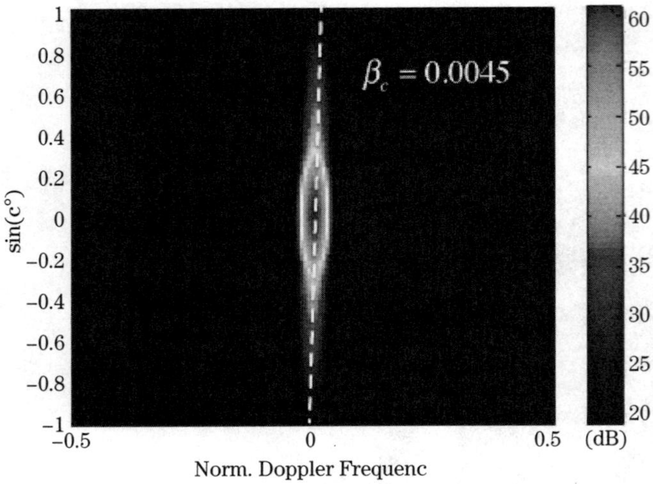

FIGURE 8–12

Clutter ridge for a 1 m/sec platform

Doppler frequency. The antenna elevation sidelobes are clearly seen along the slope of the clutter ridge of $\beta_c = 0.225$. It should be clearly noted that there is no noise simulated, so that detection is for the case of competing clutter only. By contrast, Figure 8–12 shows the clutter ridge for a 1 m/sec velocity platform, corresponding to a clutter ridge of $\beta_c = 0.005$.

The detection of targets depends on the Doppler separation from the mainbeam clutter. For targets that are outside the first null of the two-way antenna pattern, the targets are considered to be *exo-clutter*, and the detection performance can be approximated by the SNR of the return. This is strictly true if the Doppler processing provides sufficient subclutter visibility to cancel the residual sidelobe clutter. For *endo-clutter* targets that fall within the mainlobe nulls, STAP is needed to provide sufficient reduction of clutter for efficient detection. With adaptive processing, there is an attenuation of the target depending on where it lies in the normalized Doppler frequency spectrum.

8.2.3 STAP Loss

A space-time processor is defined to be a linear filter that combines all of the samples from the elements of the array with the temporal samples provided by the successive pulses of a CPI. The processor is represented by a *MN*-dimensional weight vector **w**, whose output is the inner product of the weight vector and the snapshot of the RADAR return [8]:

$$z = w^H \chi \tag{8.27}$$

The weight vector is the solution obtained from the solution of (8.27) with the undesired steering vector:

$$w = R_u^{-1}[v_t] \tag{8.28}$$

where

$$R_u = E\{\chi_u \chi_u^H\} \tag{8.29}$$

and $[v_t]$ is the steering vector towards the target. It should be noted that the covariance matrix used for calculating the adaptive weights uses only the interference-plus-noise characteristics of the environment. Significant research has been conducted to determine the methods of excluding targets from the covariance matrix to minimize attenuation of the targets. These results will not be covered in this work.

The loss of target signal can be determined as an *lower bound* by using an ideal covariance matrix, composed of the separated clutter, noise, and interference covariance matrices. In practice the losses are higher due to errors in estimation of the covariance matrix, or non-stationarity of the clutter during the coherent integration. Once R_u has been determined, the SINR level can be estimated by

$$SINR = \frac{\sigma^2 \xi_t |w^H v_t|^2}{w^H R_u w} \tag{8.30}$$

It is very important to characterize the STAP performance in terms of the RADAR performance without clutter and interference. In the noise-only case, the space-time matched filter gives the optimum processor when

$$[w] = [v_t] \tag{8.31}$$

The optimum output SNR is obtained through a gain of *MN* due to integration over *N* spatial elements and *M* pulses.

$$SNR_0 = MN\xi_t \tag{8.32}$$

The SINR loss then is the ratio of the adaptive SINR in (8.30) and the non-adapted gain [8]:

$$SINR_{Loss} = \frac{SINR(\overline{\omega})}{SINR_0} \tag{8.33}$$

The clutter covariance matrix given in (8.23) assumed a nonfluctuating clutter return. In the real world, especially in forested clutter environments, the clutter will vary temporally due to internal clutter motion. The clutter models for ICM were presented in Section 3.6 for the Billingsley clutter model. Three cases were considered in modeling the UHF GMTI system in Table 8–1 no (or zero) ICM; 1 m/sec ICM; and 5 m/sec ICM.

The SINR loss for the case of 50 m/sec platform velocity is given in Figure 8–12. Measured at the 5 dB loss point, the MDV is approximately 2 m/sec,

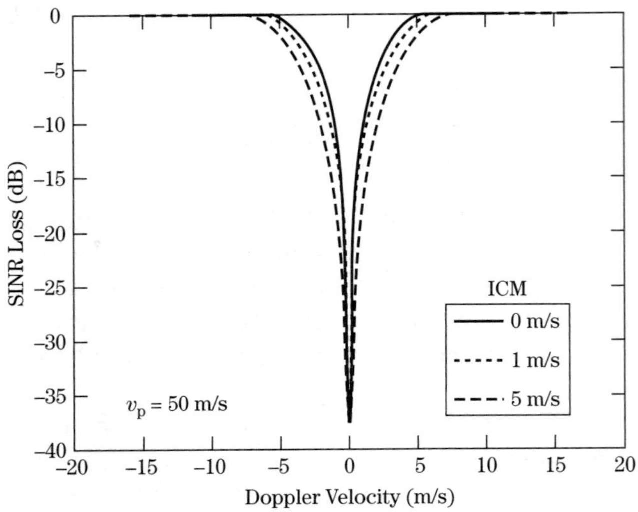

FIGURE 8–13

SINR loss for a UHF GMTI RADAR having a 50 m/sec platform velocity

which will provide excellent coverage for most vehicle traffic independent of aspect angle to the RADAR look direction. However, this MDV is optimistic, as the STAP covariance matrix was not estimated from the clutter. Estimating the R_u will normally degrade the MDV by 10 to 20% for homogeneous clutter and significantly more if the clutter is non-Gaussian. The more important aspect of Figure 8–13 is the effect of ICM on the MDV. A 1 m/sec wind characteristic (e.g., light air) can degrade the MDV by up to 0.5 m/sec. A 5 m/sec breeze has the potential to degrade the MDV to 3.5 m/sec. The amount of degradation depends on the geometry and how close to the moving target the interfering clutter resides.

The SINR loss for a 1 m/sec platform motion, representative of a hovering rotorcraft, is shown in Figure 8–14. The MDV for the no-ICM case is under 1 m/sec, sufficient for detecting dismounts under foliage. Both the light air (1 m/sec wind) and breezy (5 m/sec) conditions have an effect on the MDV. At the -10 dB SINR loss point, there is approximately a 20% decrease in the ability to detect slow-moving targets typical of dismount personnel.

8.2.4 STAP Summary

The development of STAP techniques for airborne GMTI RADAR applications is encouraging for detection of slow-moving targets at UHF, which is the most applicable frequency for foliage penetration. However, as was experienced with early FOPEN GMTI RADAR systems such as the Camp Sentinel RADAR, the proximity of the clutter location or Doppler frequency to those

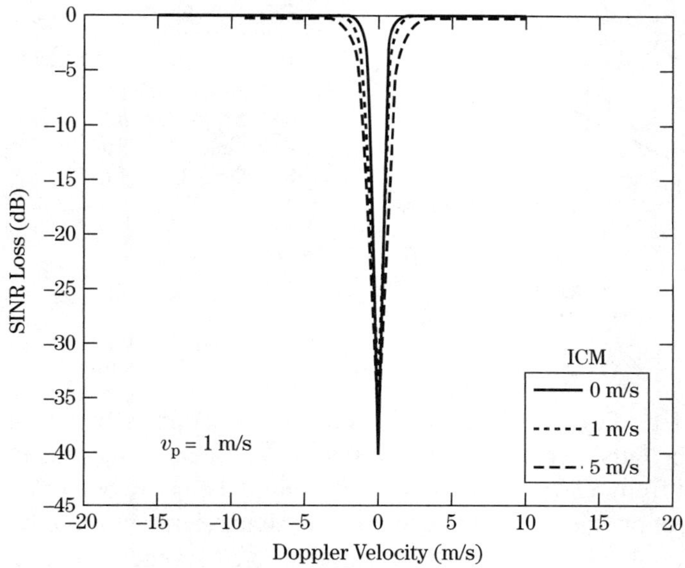

FIGURE 8–14

SINR loss for a UHF GMTI RADAR having a 1 m/sec platform velocity

of the target can have major impact on MDV. This is especially true of tree lines where the wind can modulate the clutter over a large area near targets.

STAP has the ability to cancel main beam clutter well into the main beam of the RADAR. Whereas GMTI pulse-Doppler processing is able to cancel up to 30% of the beamwidth, STAP has the potential for cancelling up to 90% of the beamwidth. This requires that the clutter environment can be measured without the targets present. It also requires that the clutter environment be stationary in terms of both spatial and temporal variations in the estimate of the interference covariance matrix. The art of STAP requires providing sufficient samples to estimate the clutter and jamming interference without the presence of targets in the training set. The dimensionality of the interference and effects of bandwidth and antenna channel match are significant challenges to real-time operation. Several development programs have been reported that use the knowledge of the terrain and lines of communications to facilitate the assumptions on target motion for training the covariance matrix [9].

8.3 Along-Track Interferometry

The design of ATI techniques has been applied for Earth resources monitoring and to a limited extent for moving target indication. This section will explore the detection of moving targets under foliage. FOPEN RADAR systems are generally designed in the VHF or UHF bands to significantly reduce the effects of foliage attenuation [10]. These longer wavelength designs are a

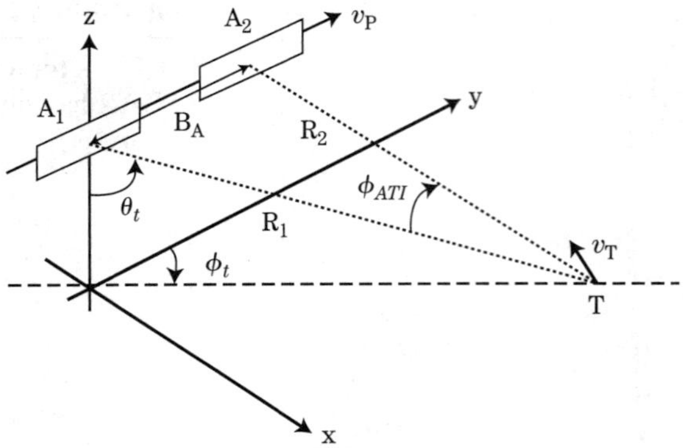

FIGURE 8–15

Along-track interferometry coordinate system
Source: NASA Jet Propulsion Laboratory
© 2000 IEEE [12]

problem for GMTI RADAR applications, where the detection is based on target Doppler being outside of mainbeam clutter (exo-clutter) or significant clutter cancellation from STAP (endo-clutter). To limit the clutter competing with the target detection, the resolution cell must be significantly reduced by high range resolution (large signal bandwidth) and fine cross-range resolution (long coherent integration times). These two criteria are synonymous with a SAR implementation.

ATI RADARs have been proposed for both space [11–13] and airborne applications [14] to detect very slow-moving vehicles and surface phenomena. Two SAR images are simultaneously formed with a baseline between the transmit/receive aperture and a second receive-only aperture, as shown in Figure 8–15. By measuring the phase of each scattering cell, the fine Doppler content can be used to discriminate motion with respect to the background clutter. This approach has been shown to depend on the strong correlation of the return to the two receive apertures. When there is an attenuation of signal due to foliage volumetric scattering or the target signature varies as a function of the angle to each receiver, this correlation criterion affects both the target detection and its localization.

In this section, the design of a UHF ATI RADAR is considered from several important aspects, including the following:

- Antenna baseline separation for discriminating slow-moving vehicles from ground clutter
- Effect of SCR on vehicle MDV
- Effect of ICM and foliage attenuation on ATI ability to discriminate target from clutter and to determine the target position and velocity

First, the concept of ATI will be summarized, and its potential for detecting moving targets will be examined. As we have shown, the impact of moving platforms and volumetric foliage clutter on SNR and SCR are complex. A complex scenario has been modeled where both fixed foliage clutter and moving targets exist in a distributed scene. Based on these simulation results, several trades on antenna and signal processing architecture will be explored. Finally, the effects of ICM on the MDV will be presented.

8.3.1 ATI Theory

ATI is the process of combining the return from a transmitted SAR signal into two or more receive apertures. If those apertures are along the velocity vector of the platform or platforms, the background clutter can be canceled. Previous displaced phase center antenna systems have canceled the clutter magnitude, counting on the moving target to be highly correlated between the looks. However, this "simple" subtraction also reduces the target SNR, limiting the probability of detection.

ATI determines the phase information on each resolution cell and uses this phase to detect Doppler motion with respect to the background. As was the historical case with RADAR when going from noncoherent to coherent processing, there is a significant advantage to using phase as a target discriminant.

The ATI phase for a moving object in the scene is dependent on the difference in separation between the two apertures and the SNR of the signal. For small separation between the antennas, higher speeds are needed to discriminate the phase from the background. At longer separations, the phase measurement is more accurate, but there appears an ambiguity in the phase. These ambiguity effects are well known in the application of interferometric SAR (INSAR) techniques for terrain mapping [12].

Along-track interferometry is an alternate GMTI RADAR architecture for detection of very slow-moving targets. The concept uses the along-track separation between antenna centers to measure the phase difference of a target between two different signal paths. Consider two independent receive signals S_1 and S_2 containing only a target signal. ATI processing on these signals yields [14]

$$\gamma_{tgt} \exp(j\phi_{tgt}) = \frac{\langle S_1 S_2^* \rangle}{\sqrt{\langle S_1 S_1^* \rangle \langle S_2 S_2^* \rangle}} \tag{8.34}$$

where <.> denotes the expected value, and * denotes the complex conjugate. ATI processing for detecting slow-moving targets with radial velocity v_T relies on a high correlation of the target. Both the target amplitude γ_{tgt} and phase ϕ_{tgt} returns are then distinguishable from the background clutter. If the corresponding pixel in each image is highly correlated and stationary, there will be no phase difference other than a constant phase across the motion compensated image. The target Doppler frequency creates a phase difference

φ_{tgt} between the two paths. From that phase difference, the target velocity v_T can be determined from the platform *velocity* v_P and baseline separation B_A:

$$v_T = \frac{\phi_{tgt}\,\lambda_0\,v_P}{2\pi\,B_A} \tag{8.35}$$

Because the target phase is modulo 2π, a Doppler frequency ambiguity exists, which is given by

$$v_{ambig} = \frac{\lambda_0\,v_P}{B_A} \tag{8.36}$$

For a target that has radial velocity different from the background clutter, the target will have a velocity measurement error given by [13]

$$\sigma_v = \sigma\frac{\lambda}{2\pi}\frac{v_T}{B_A\sin(\gamma_g)} \tag{8.37}$$

where γ_g is the nominal incidence angle, and v_T is the radial velocity of the target.

The MDV is then related to the target radial velocity, the separation B_A between the phase centers, and the additive noise of the receivers. This additive noise σ in (8.37) is related to the SNR of the detection cell, and the number N_I of independent cells averaged by

$$\sigma = \frac{1}{\sqrt{2N_t}}\frac{\sqrt{1-\rho^2}}{\rho} \tag{8.38}$$

where

$$\rho = \frac{1}{(1+1/SNR)} \tag{8.39}$$

When there is a large separation B_A between antenna phase centers, (8.37) indicates there is potential for detecting targets having small radial velocities. This detection provides a significant improvement in MDV over the conventional GMTI process. However, an attendant Doppler frequency ambiguity condition needs to be analyzed in the processing.

Figure 8–16 parametrically shows the MDV for two UHF array antennas as a function of the separation of the phase centers and of the SNR [16]. The figure evaluates MDV as a function of baseline separation for SNR values of 5, 10, and 20 dB and for none independently averaged cells.

Figure 8–17 shows the velocity ambiguity for the same baseline separation range, evaluated at 400 MHz. With a large separation between phase centers of a UHF RADAR, the potential exists for detecting targets having small radial velocities, with a significant improvement in MDV over the conventional GMTI process. However, the analysis assumes that the signal return is noise

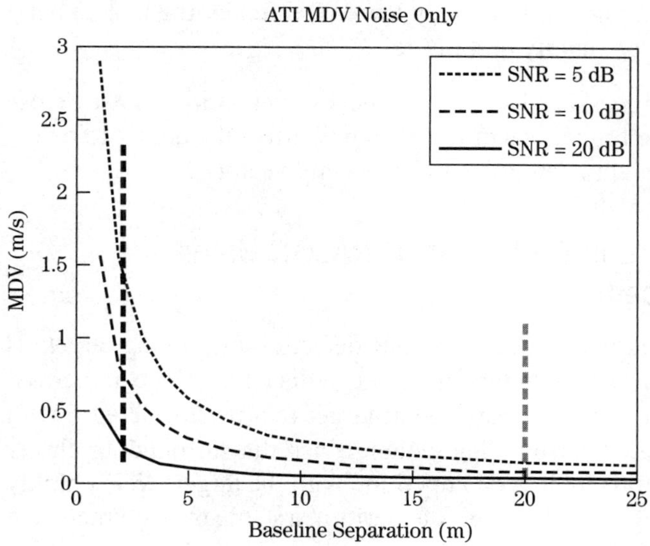

FIGURE 8–16

ATI MDV with noise only at 400 MHz © 2008 IEEE [16]

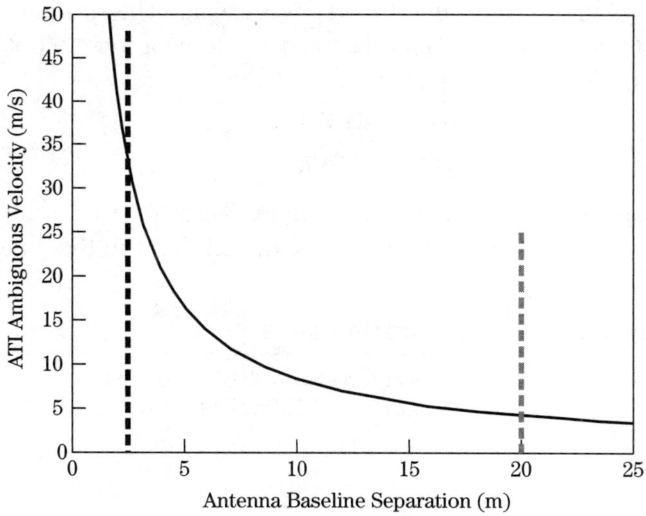

FIGURE 8–17

Velocity ambiguity at 400 MHz © 2008 IEEE [16]

limited, not clutter limited, and does not include the effects of internal clutter motion or target decorrelation. Each of these effects must be included in a more analytic approach to determine a practical MDV.

Even at a low SNR level of 5 dB, there is a significant improvement in MDV over a real aperture RADAR. Two antenna separations will be examined in greater detail:

- A 2 meter separation that has a modest MDV of 0.75 m/sec and large unambiguous velocity interval (0 to 30 m/sec)

- A 20 meter separation that has a very low MDV (under 0.2 m/sec) but a very low velocity ambiguity of 4 m/sec

The complexity of processing Doppler frequency ambiguities must be explored. In addition the limitations of the SCR will affect the ability to detect slow-moving targets and to determine their velocity vector.

8.3.2 SCR Effects in FOPEN GMTI RADAR Using ATI Processing

The predominant issue with GMTI RADAR detection of moving targets is target-to-clutter ratio. As the clutter cell area is reduced by range and cross-range resolution, it is expected that a point target return will increase with respect to the distributed surrounding clutter. Since the surrounding clutter within a resolution cell is in direct competition with the target, ATI is highly dependent upon having a sufficient SCR to perform a stable measurement. An insufficient SCR would decrease correlation, and would produce a nonlinear target phase–velocity relationship during the ATI processing. The resultant target correlation amplitude from (8.33) has been shown to be a function of both the SCR and the CNR, as given by [13]

$$\gamma_{ATI} = \frac{SCR \exp(j\phi_{tgt}) + 1}{SCR + 1 + 1/CNR} \tag{8.40}$$

The measured ATI target correlation, shown in Figure 8–18 for a baseline separation of 2 meters, suffers when there is an insufficient SCR in the ATI

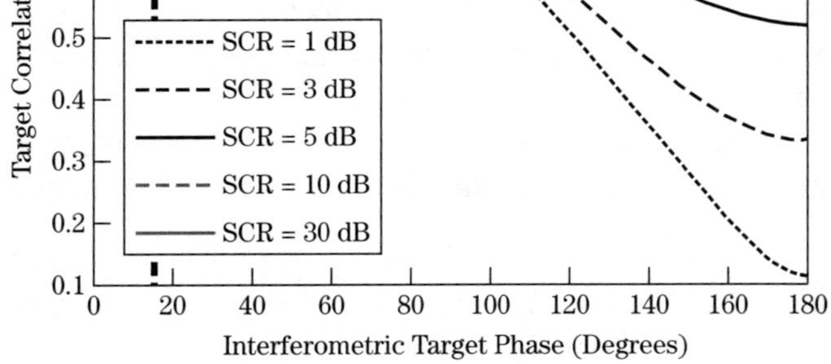

FIGURE 8–18

ATI target correlation—2 meter baseline © 2008 IEEE [16]

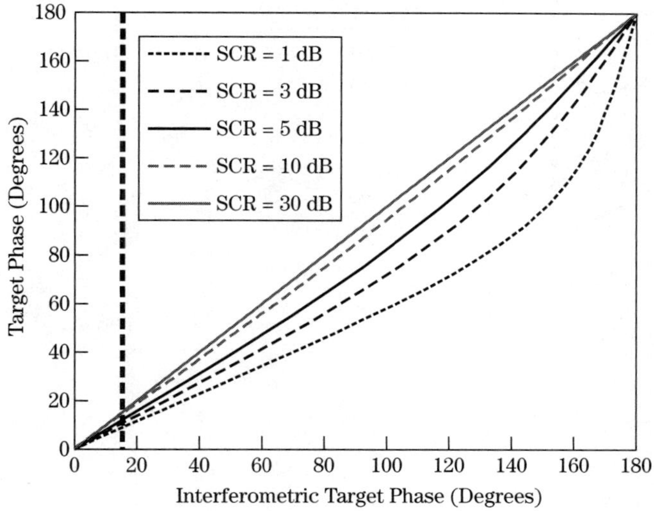

FIGURE 8–19

ATI target phase—2 meter baseline

processing. However, the figure shows that acceptable target correlation can be maintained for SCRs greater than 5 dB.

The ability to detect and locate the targets in the ATI images depends on an accurate measurement of the ATI phase ϕ_{ATI}. This phase is also affected by the SCR, as given by

$$\phi_{ATI} = \frac{SCR}{SCR + 1}\phi_{tgt} \tag{8.41}$$

The nonlinear phase relationship as a function of the SCR is illustrated for the 2 meter baseline separation in Figure 8–19. When the SCR is high, a linear relation is preserved between the target and ATI phases. However, as the SCR is reduced, more or stronger clutter competes with the target phase, contributing to a measurement error in ϕ_{ATI}.

The analysis of ATI performance in foliage will focus on slow-moving targets, such as vehicles perpendicular to the RADAR look direction or dismounts. The region for target velocity below 3.1 m/sec is indicated by the region to the left of the vertical dotted line in both Figure 8–18 and Figure 8–19. For this small separation between ATI subarrays, the measurement is well within the high correlation region of the predicted performance. However, because of the small difference in the baseline, there will be little precision in determining the target velocity [16].

When the baseline is increased to 20 meters, there is significant increase in ATI phase sensitivity to target velocity with the applicable range of phase for the ±3 m/sec targets increasing to 160 degrees. In Figure 8–20 it is shown that with a large baseline the targets will decorrelate significantly with lower SCR. A normal metric for the target correlation coefficient is 80%. So for targets

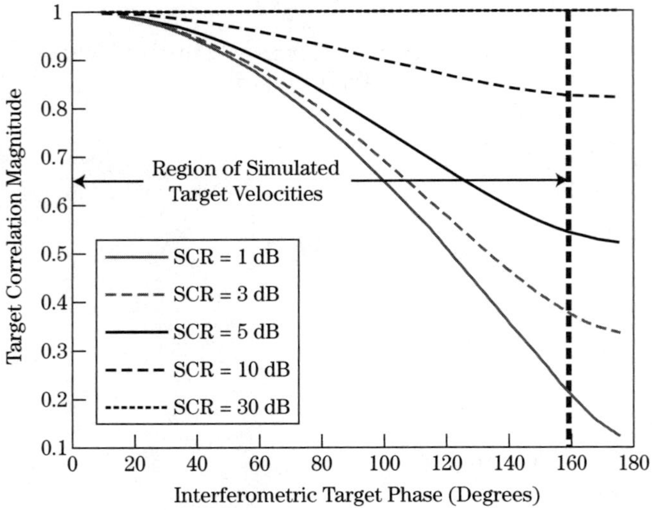

FIGURE 8–20

ATI target correlation for a 20 meter baseline © 2008 IEEE [16]

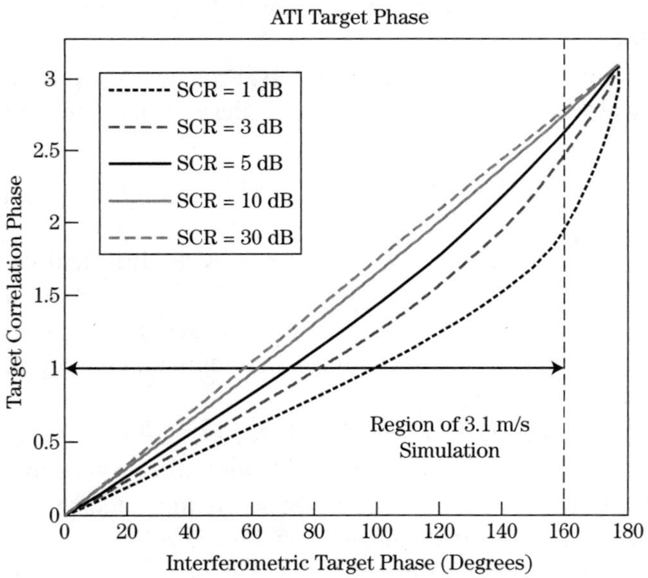

FIGURE 8–21

ATI target phase for a 20 meter baseline © 2008 IEEE [16]

having higher Doppler frequencies and lower SCRs, as shown in Figure 8–21, the errors in measuring the velocity component are significantly increased.

In the case of wider range of target radial velocity, the targets wrap in ATI phase, and the resultant ambiguity needs to be resolved. One applicable technique for resolving this ambiguity is to employ a first-order target tracking within the individual SAR subapertures to estimate the target coarse range rate. By using predicted target motion, it is possible to estimate the order of ambiguity that is associated with a particular target and to provide a correction into the estimation of target velocity and displacement due to radial motion. However, the performance of this technique is also susceptible to low SNR and CNR factors.

8.3.3 ATI Modeling and Simulation

To evaluate the SCR effects on ATI detections of slow-moving targets under foliage, a high-fidelity modeling program was created. The evaluation of the ATI performance of a multiple subaperture RADAR must include the waveform generation from the transmit aperture, the coherence and propagation of energy from the RADAR to the earth's surface, along with the clutter and target return to each receiver channel. This process has been carried out using RLSTAP, a modeling environment developed to analyze STAP performance in many airborne and space-borne RADAR applications [16,17].

RLSTAP has the capability to perform a pulse-by-pulse simulation with the following characteristics needed for FOPEN ATI evaluation:

- Modeling of the terrain type, terrain height, and foliage clutter to obtain representative SCR statistics [18]
- Distributed transmit and receive apertures, with independent location and motion
- Modeling multiple targets, with details of target motion for signal decorrelation studies
- Statistical ICM to evaluate clutter motion effects on MDV

Figure 8–22 shows the RLSTAP data flow method for independent transmit and receive operation, used for parametric evaluation of FOPEN ATI operation. Models for the clutter and foliage between the RADAR and the targets are provided to directly characterize the SCR and CNR as a function of the simulation geometry. The clutter model and foliage loss has been validated with recorded flight data, such that the target returns will have statistical fluctuation representative of a realistic airborne RADAR.

Figure 8–23 illustrates the scenario modeled in an evaluation of ATI for detecting slow-moving targets under foliage, with key parameters summarized in Table 8–2. In understanding FOPEN GMTI RADAR, it is particularly important to assess the effect of foliage attenuation and ICM on the ATI signal correlation. The simulation has the ability to selectively use foliage model types and employ ICM.

It is important to understand the impact of FOPEN propagation and losses on the ability to detect targets moving under foliage. To evaluate the effects of clutter, foliage attenuation, and ICM, a detailed scenario was developed. An airborne platform with modest speed (120 m/sec) and altitude (7 Km) was

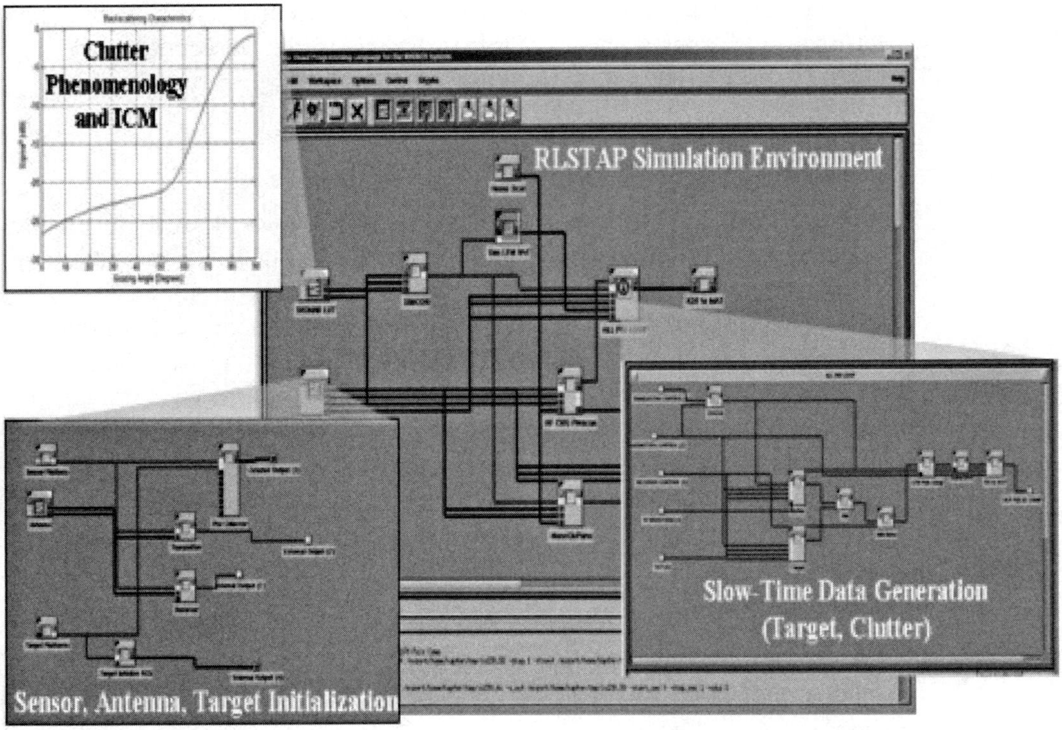

FIGURE 8–22

RLSTAP simulation environment for ATI evaluation © 2008 IEEE [16]

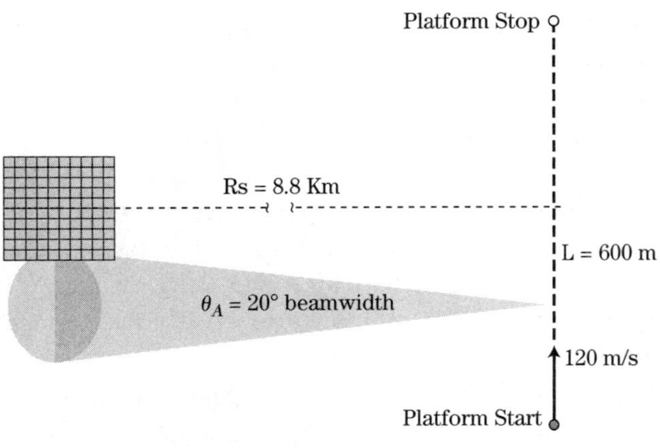

FIGURE 8–23

ATI modeling scenario © 2008 IEEE [16]

Table 8–2 ATI modeling parameters

Altitude	7 km	Subarray	2 m × 1 m
Velocity	120 m/s	Subarray Weights	Uniform
Frequency	360 MHz	ATI Baseline	[2, 20] meters
Bandwidth	160 MHz	Target RCS	5 m^2
SAR Length	600 meters	Target Radial Velocity	[−3.1:0.37:3.1] m/s
Peak Power	1 Kilowatt	Foliage Height	20 meters
Duty Factor	0.1	Slant Range	8.8 Km

used. The ground moving targets were simulated in 20 meter high, single-canopy forest moving along narrow roads at small velocities. With a synthetic aperture length of 600 meters at 8.8 Km slant range, the cross-range resolution is 6 meters, based on (4.1).

These simulation conditions are at higher grazing angles than expected for an operational system, thereby yielding lower than reasonable foliage attenuation. However, the independent signal, noise, and clutter data cubes were combined with the appropriate attenuation to simulate several SCR conditions. These approximations were employed because of the very high fidelity of the RLSTAP simulation, resulting in very long run times. For the design of a more realistic system, the scenario would be modified to have appropriate grazing angles to get target loss through the foliage as well as clutter cell sizes.

In the scenario shown in Figure 8–24 19 targets with 5.0 m^2 RADAR cross section (RCS) each are modeled, along with ground velocities ranging over ±3.3 m/sec. Each target is modeled on a path within a forest, so that the moving objects will be obscured to a sensor without foliage penetration qualities. The attenuation of the targets is based on the grazing angle and frequency, employing the model in (3.33). Each target has a unique radial velocity to assess the measurement accuracy on ATI phase. These accuracies will also affect the ability to reposition the targets at the correct geospatial position.

Figure 8–25 illustrates the ATI image created from the two SAR images and evaluated for correlation from (8.35) with respect to the fast-time (range) and the slow-time (cross-range) dimensions. It should be noted that the targets exhibit the customary SAR displacement in cross-range by the amount

$$Y_{disp} = R_s \frac{v_{tgt}}{v_{plat}} \tag{8.42}$$

This displacement is the result of the target Doppler frequency component in the image formation process. The individual target locations in the image magnitude are isolated so that they are not in the same range or cross-range

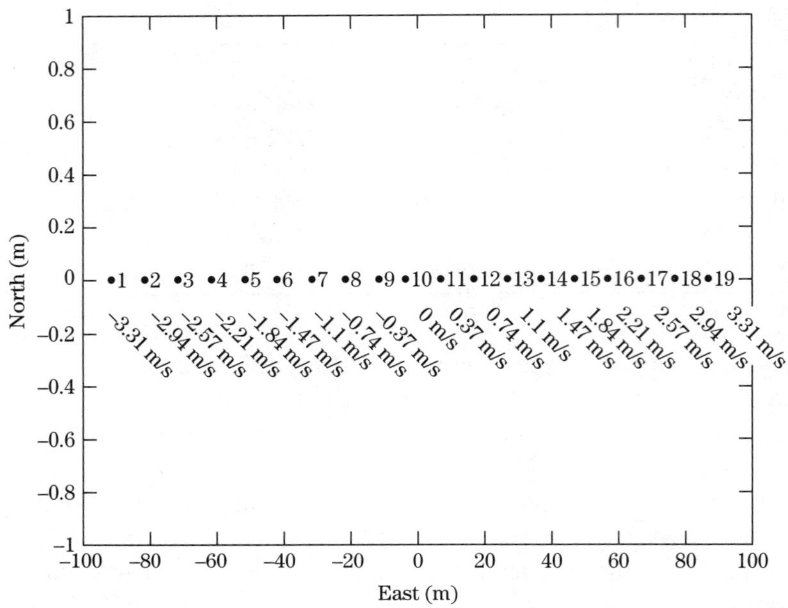

FIGURE 8–24 Physical location and velocity for moving target modeling © 2008 IEEE [16]

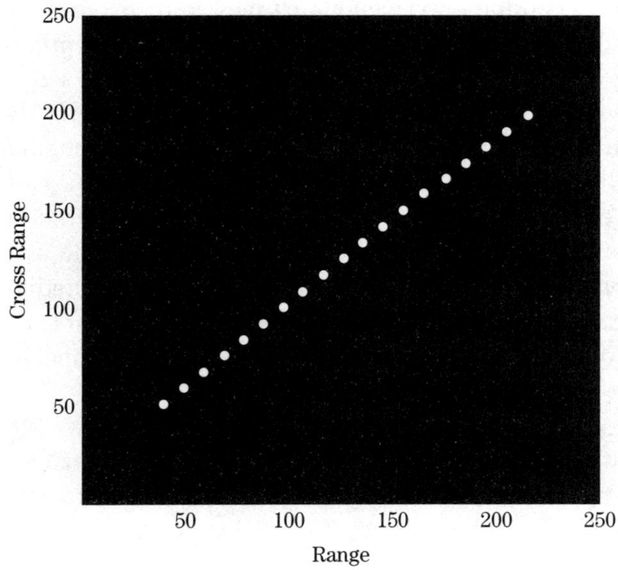

FIGURE 8–25 SAR location of moving targets due to radial velocity © 2008 IEEE [16]

resolution. However, the phase is smeared due to the physical motion of the target during the integration time. A multilook ATI image was created by applying a 3-by-3 moving average window to the SAR image. The multilook process is an effective way to reduce unwanted speckle in SAR images.

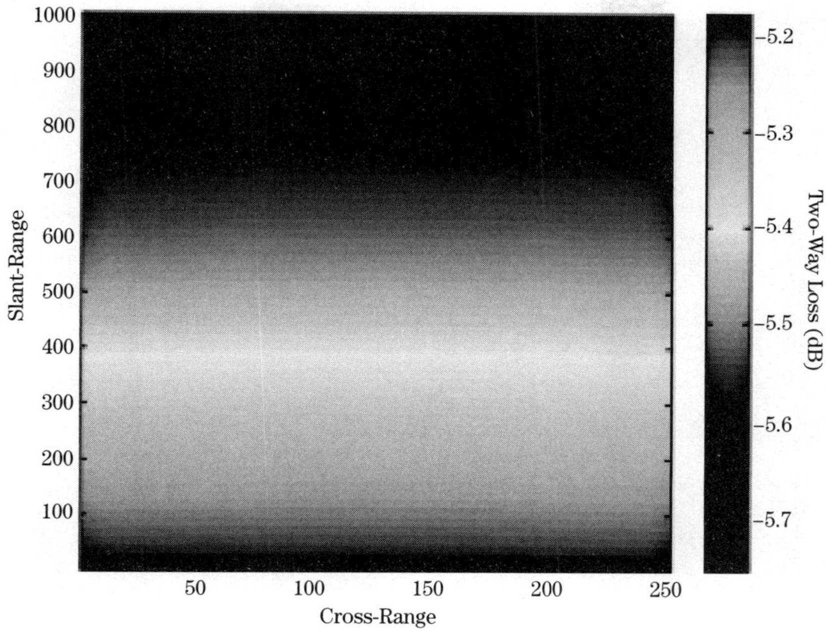

FIGURE 8–26

Attenuation of targets within scenario © 2008 IEEE [16]

The simulation includes detailed synthesis of the antenna, transmitter, and receive chain, including a LFM waveform characteristic for the FOPEN SAR. The slant range from the RADAR to each clutter cell is evaluated, and the foliage loss is applied to any targets under the foliage, as shown in Figure 8–26 [19].

By having the combination of platform motion and LFM waveform, a realistic synthesis of the range and Doppler walk of each target is possible. However, for this example, no RFI or system errors were included. A parametric evaluation was conducted to assess ATI's ability to detect targets in forest clutter. Two baseline separations 2 and 20 meters were modeled, corresponding to the analysis in the previous section. For each of these baseline separations, three SCR conditions were evaluated: 5, 10, and 20 dB. For this experiment the SCR was measured based on the root mean square (RMS) of the clutter return to the peak of the target return in the SAR image. Finally, the impacts of windblown clutter and attendant foliage ICM were simulated with RLSTAP.

Each antenna had an area of 2 square meters and was formed with a representative aperture design of six elements in azimuth and three elements in elevation. Uniform transmit weighting was applied in both azimuth and elevation, providing 13 dB sidelobe levels. The antenna azimuthal pattern for a 2 meter baseline separation is depicted in Figure 8–27, and results in representative elevation and azimuth sidelobes.

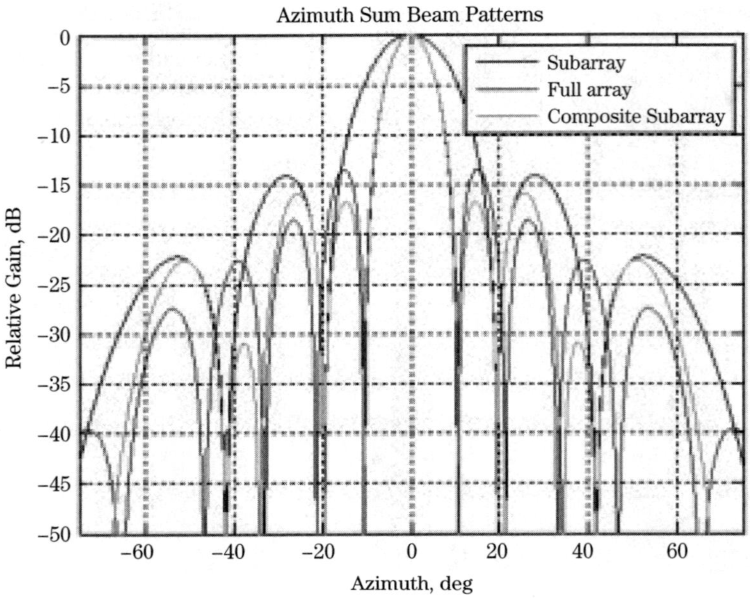

FIGURE 8–27

ATI antenna pattern performance—2 meter baseline © 2008 IEEE [16]

The individual subapertures provide an illumination gain for ATI SAR image collection, given by the outer curve. However, the composite pattern is shown as if the two subapertures were coherently combined from individual array patterns and forms a composite pattern with grating lobes. In the ATI implementation, two separate antennas are used to form simultaneous SAR images. The baseline separation distance causes phase ambiguities during the ATI processing. These ambiguities are analogous to the grating lobes in the patterns, as expected from sparse array theory. The azimuth angle for the n-th grating lobes θ_{GL} with respect to the scan angle is θ_0 given by [20]

$$\sin\theta_0 - \sin\theta_{GL} = \frac{n\lambda}{B_A} \tag{8.43}$$

For the 20 meter baseline, the increased sensitivity in phase provides a method of evaluating target phase sensitivity and location accuracy. The composite ATI pattern in Figure 8–28 shows the grating lobe effect and illustrates the potential for improving target location accuracy, if these grating lobes could be resolved. Again, the spatial grating lobes are not realized in the ATI processing but are indicative of spatial variation of target phase ambiguities.

8.3.4 ATI Signal Processing

The ATI signal processing steps are illustrated in Figure 8–29. Data cubes were generated from RLSTAP to provide signal phase history for SAR and

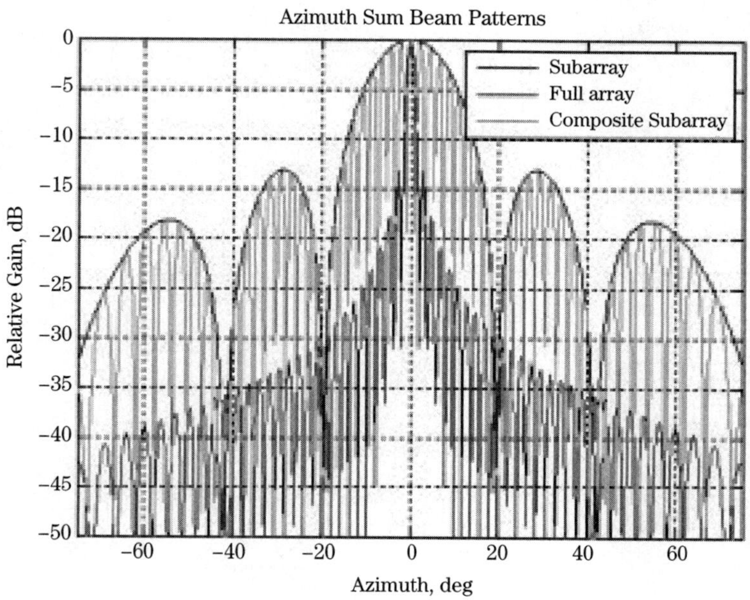

FIGURE 8–28

ATI antenna pattern performance—20 meter baseline © 2008 IEEE [16]

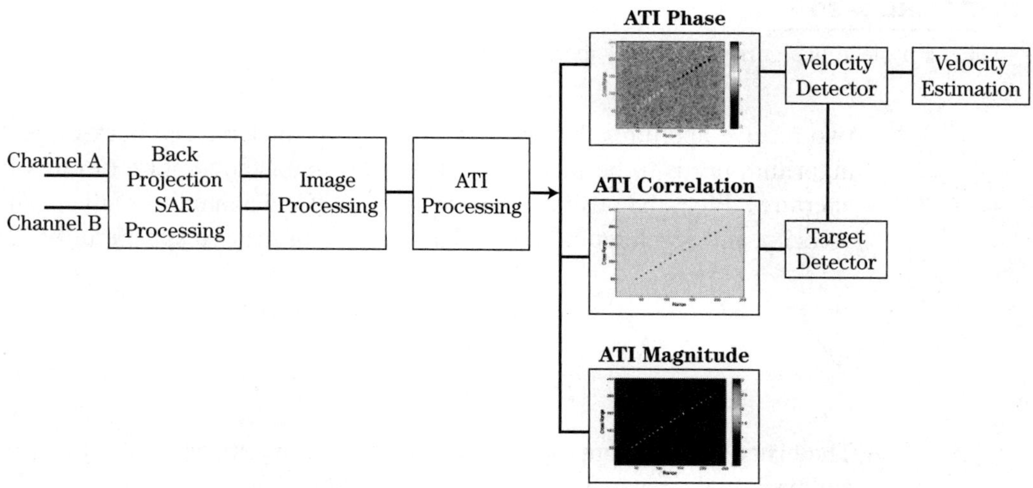

FIGURE 8–29

ATI signal processing flow © 2008 IEEE [16]

ATI image formation. Once the data cubes for the two channels were created, MATLAB code was used to form the ATI images.

The first step was to form SAR images for each of the subapertures, as illustrated in Figure 8–30. Each individual SAR image was created by back projection image formation processing into the ground plane, formed at the

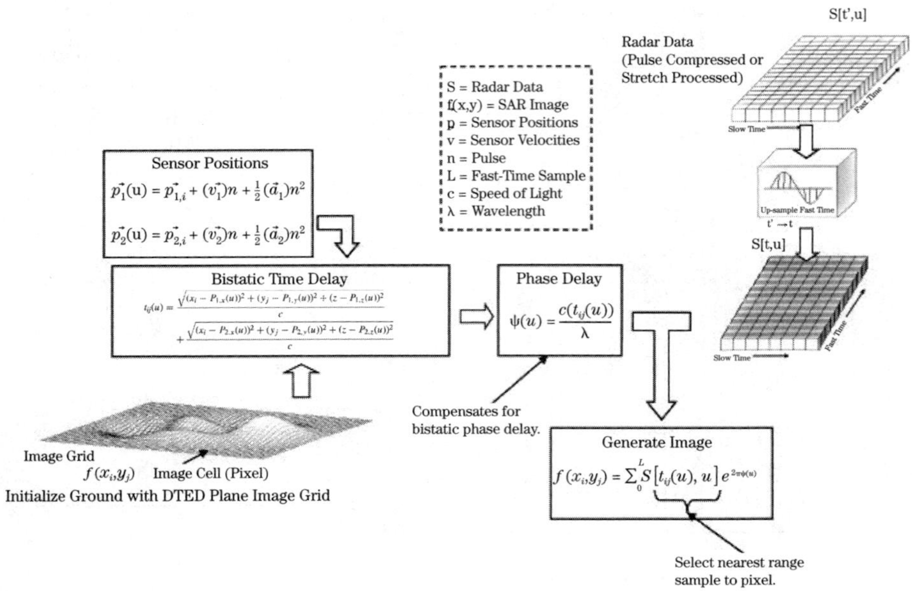

FIGURE 8–30

Back projection algorithm for dual aperture ATI [23]

two receive apertures that are separated by B_A meters. The back projection algorithm needs to be modified because of the bistatic nature of one of the apertures. First, the position of each aperture is calculated from the position, velocity, and acceleration of the phase centers of the two subapertures [21]:

$$\begin{aligned}\vec{p}_1(u) &= \vec{p}_{1,i} + \vec{v}_1 n + \frac{1}{2}\vec{a}_1 n^2 \\ \vec{p}_2(u) &= \vec{p}_{2,i} + \vec{v}_2 n + \frac{1}{2}\vec{a}_2 n^2\end{aligned} \tag{8.44}$$

The bistatic delay from the transmit antenna to the image cell on the ground and back to the receive antenna is then calculated as

$$\begin{aligned}t_{ij}(u) = &\frac{\sqrt{(x_i - P_{1,x}(u))^2 + (y_j - P_{1,y}(u))^2 + (z - P_{1,z}(u))^2}}{c} \\ &+ \frac{\sqrt{(x_i - P_{2,x}(u))^2 + (y_j - P_{2,y}(u))^2 + (z - P_{2,z}(u))^2}}{c}\end{aligned} \tag{8.45}$$

Next, the bistatic phase delay needs compensation:

$$\psi_A(u) = \frac{ct_{ij}(u)}{\lambda} \qquad \psi_A(u) = \frac{ct_{ij}(u)}{\lambda} \tag{8.46}$$

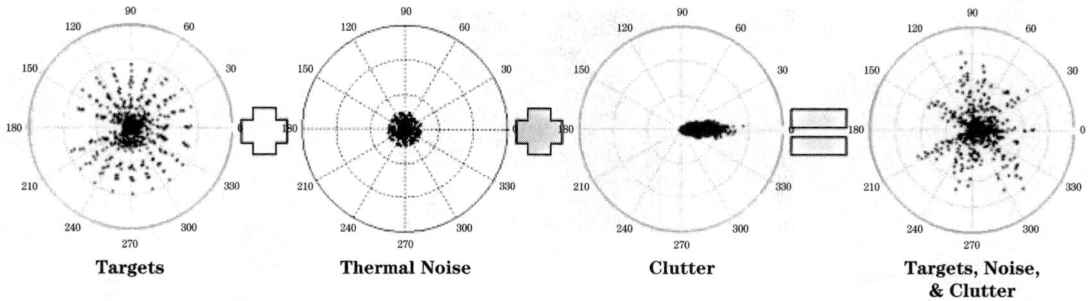

FIGURE 8–31

Summation of target, clutter and noise in ATI simulation © 2008 IEEE [16]

Finally, the image is formed by appropriate delay of the RADAR return from each pixel in the SAR image function $S(t, u)$:

$$f(x_i, y_j) = \sum S(t_{ij}(u), u) \exp[j2\pi \psi_A(u)] \tag{8.47}$$

A 3×3 boxcar filter was then passed over the image to reduce the speckle [22]. Finally, the ATI image was formed with knowledge of the subarray separation and of a common image focal point on the ground. Within the ATI image, three factors were calculated: ATI phase, pixel correlation between the two images, and ATI magnitude. These three factors were used to detect the targets and to estimate the location and velocity.

The simulation results are summarized in Figure 8–31 with the different phenomena for targets, thermal noise, and distributed clutter. The 19 targets are manifest as radial lines in the polar plot. There are multiple signals shown for each target, since there will be a finite target motion in range and Doppler during the SAR image formation. With the wide separation in targets, it is reasonable to expect that the peak detection is the center of the dominant range or Doppler resolution cell.

Noise will be randomly distributed in angle due to its statistical nature in the RADAR. RLSTAP imparts a random phase and amplitude on each clutter cell that is distributed in amplitude and phase. The phase distribution will be small, while the amplitude has imparted the Weibull statistics of the clutter power spectral density. Note that the ground clutter is predominantly distributed along the ATI's zero phase line. This is the effect of ATI motion compensation on each subarray to the scene focal point.

The target detection processing, which is accomplished in the SAR image plane, is shown in Figure 8–32. The left plot is the ATI phase around the area of moving targets, with image intensity showing the phase for the targets that are either moving away from the RADAR (black) or toward the RADAR

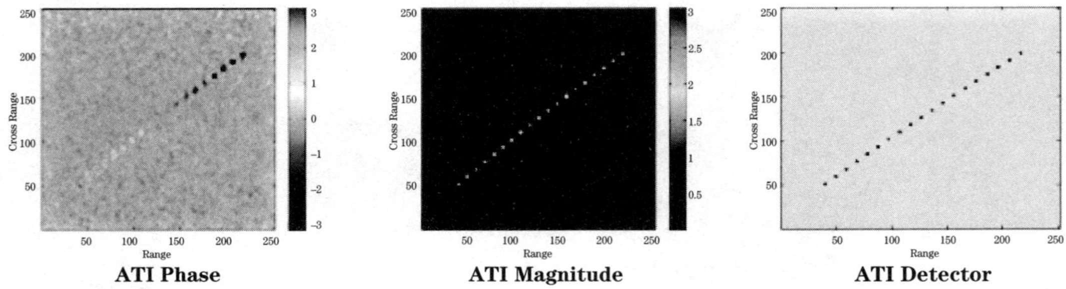

FIGURE 8–32

Three components of ATI target signal © 2008 IEEE [16]

(white). The gray background is the variation in motion compensated ground clutter.

The second image plane in Figure 8–32 shows the magnitude of the ATI pixel returns. In the example, with its high signal to clutter, the individual targets are evident. A combination of amplitude and phase detection averaging and CFAR comparison to a local mean threshold was used to declare a target, as is shown in the ATI detector panel.

Figure 8–33 provides the ATI polar results of all pixels over a low threshold, for the baseline separation of 2 meters and for three different SCR conditions. The targets are separated in phase related to their radial velocities. Moreover, the ATI detector shows the targets in the SAR coordinates and the proportional azimuthal offset. It should be noted again that the clutter is distributed predominantly along the zero ATI phase.

The polar plot illustrates the ATI angle effects due to the target Doppler frequency. The magnitude of the ATI is a measure of the SNR or CNR. A high detection probability is illustrated for both the 20 dB and 10 dB SCR plots. On the 5 dB SCR, there is a scatter in the Doppler location of the targets caused by the low correlation of the ATI pixels. This will be examined below in greater detail.

The angular separation between the targets is scaled by the effects of the ± 22.5 m/sec Doppler ambiguity, as calculated from (8.36). In this case there is a high degree of compression of the slow-moving targets in ATI phase, directly competing with ground clutter.

These results reinforce the theory by illustrating the demands on phase resolution for slowly moving target velocity estimation. The target correlation for the 2 meter baseline can be seen in Figure 8–35a. The high correlation for each of the SCRs is as expected due to the minimal baseline separation between ATI processes apertures. The estimated ATI phase for the 2 meter baseline is shown in Figure 8–35b.

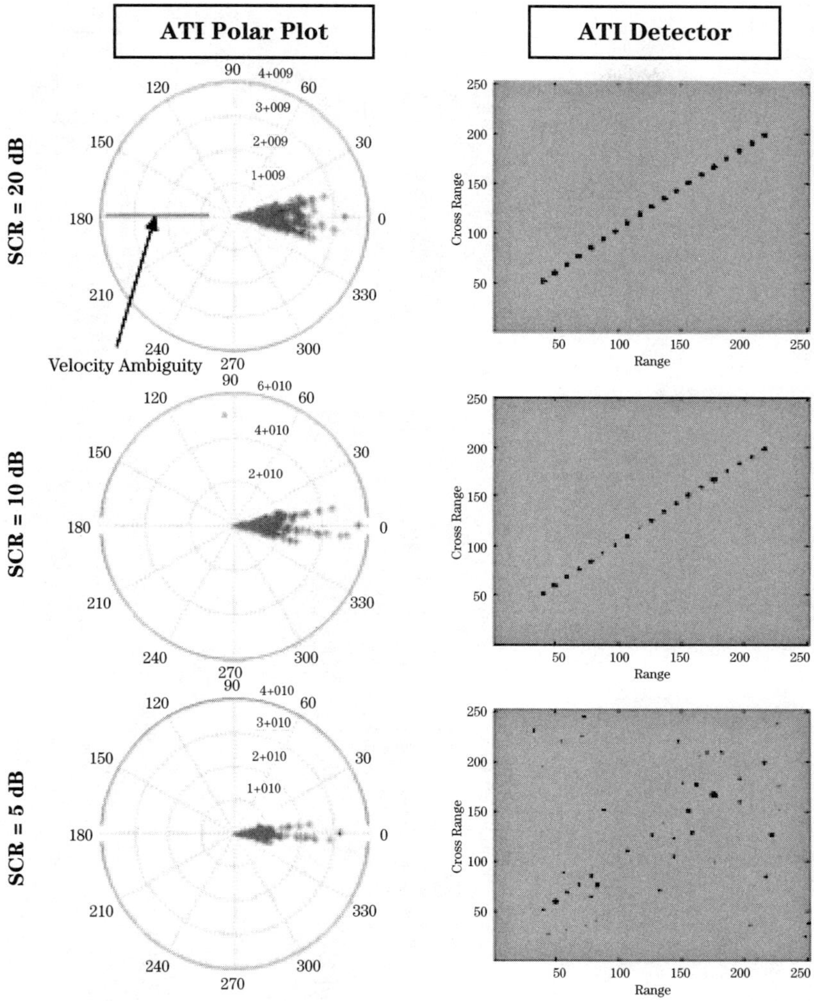

FIGURE 8–33

Impact of signal to clutter on ATI phase determination—2 meter baseline © 2008 IEEE [16]

Figure 8–34 depicts the results for ATI detection of the 19 targets with the wide antenna phase center separation of 20 meters. For the 20 meter antenna phase center separation case the slow-moving targets occupy the entire 2π phase domain due to the ± 2.25 m/sec velocity ambiguity introduced through ATI processing. The clutter at the middle represents the foliage scene and the competing signal for both detection and target location accuracy. For the 20 dB SCR, the scattering of the targets is tight and well separated from the ground clutter. For the 5 dB SCR, the ground clutter is a large percentage of the image and provides only a small separation from the detected ATI phase. More importantly, the variance of the angular accuracy of ATI phase

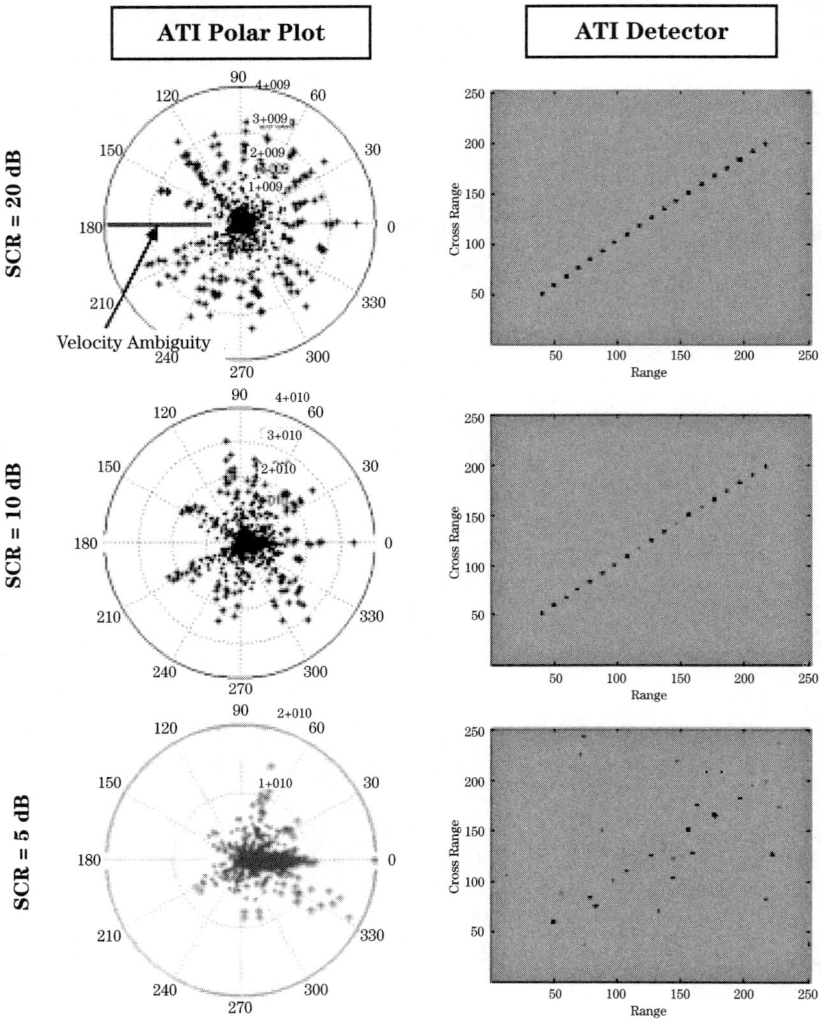

FIGURE 8–34

Impact of SCR on minimum discernable velocity—20 meter baseline © 2008 IEEE [16]

is observed, which directly affects the ability of the algorithm to relocate the targets in azimuth.

The target's radial velocity can be estimated from Figure 8–35 and Figure 8–36 using (8.35). These results are well correlated to the theory in (8.38), supporting the concept that SCR directly affects the velocity estimation accuracy in ATI processing.

The correlation between the pixels in the 2 meter ATI example is shown Figure 8–35. The estimated curves on the bottom are derived by a curve fit of the data. There is not a separation of the correlation as a function of phase

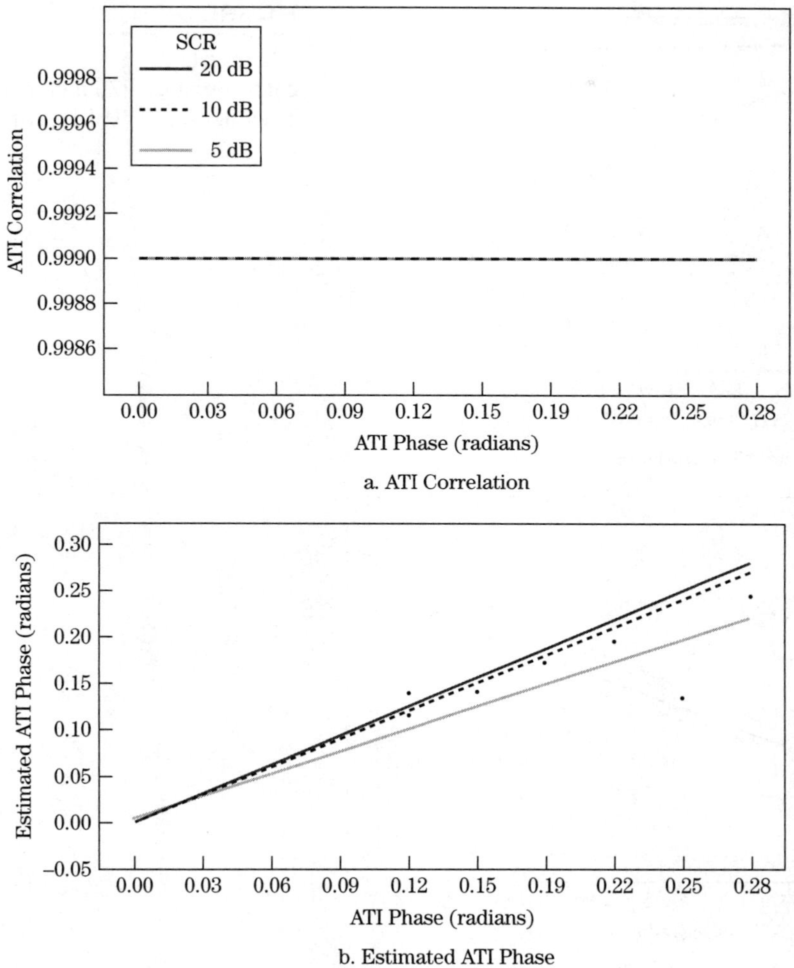

FIGURE 8–35

Simulation of target ATI versus signal to clutter ratio—2 meter baseline © 2008 IEEE [16]

with this small separation. This is attributed to two factors:

- The theoretical analysis assumes that the targets are point targets.
- The simulated targets do not occupy single pixels because of the potential for targets occurring in adjacent range or Doppler filters.
- The simulated of the clutter, without ICM, is highly correlated. Because this effect is additive to the target return, the estimation will be biased.

The comparison of the theory and the simulation results in the ATI phase is much closer in agreement. Again the estimated ATI phase is curve fit to the detected data in the three simulations of SCR conditions. At the lower SCR, the clutter in the resolution cell modifies the phase; hence, the phase error is difficult to predict. The magnitude of the dispersion of the phase is close to

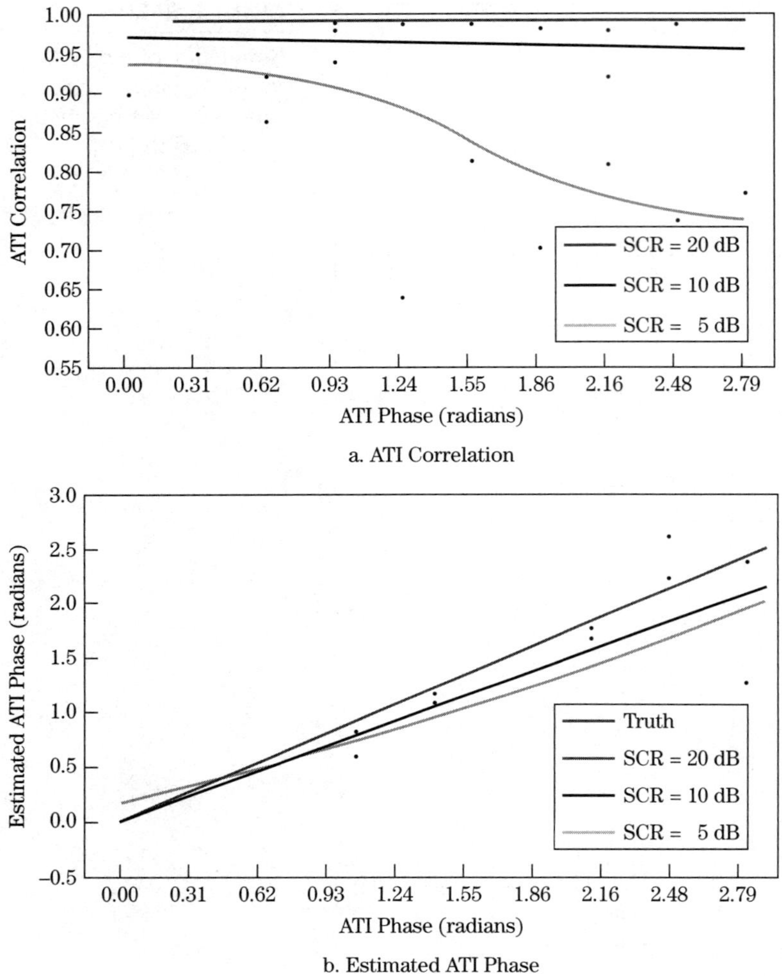

FIGURE 8–36

ATI target characteristics—20 meter baseline © 2008 IEEE [16]

the theory. However, as stated already, the estimate is expected to be biased, due to the lack of clutter decorrelation in the simulation model.

ATI performance for the 20 meter wide baseline is shown in Figure 8–36. The sensitivity of ATI phase to target velocity is much more pronounces than the 2 meter case just examined. The clutter occupies the center of the polar diagrams, with some extent in the zero-phase region. Because the targets can straddle range gates and Doppler filters, there is a large number of data points clustered around the detected target. By centroiding the data points and applying detection thresholds, the appearance of distinct targets result.

After ATI detection, the 19 targets are seen to follow the expected location in range and Doppler frequency for high SCR values. In the 5 dB SCR case, the detection points show some geolocation scatter, as is expected due to the dispersion of measuring the ATI phase.

By curve fitting the data for the ATI correlation, curves over the wide target velocity phase extent are developed. Compare the *estimated* dispersion with the *theoretical*. Good quantitative agreement exists for ATI phase dispersion for the high SCR case. For the lower SCR case, the simulated curves are biased again by the strong correlation of the targets to the background clutter. This is thought to be an artifact of the simulation environment. The ATI phase exhibits excellent agreement between the *estimated* and *theoretical* curves for all three SCR levels. This verifies the viability of using ATI phase for performing target velocity estimation.

The geolocation errors in this wide baseline case will be minimized if the target can be detected. In fact, it is postulated that the error could be corrected by estimating the SCR in the region of the target detection.

Finally, the effects of windblown clutter and attendant foliage ICM were analyzed for three different baselines: 2 meters, 10 meters, and 20 meters. The foliage ICM was modeled using a SCR of 10 dB along with the Billingsley clutter model with a 5 mph wind velocity [16]. The effects of ICM can be seen in Figure 8–37, where there is a widening of the clutter phase due to its nonstationary characteristics.

The estimated MDV was compared with the theoretical MDV with no ICM and with 5 mph wind velocity. Figure 8–38a illustrates the computed multilook ATI MDV as a function of baseline separation for distance SCRs of 5, 10, and

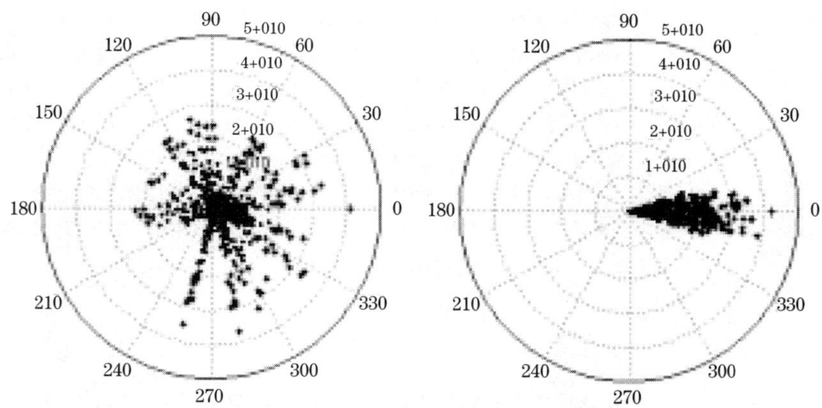

a. B = 20 m b. B = 2 m

FIGURE 8–37

Effect of internal clutter motion (5 mph) on ATI phase © 2008 IEEE [16]

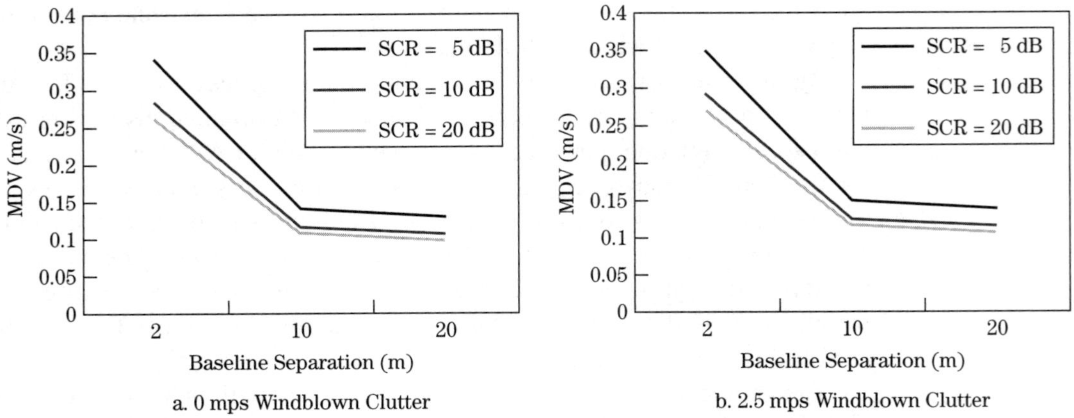

FIGURE 8–38

Effect of billingsley windblown clutter model

Table 8–3 Baseline separation and clutter model effects on MDV © 2008 IEEE [16]

ATI Baseline [m]	MDV Theory Limit [m/s]	Simulated ATI MDV [m/s]	Simulated ATI MDV with ICM [m/s]
2	0.25	0.28	0.30
10	0.05	0.12	0.13
20	0.03	0.11	0.12

20 dB without ICM. Figure 8–38b shows the computed multilook ATI MDV when ICM for a 5 mph surface wind speed is included. The increase in MDV for clutter with ICM is only marginal over the simulated clutter without ICM. Additional analysis and measurements are required to determine the effects of wind velocity and the resultant impact on the ATI velocity error.

Table 8–3 lists the ATI MDV results for the cases depicted in Figure 8–38 along with a 10-meter baseline separation case. The theoretical MDV limit is also included in the table.

8.3.5 ATI Conclusions

In summary, ATI provides an approach for performing ground moving target detection from a moving platform. By employing two separated apertures, the correlation of small ground segments can be processed for detecting targets with radial velocities significantly lower than classical GMTI RADAR systems with STAP. The approach can also provide SAR images for fixed target detection with additional signal processing. However, it will be important to have very fine resolution and polarization agility for fixed target detection and classification. This may increase the cost and complexity of the system beyond practical utility.

More analysis and data collections are needed to verify the foliage ICM impacts on ATI performance. The RLSTAP simulation has indicated a first-order impact on the ATI correlation and phase measurement. But there are details of short- and long-term correlation of the foliage motion that can be answered only with experimental measurements.

In this section, we outlined a design approach that provides for detecting slow-moving ground vehicles in the presence of foliage. By modeling the phenomenology and sources of signal decorrelation we were able to determine the effects of baseline separation, SCR, and ICM on ATI's ability to detect slow-moving targets and estimate their Doppler.

The modeling results are well aligned with the theory. It was possible to illustrate through a high-fidelity RADAR simulation that target detection relies on both signal correlation and phase separation from clutter. It was also shown that SCR, baseline separation, and ICM all reduce target correlation. The target Doppler frequency estimation is bounded by baseline separation and limited by SCR and ICM. These effects provide realistic bounds on the detection and target location accuracy of ATI, as an emerging FOPEN GMTI RADAR systems approach.

8.4 References

[1] Robinson, C. A., "RADAR Counters Camouflage," *AFCEA SIGNAL Magazine,* Alexandria, VA, June 2007.

[2] Staudaher, F. M., "Airborne MTI," in *RADAR Handbook* (2d ed.), Ed. Skolnik, M., McGraw Hill, New York, 1990, Chapt. 16.

[3] Sullivan, R. J., *Microwave RADAR—Imaging and Advanced Concepts*, Artech House, Norwood, MA, 2000, Chapt. 11.

[4] Entzminger, J. N., Fowler, C. A., and Kenneally, W. J., "Joint STARS and GMTI: Past Present and Future," *IEEE Trans. AESS*, Vol. 35, No. 2, April 1999, pp. 748–761.

[5] Norris, G., "Boeing Rotary UAV Aims to Set Records," *Aviation Week and Space Technology,* May 30, 2008.

[6] Thunderbolt, "A160T Unmanned Helicopter Flies with Foliage-Penetrating RADAR in DARPA Tests," *Military Technology,* October 2009, http://www.armybase.us/2009/10

[7] Jao, J. K., "FOPEN GMTI Study Summary," Unpublished FOPEN Project Report FPR-6, MIT Lincoln Laboratory, Lexington MA, Prepared for DARPA Information Systems Office, May 2000.

[8] Ward, J., "Space-Time Adaptive Processing for Airborne RADAR," MIT Lincoln Laboratory Technical Report 1015, Lexington, MA, publically released December 1994.

[9] Guerci, J. R., *Space-Time Adaptive Processing for Radar,* Artech House, Norwood, MA, 2003.

[10] Davis, M. E., Tomlinson, P. G., and Maloney, R. P., "Technical Challenges in Ultra-Wideband RADAR Development for Target Detection and Terrain Mapping," *Proc 1999 IEEE RADAR Conference,* Boston, MA, April 1999, p. 1.

[11] Thompson, A. A. and Livingstone, C. E., "Moving Target Performance for RADARSAT-2," *Proc. 2000 Aerospace Conference,* Big Sky, MT, March 2000, p. 2599.

[12] Rosen, P. A., et al., "Synthetic Aperture RADAR Interferometry," *Proceedings IEEE* Vol. 88, No. 3, March 2000, p. 333.

[13] Rodriguez, E. and Martin, J. M., "Theory and Design of Interferometric Synthetic Aperture Radars," *IEE Proceedings-F* Vol. 139, No. 2, April 1992, pp. 147–59.

[14] Chen, C. W., "Performance Assessment of Along-Track Interferometry for Detecting Ground Moving Targets," *Proc. 2004 IEEE RADAR Conference,* Philadelphia, PA, April 2004.

[15] Livingstone, C. E., et al., "An Airborne Synthetic Aperture RADAR (SAR) Experiment to Support RADARSAT-2 Ground Moving Target Indication (GMTI)," *Can. J. Remote Sensing* Vol. 28, No. 6, 2002, p. 794.

[16] Kapfer, R. and Davis, M. E., "Along Track Interferometry for Foliage Penetration Moving Target Indication," *Proc. 2008 IEEE RADAR Conference,* Boston, MA, April 2008.

[17] Davis, M. E., Maher, J., Hancock, R. J., and Theis, S., "High Fidelity Modeling of Space-Based RADAR," *Proc. of the 2003 IEEE RADAR Conference,* Huntsville, AL, May 2003.

[18] Hancock, R. J., "AFRL Research Laboratory Space Time Adaptive Processing (RL-STAP) Simulation Tool," KASSPER Technical Meeting, DARPA, Arlington, VA, April 2002, http://www.darpa.mil/sto/space/pdf/KASSPER02_Hancock.pdf

[19] Bessette, L. A., Crooks, S. M., and Ayasli, S., "P-3 Ultra-wideband SAR, Grayling Michigan, Target and Clutter Phenomenology," *Proc 1999 IEEE RADAR Conference* Boston, MA, April 1999, p. 125.

[20] Hansen, R. C., *Phased Array Antennas,* Wiley Interscience, New York, 1998, p. 13.

[21] Soumekh, M., *Synthetic Aperture RADAR Signal Processing*, Wiley, New York, 1999.

[22] Gierull, C.H., and Sikaneta, I.C., "Estimating the Effective Number of Looks in Interfereometric SAR Data", IEEE Trans. On Geoscience and Remote Sensing, Vol. 40, No. 8, August 2002, pp. 1733-42.

[23] Kapfer R. Black River Systems, Utica NY, Private communications, 2008.

CHAPTER 9

Bistatic FOPEN SAR

Airborne RADARs have been developed for detecting slow-moving targets in a large number of terrain and clutter environments, as discussed in Chapter 8. Reliable detection of slow-moving targets is enabled only by very large phased array RADARs and space-time adaptive processing to limit or remove the effects of ground clutter due to the motion of the RADAR system. When the targets stop, synthetic aperture RADAR (SAR) modes need to be scheduled to discriminate the stationary targets or cultural features for improved situational information. For targets moving under foliage, a very high frequency (VHF) or ultra high frequency (UHF) is used for reduced foliage attenuation [1]. However, large arrays at signal wavelengths from 0.5 to 6 meters are not typically feasible for installation on most aircraft. Moreover, platform and clutter motion severely limits the minimum discernable velocity (MDV) to several meters per second.

A typical platform for performing persistent long-range detection of moving targets approaching a fixed geographic location is the tethered airborne RADAR system (TARS) shown in Figure 9–1 [2,3]. An alternative platform for foliage penetration (FOPEN) ground moving target indicator (GMTI) RADAR is the A-160 Hummingbird unmanned rotorcraft, previously shown in Figure 8–4 [4]. Both platforms have the ability to carry the large antenna required for low MDV detection of ground targets. However, if the targets stop, the RADAR would be unable to detect the targets under the foliage. If the object must be detected again, the system needs to schedule a fixed target detection mode such as SAR.

Unfortunately, if the platform moves to collect SAR images, there is a severe penalty in the ability to maintain slow-moving target detection. Moreover,

FIGURE 9–1

Tethered airborne RADAR system [4]

a TARS system does not have the ability to move and has no capability for detecting small stationary targets in severe foliage clutter.

Bistatic GMTI RADAR has been demonstrated using TARS as the illuminator and an airborne receiver as the receiver [5]. There have been several discussions of the need for bistatic FOPEN SAR for improved detection of fixed targets [6]. The proposed solution to obtaining both fixed and moving target surveillance of an area is to use a bistatic RADAR adjunct to the GMTI RADAR, as depicted in Figure 9–2 [7]. This approach maintains the surveillance coverage while providing for simultaneous imaging of fixed targets and terrain features.

The motion of the passive receiver can form a SAR image, whereas the stationary GMTI system continues to detect very slow-moving targets. For a receiver flying at an appropriate distance, a SAR image can be generated. This image has the potential for providing cross-range resolution significantly better than a monostatic image. More importantly, the signal-to-noise ratio (SNR) and signal-to-clutter ratio (SCR) can be enhanced when the receiver is sufficiently close to the target. The most critical RADAR requirement is bistatic compatibility between the primary GMTI RADAR waveform, and the signal fidelity for focusing SAR images at the bistatic receiver. This chapter will explore a hybrid system that enables maintenance of GMTI while obtaining moderate to fine SAR images on targets under foliage. This hybrid architecture employs a small unmanned aerial vehicle (UAV) as a bistatic

FIGURE 9–2

Bistatic scenario for simultaneous GMTI and SAR

adjunct receiver. Moreover, not having a transmitter onboard makes it feasible to integrate this function in a smaller and lower-cost unmanned aerial system (UAS).

The need for UHF operation to see through trees and the need for high grazing angles to minimize the foliage attenuation of small or slow targets has already been established in Chapter 3. This chapter will discuss the method to obtain the SAR image resolution as well as the advantage in resolution and SNR for bistatic operation. The predicted results are examined with a detailed modeling and simulation of the scenario [7].

9.1 Bistatic RADAR

Bistatic RADARs, which have existed since the invention of RADAR, have provided incentives for improvement in target detection and localization [8]. However, the fact that the transmitter and receiver are on separate platforms has been a challenge for coherent signal processing. For any coherent RADAR to operate, the receiver needs to know the location, velocity, and time reference

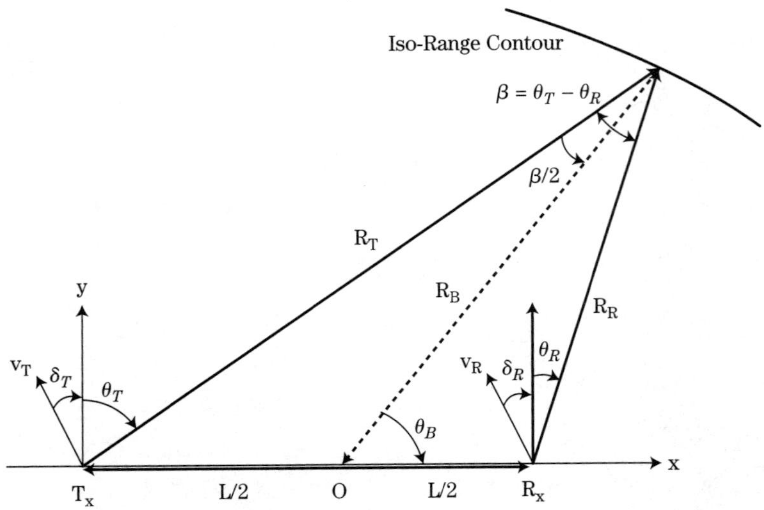

FIGURE 9–3

Bistatic RADAR coordinate system in two dimensions

of the transmitter. With the development of the global positioning system (GPS), accurate clocks, and high-speed digital signal processing, bistatic and multistatic RADAR systems are becoming more popular and practical.

The geometry for bistatic RADAR operation is given in Figure 9–3. Four geometric factors are important in the definition and performance of a bistatic RADAR:

- The slant range from the transmitter to the target or clutter, R_T
- The slant range from the receiver to the target or clutter, R_R
- The baseline between transmitter and receiver, L
- Bistatic angle between transmitter and receiver range vectors, β

GPS enables the three-dimensional location of the transmitter and receiver platforms to enable calculation of the baseline and bistatic angle. However, it is just as important that the onboard inertial navigation system (INS) provide the attitude and heading of the platform and an accurate clock provide the absolute timing.

The coordinate system forms elliptic contours, referred to as the iso-range contours, for the same range sum $(R_T + R_R)$ between transmit and receive vectors. The angle between transmit and receive vectors to the target location is the bistatic angle β. An important relation is that the bisector of the bistatic angle is orthogonal to the tangent of the iso-range contour at any point on the iso-range ellipse [8]. This bistatic bisector helps define many of the important parameters in RADAR coordinates, including the iso-range contours and both the range and Doppler resolutions. These geometric factors are described subsequently.

The RADAR range equation for bistatic operation is derived analogous to the monostatic range equation. The SNR to a point is given by

$$SNR_B = \frac{P_T G_T G_R \lambda_0^2 \sigma_B}{(4\pi)^3 k T_s B_n R_T^2 R_R^2 F_n L_T L_R L_{fol}} \tag{9.1}$$

where

P_T	Transmitter output power
G_T	Transmitter gain in the direction of the target
G_R	Receiver gain in the direction of the target
λ_0	Wavelength of waveform
σ_B	Bistatic RADAR cross section of the target or clutter return
L_T	Transmitter losses to the target
L_R	Receiver losses from the target
L_{fol}	Loss due to propagation through foliage
kT_sB_n	System noise temperature bandwidth product
F_n	Receiver noise figure

It should be noted that the foliage loss L_{fol} depends on both the transmit and receive path. The bistatic RADAR range equation is related to the monostatic equation, when the target RADAR cross section (RCS) values are equivalent ($\sigma_0 = \sigma_B$), and the range product is given by

$$R_T R_R = R_0^2 \tag{9.2}$$

The maximum range product is defined in terms of the minimum SNR to detect a target as κ:

$$\kappa \triangleq (R_T R_R)_{\max} \quad \text{at} \quad SNR_{\min} \tag{9.3}$$

It is well known that the geometry of bistatic RADAR system differs from that of monostatic RADARs due to the effects of the baseline distance L. First, the iso-range contours are ellipses instead of circles in a two-dimension plane and ellipsoids instead of spheres in three dimensions. Second, the contours of constant SNR are not coincident with the iso-range contours, but follow *ovals of Cassini* with the relationship [8]:

$$SNR_B = \frac{K}{R_T^2 R_R^2} \tag{9.4}$$

The term K is the bistatic RADAR constant, which is related to the bistatic maximum range product κ as

$$K = \kappa^2 SNR_{\min} = \frac{P_T G_T G_R \lambda_0^2 \sigma_B}{(4\pi)^3 k T_s B_n F_n L_T L_R L_{fol}} \tag{9.5}$$

A convenient method to analyze the constant SNR profiles is to convert to polar coordinate (R_B, θ_B) with the transformation

$$R_T^2 R_R^2 = (R_B^2 + L^2/4)^2 - R_B^2 L^2 \cos^2\theta_B \tag{9.6}$$

The range resolution of bistatic waveforms has an analogous behavior in terms of compressed pulse width τ_M from the monostatic waveform. However, because of the geometries between iso-range contours, the resolution is "diluted" by the bistatic angle β. This geometric dilution of range resolution δR_B is expressed by

$$\delta R_B = \frac{c\tau_M}{2\cos(\beta/2)} \tag{9.7}$$

The consideration of the range ambiguity is as important to bistatic operation as in a monostatic RADAR system. The unambiguous iso-range contour is related to the PRF of the waveform by

$$(R_T + R_R)_u = \frac{c}{PRF} \tag{9.8}$$

This is the equation of an ellipse with major axis of length c/PRF.

The Doppler frequency is derived from the relationship of the radial velocity of the target with respect to both the transmit and receive sites. If the two sites are stationary, then the target Doppler frequency depends on the angle δ that the velocity vector makes with bistatic angle β.

$$f_B = -\left(\frac{2v_T}{\lambda_0}\cos\delta\cos(\beta/2)\right) \tag{9.9}$$

Thus, the magnitude of the target Doppler frequency will be a maximum when it is traveling toward or away from the bistatic angle bisector. It also depends on the bistatic angle. For the monostatic case in which $\beta = 0$ degrees, one obtains the familiar target Doppler frequency relationship. As the bistatic angle increases, the target Doppler frequency is reduced. At a forward scatter relationship, $\beta = 180$ degrees, and the target Doppler frequency is zero.

When the target is stationary and the transmitter and receiver move with velocity v_T and v_R, respectively, the Doppler at the target site can be determined as [9]

$$f_B = -\frac{v_T}{\lambda_0}\cos(\delta_T - \theta_T)\cos\gamma_T - \frac{v_R}{\lambda_0}\cos(\delta_R - \theta_R)\cos\gamma_R \tag{9.10}$$

The angles δ_T and δ_R are the azimuth angles of the target velocity vector with respect to the respective transmitter and receiver platform coordinate angles. In addition, the grazing angles γ_T and γ_R between the two platforms and the earth's surface must be included. In the following analyses, the transmitter and

Table 9–1 Parameters for bistatic RADAR example

Transmitter Altitude	7 Km AGL	Receiver Altitude	1 Km AGL
Transmitter Velocity	120 m/sec N	Receiver Velocity	50 m/sec N
Transmitter Position	X = −10 Km Y = 0 Km	Receiver Position	X = 0 Km Y = 0 Km
Frequency	450 MHz	PRF	1000 Hz
Iso-Range Contour Spacing	50 μsec	Iso-Doppler Contour Spacing	PRF/64

receiver altitudes are all considered in terms of above ground level (AGL). A more accurate representation would be to include the earth's radius and local terrain altitude in the formulation.

Iso-Doppler contours are not an easy to plot as the ellipses for iso-range contours. Each point on the surface of the earth has a unique Doppler frequency depending on the velocity vector of the transmitter and receiver and the geometric projection on the surface. A scenario and relevant parameters applicable to FOPEN GMTI RADAR are summarized in Table 9–1. The scenario is illustrated in Figure 9–4, with both transmitter and receiver moving north and iso-range and iso-Doppler contours. The baseline dimension L_B is 10 Km, providing a relatively long separation between transmitter and

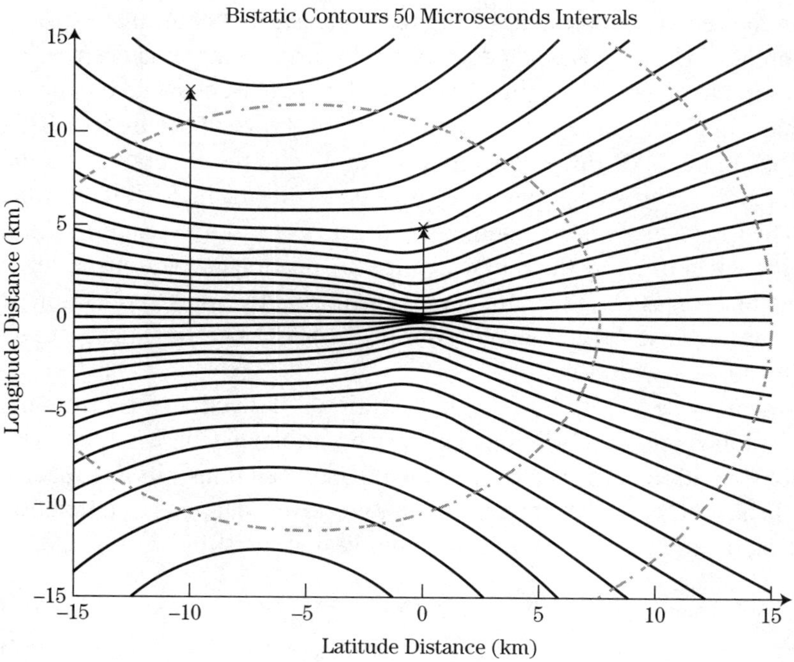

FIGURE 9–4

Bistatic Iso-range and Iso-Doppler contours

receiver. This separation will serve as the basis for showing the bistatic RADAR advantage over a monostatic RADAR.

The length of the lines on the graph represents the relative velocity magnitude and orientation. The iso-range contours, which are drawn 50 μsec intervals, are unambiguous at the 1000 Hz PRF (which is 150 Km in the pseudo-monostatic case). The iso-Doppler contours are drawn at PRF/64, which might correspond to the Doppler filter width for a GMTI RADAR signal with a 64 msec CPI. These parameters were chosen to easily show the clutter Doppler frequency contours on the surface of a smooth earth. Specific range and Doppler spectra will be illustrated with the SAR development in the next section.

9.2 Bistatic SAR Signal Geometry

Signal processing for bistatic RADAR systems is complicated by the remote location of the transmitter from the receiver. There are several conventions as to where to place the origin of the coordinate system to facilitate the signal processing. For example, the coordinate system is located at the midpoint of the baseline vector in Figure 9–3. This simplifies the description of the elliptic function for iso-range contour. An alternative reference point is at the receiver, where the signal processing and time reference can be conveniently coordinated. For this case, the baseline orientation and length will be appropriately located in the receiver coordinates. The receiver-oriented coordinate system is common in MTI systems when detection of a discrete target is desired.

However, the bistatic SAR processing can be greatly enhanced by choosing an image plane centered coordinate system, as shown in Figure 9–5. This is analogous to the SAR imaging from Chapter 4, with the exception that the transmitter and receiver will traverse separate paths during the SAR image formation. The image plane is represented by an array of discrete scatters whose location is each defined by the $\vec{r}_0(\tau)$ vector. The location of the transmitter is $\vec{R}_T(\tau)$, the receiver is $\vec{R}_R(\tau)$, and τ is the time during the image collection. It is evident that because R_T and R_R vary during the SAR operation, the baseline $\vec{L}(\tau)$ will also vary with time.

The signal phase history for bistatic operation is derived directly from the analogous monostatic RADAR signal history described in Chapter 4, but with careful consideration of the vector relationship between transmits and receive range to the scene center. The transmitter and receiver locations as a function of slow-time τ are given by the time-varying vectors [9,10]

$$\vec{R}_T(\tau) = [x_T(\tau), y_T(\tau), z_T(\tau)] \tag{9.11}$$

$$\vec{R}_R(\tau) = [x_R(\tau), y_R(\tau), z_R(\tau)] \tag{9.12}$$

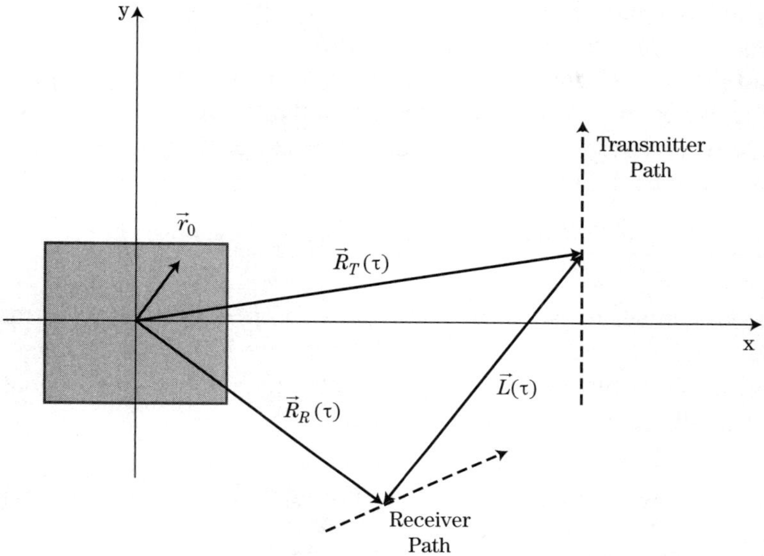

FIGURE 9–5

Coordinate system for bistatic SAR image processing

If the platforms are assumed to be stationary during the fast-time pulse propagation, the signal phase history can be expressed in terms of the location of the two platforms in slow-time only. Thus, in a slow-time coordinate system, the actual distance from the transmitter to a scatterer located at $\vec{r}_0(\tau)$ in the scene is given by

$$d_{T0}(\tau) = \|\vec{R}_T(\tau) - \vec{r}_0(\tau)\| \tag{9.13}$$

and the actual distance from the receiver to the scatterer is:

$$d_{R0}(\tau) = \|\vec{R}_R(\tau) - \vec{r}_0(\tau)\| \tag{9.14}$$

The fast-time delay of the receive pulse, relative to the travel time to and from the scene origin, is given by

$$t(\tau) = \frac{d_{R0}(\tau) + d_{T0}(\tau)}{c} = \frac{\Delta R(\tau)}{c} \tag{9.15}$$

The term $\Delta R(\tau)$ is commonly referred to as the differential range. It should be noted that equations (9.13) through (9.15) assume perfect measurement of the absolute time, and of the positions of the transmit and receive platforms. This formulation provides a simplistic assessment of the image processing approach. Extensive research has gone into the effects of measurement errors and their impact on the location and image quality functions of bistatic SAR. These factors will not be considered in this introductory treatment of the subject [11,12].

The transmitted pulses have a bandwidth B of frequency around the center frequency f_0. Over this bandwidth, the waveform is considered to be composed of multiple radian frequencies $\omega_i = 2\pi f_i$ within the bandwidth. Given the differential range in (9.15) to the m-th cell on the surface, a linear phase function of the received radian frequency samples corresponding to a scatterer located at $r_0(\tau)$ can be defined as

$$\Phi_m(\omega_i, \tau_k) = -\frac{\omega_i \Delta R_m(\tau_k)}{c} \qquad i = 1 \ldots N_f;\ k = 1 \ldots N_T \tag{9.16}$$

where N_f is the number of radian frequencies ω_i sampled in the waveform, and N_T is the number of transmit pulses used to form the bistatic SAR image. Given the linear phase function for a single scatter, the total received signal is the sum from all of the scatterers in the scene

$$s(\omega_i, \tau_k) = \sum_{m=1}^{M} a_m \exp[j\Phi_m(\omega_i, \tau_k)] \tag{9.17}$$

The image is then formed by correlation of the phase history $s(\omega_i, \tau_k)$ with the reference phase function $\Phi_m(\omega_i, \tau_k)$ [9]

$$P(x, y, z) = \frac{1}{N_f N_T} \sum_{i=1}^{N_f} \sum_{k=1}^{N_T} s(\omega_i, \tau_k) \exp[-j\Phi_{xyz}(\omega_i, \tau_k)] \tag{9.18}$$

where Φ_{xyz} is the reference phase evaluated for each (x, y, z) location in the scene. This is the matched filter from of image construction, assuming perfect knowledge of time reference, and accurate knowledge of the location, orientation, and speed of on both the transmit and receive platforms. The number of operations to form the scene scales with the number of pixels in the scene, the number of frequency samples N_f, and the number of pulses transmitted N_T. Assuming N^2 pixels in the scene, and $N_T N_f \alpha N^2$, the total number of operations to form a bistatic SAR image using a matched filter is $O(N^4)$.

Multiple techniques for image formation have been developed that are more computationally efficient than the matched filter. It should be noted that polar format algorithm (PFA) and range migration algorithm (RMA) techniques have been developed for bistatic SAR. However, the ultra wideband (UWB) waveform, wide integration angles, and varying baseline violate many of the approximations for PFA and RMA processing [13]. The backprojection algorithm (BPA) provides the best image focus results for bistatic SAR as it did for monostatic SAR, but with a computational load is order of $O(N^3)$. However, because of the varying baseline, BPA will place significant demands on precise motion compensation (MOCOMP) and will limit the size of the image. Nevertheless, BPA is the best current algorithm choice for UWB SAR, since no approximations have been made in the image formation processing.

9.3 Bistatic SAR Resolution

The general case of three-dimensional bistatic RADAR operation with separate transmitter and receiver platforms is illustrated in Figure 9–6. The transmitter is flying with velocity $\vec{v}_T$ at a position vector $\vec{R}_T$ and altitude h_T above the earth. Similarly, the receiver is flying with velocity $\vec{v}_R$ at a position vector $\vec{R}_R$ and h_R altitude above the earth. An imaged area is described by each point of vector $r_0(\tau)$ around the origin of the coordinate system.

Several key system attributes need to be analyzed for SAR operation. First, the surface clutter return needs to be evaluated in terms of the Doppler frequency, which is the primary characteristic exploited to image the terrain. Because there are separate transmit and receive platforms, the bistatic Doppler frequency will be a function that is dependent on the motions of the two. Next, the range resolution needs to be determined by the transmit bandwidth and the geometry of the transmit-surface-receiver spatial vectors. Finally, the cross-range resolution will be determined by the angular velocities for the imaged cell.

This section will explore the general case of the bistatic Doppler frequency and its effect on resolution. Next, the analysis will consider the special case of a tethered transmitter and the surface Doppler frequency due to the receiver motion. Finally, the advantage of bistatic RADAR operation will be explored when the receiver is closer to the image plane than the transmitter.

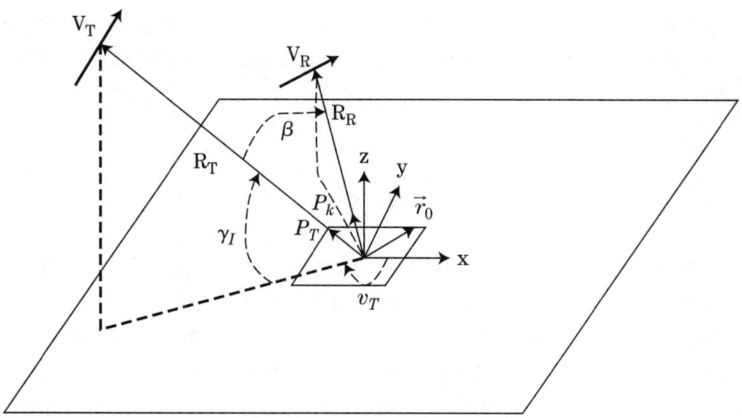

FIGURE 9–6

Bistatic RADAR coordinate system © 2009 IEEE [7]

9.3.1 Bistatic Doppler Frequency

We first look at the three-dimensional coordinate system and contributions to bistatic SAR. If a platform is moving, the surface Doppler frequency is a combination of the motion of both the transmitter and receiver, projected on the surface described by the vector $\vec{r}_0$, is derived from the angular speed of the two platforms. Using a Taylor expansion for a small target area $\vec{r}_0$ around the origin, the Doppler frequency is given by [14]

$$f_D \cong -\frac{1}{\lambda_0}\left[\omega_T + \omega_R\right]^T \left[r_0\right] \tag{9.19}$$

where λ_0 is the wavelength, and ω_T and ω_R are the transmitter and receiver angular speeds with respect to the origin. These variables are determined using:

$$\omega_T = \left|\frac{(I - \hat{\rho}_T\hat{\rho}_T^T)v_T}{\|R_T\|}\right| \tag{9.20}$$

$$\omega_R = \left|\frac{(I - \hat{\rho}_R\hat{\rho}_R^T)v_R}{\|R_R\|}\right| \tag{9.21}$$

In (9.2) and (9.3), $\mathbf{I}$ is the identity matrix, and $\hat{\boldsymbol{\rho}}_{\mathbf{T}}$ and $\hat{\boldsymbol{\rho}}_{\mathbf{R}}$ are the unit vectors from the origin to the transmitter and receiver respectively.

It is conventional to use the *bistatic equivalence theorem* in evaluating metrics of resolution. When the transmitter has a bandwidth of B, a bistatic angle of β, and a bistatic grazing angle of γ_B, the range resolution is measured along the bistatic angle bisector β between the transmitter and receiver vectors in Figure 9–6. In the ground plane that resolution is the monostatic range resolution $c/(2B\cos(\gamma_g))$ but is modified by the bistatic angle bisector. This bistatic range resolution is given by

$$\delta_R = \frac{c}{2B\cos(\beta/2)\cos(\gamma_B)} \tag{9.22}$$

The angular motion of a platform around the image plane creates the Doppler frequency variation with time required to form a SAR image.

The SAR cross-range resolution δ_{CR} can be expressed inversely proportional to the summation of the angular velocity around the center times the integration time τ_I. Following the same convention, the bistatic SAR cross-range resolution is defined along the bistatic angle bisector between $\vec{R}_T$ and $\vec{R}_R$ by [14]

$$\delta_{CR} = \frac{\lambda_0}{\tau_I\sqrt{\omega_T^2 + \omega_R^2 + 2\omega_T\omega_R\cos\beta}} \tag{9.23}$$

The term δ_{CR} is strictly speaking valid only for a circular SAR operation. However, linear paths can be motion compensated to the scene center to approximate the relationship given above. The cross-range resolution can be solved by Taylor series expansion of the geometry at the center of the SAR image collection.

Note that if one of the platforms is stationary the bistatic SAR cross-range resolution simplifies into twice the monostatic SAR cross-range resolution (i.e., it is coarser by a factor of 2). This is important, as we will see in the next section. The integration time can be determined by the speed of the platform and the integration angle required for a specified cross-range resolution, as shown in Figure 9–7. The dotted line above is the integration time for a monostatic UHF platform at a 20 Km slant range. At these very slow speeds and long-range, achieving fine cross-range resolution SAR is very difficult.

In the bistatic SAR case, when the transmitter is at the same 20 Km range and is stationary, the cross-range resolution can be derived for various receiver radii around the scene. For an integration time of 10 seconds, resolutions

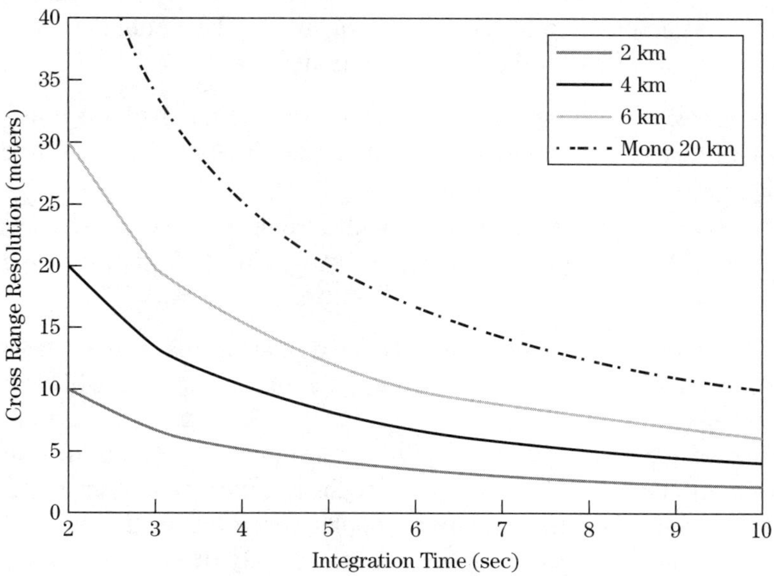

FIGURE 9–7

Integration time for bistatic SAR cross-range resolution © 2009 IEEE [7]

Table 9–2 Stationary transmitter bistatic RADAR parameters

Transmitter	Value	Units	Receiver	Value	Units
Slant Range	20	[Km]	Slant Range	2	[Km]
Velocity	0	[m/sec]	Velocity	75	[m/sec]
Altitude	7	[Km]	Altitude	1	[Km]
Grazing Angle	20.5	[deg]	Grazing Angle	30.0	[deg]
Orientation φ_T	180	[deg]	Orientation φ_R	90, 180	[deg]
Antenna Area	6	[m^2]	Antenna Area	1	[m^2]
PRF	100	[Hz]	Target RCS	10	[dBsm]
Frequency	400	[MHz]	Bandwidth	30	[MHz]
			Integration Time	10	[s]

approaching 2 meters can be obtained at ranges of 2 Km from the scene. This cross-range resolution is sufficiently fine so that the observation of cultural details and the verification of vehicles under the forest canopy are possible.

9.3.2 Stationary Bistatic SAR Transmitter

The general case of a stationary UHF transmitter and moving bistatic RADAR receiver is considered next. Table 9–2 summarizes the parameters used in the analysis and serves as the starting point for a detailed simulation of the scene.

The transmitter, which is assumed to support a GMTI RADAR system, is assumed to be stationary at 20 Km slant range from the area under surveillance, and the receiver is assumed to be moving at a radius of 2 Km from the scene center. There are two cases for the receiver motion:

- *Pseudo-monostatic*: Defined as having the receiver in line with the transmitter illumination path, in this case, moving north along the radial line from the transmitter to the scene center
- *90 degree bistatic*: Defined with a flight path at right angles to the transmit illumination; this example is moving east at a point 2 Km north of the scene center

For simplicity, this motion can be treated as in the plane of the image. These features will be exploited to determine the image quality of a candidate bistatic SAR system.

With a stationary transmitter ($\omega_T = 0$), the cross-range resolution from (9.23) simplifies to be inversely proportional to the receiver integration angle. Figure 9–8 illustrates the surface clutter Doppler spectrum for the pseudo-monostatic case and in terms of bistatic receiver velocity of $v_R = 75$ m/sec. The angular spread of the Doppler frequency is scaled by v_R, and is

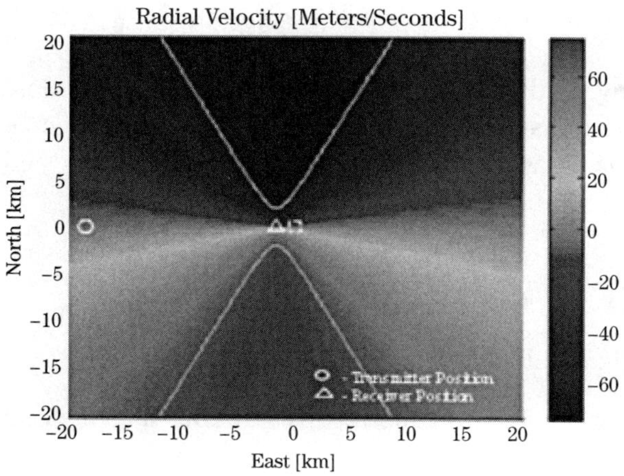

FIGURE 9–8

Pseudo-monostatic surface Doppler frequency within 1 kilometer square image

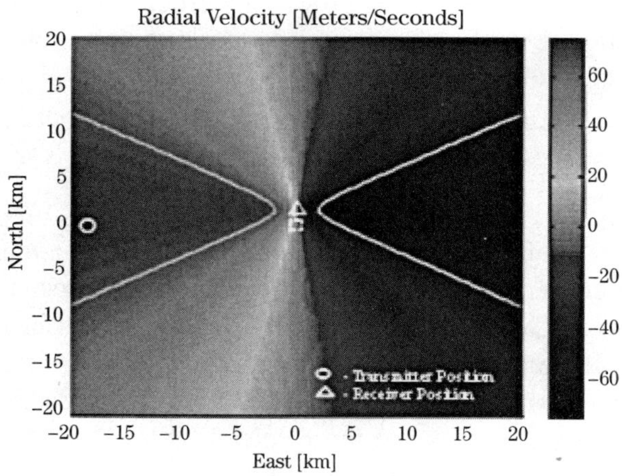

FIGURE 9–9

90-degree bistatic surface Doppler frequency within 1 kilometer square image

perpendicular to the velocity vector. The transmitter and receiver are indicated by a circle and triangle, respectively, and the first velocity ambiguity is shown by the grey hyperbola. Figure 9–9 shows the surface clutter Doppler for the 90 degree bistatic angle scenario. Because the transmitter is stationary, the clutter spectrum is identical to the pseudo-monostatic case but rotated by 90 degrees around scene center.

9.3.3 Bistatic SAR Advantage

Bistatic operation is difficult, due principally to the need for synchronizing the receiver sampling to a transmitter source. Thus, there needs to be an advantage to bistatic operation that warrants this increase in system complexity (and hence cost). Principal tactical advantage is that the receiver is silent and can generally be lower in power that a monostatic RADAR. One major technical benefit that has been examined for several decades is known as the *bistatic advantage*.

Because the receiver range to the target is significantly less than the transmit range, there is an SNR advantage to bistatic operation. This advantage I_B can be related to the antenna areas (A_T and A_R), the target ranges (R_T and R_R), and the bistatic loss L_B as [15]

$$I_B = \left[\frac{A_R}{A_T}\right] \cdot \left[\frac{R_T^2}{R_R^2}\right] L_B \tag{9.24}$$

The bistatic loss L_B needs to include several important factors:

- *Beam-shape loss*: gain of the transmit and receive patterns at each point on the surface of the scene
- *Waveform correlation loss*: losses in signal processing efficiency due to timing, and relative motion (Doppler effects) in the correlation of a point scatterer
- *Propagation loss*: difference in loss from the transmitter to the target or clutter patch to the receiver; especially relevant for bistatic FOPEN SAR operation

Thus, the technical advantages afforded by the bistatic RADAR operation are concerned by the time and energy needed to collect the SAR. First, it takes less time to fly a given angle at the shorter range for SAR resolution, even though the integration angle must be twice as large to achieve the same cross-range resolution as in a monostatic SAR. Second, the bistatic RADAR range equation provides a $1/R_R^2$ advantage to the received SNR. Third, an advantage in SCR is realized through the reduction of both the range and the size of a clutter cell on the ground. The bistatic advantages considered here are under assumption of constant integration time and can be approximated by [7]

$$SCR_{adv} \approx \frac{\sigma_m^0 \rho_{Rm} \rho_{CRm}}{\sigma_b^0 \rho_{Rb} \rho_{CRb}} L_\delta \tag{9.25}$$

where ρ_R and ρ_{CR} are the range and cross-range resolutions and the σ^0 values are the RADAR backscatter coefficient of the monostatic and bistatic orientations, respectively, and L_δ is the bistatic alignment loss due to timing and antenna directivity. Fourth, two-way monostatic foliage attenuation signal degradation can be avoided or minimized with different receiver geometries and grazing angles. Finally, with a passive receiver it is easier to fly a shorter slant range to the targets, albeit at a higher grazing angle. These bistatic advantages have been evaluated for several scenarios.

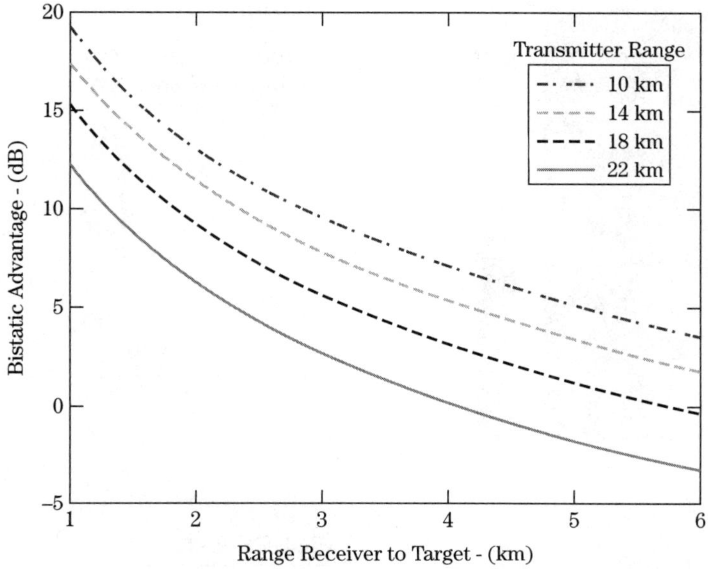

FIGURE 9–10

Bistatic signal-to-noise advantage

The analytic bistatic SNR advantage from (9.24) is illustrated in Figure 9–10 for several transmitter and receiver ranges to the scene center. A clear enhancement in the SNR is indicated for a wide range of operating conditions, so long as the operation synchronizes the transmit and receive antennas to observe the same SAR area in all these scenarios. The Doppler frequency spread, and hence the cross-range resolution, is scaled by the receiver range and velocity, thus providing an improvement over a monostatic SAR operating with the same parameters.

Figure 9–11 illustrates the achievable cross-range resolution verses integration time for the 75 m/sec receiver platform velocity and pseudo-monostatic operation. The principal scenario parameter is the slant range from the receiver to the scene center, illustrating the integration angle dependence of (9.23). This is compared with monostatic SAR resolution (dashed line) possible with the transmitter $R_T = 20$ Km, and $v_T = 75$ m/sec. This clearly illustrates the cross-range resolution advantage of bistatic operation.

The improvement in cross-range resolution between the monostatic and pseudo-monostatic case is shown in Figure 9–11. Each case examines the cross-range resolution in meters of a 1 Km square area. The cross-range resolution for the monostatic case as calculated in Figure 9–7, considers a common transmitting and receiving platform at a range of 20 Km moving at 75 m/sec. The small-angle bistatic case (pseudo-monostatic), which is shown in Figure 9–12, is consistent with the parameters in (9.23).

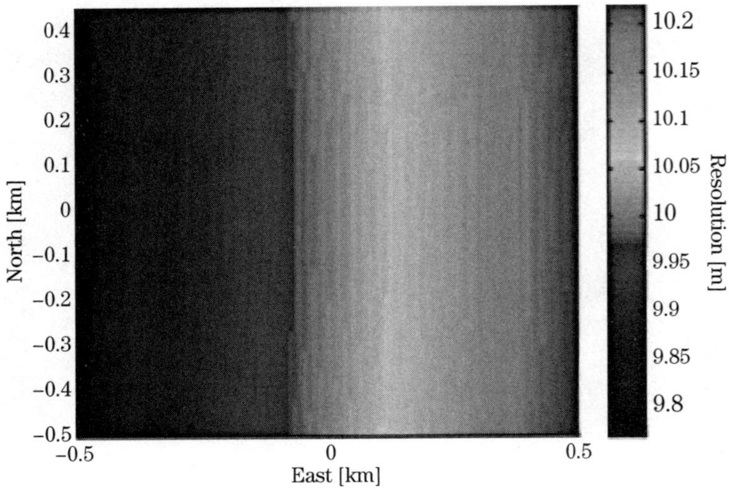

FIGURE 9–11

Psuedomonostatic resolution within 1 Km^2 area

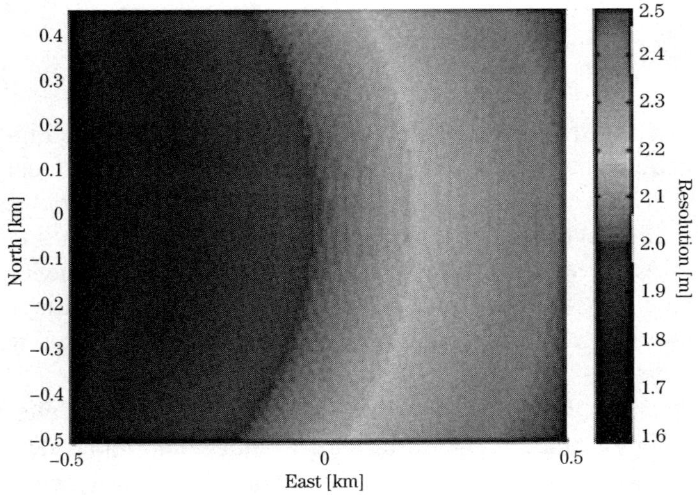

FIGURE 9–12

90 degree bistatic resolution within 1 Km^2 area

We also expect an advantage in cross-range resolution for a shorter-range receiver. This is caused by the higher angular rate of synthetic aperture formation. The cross-range resolution for the bistatic receiver is a 5-to-1 improvement over the monostatic SAR given the scenario parameters and a constant 10 second integration angle. We also notice that this resolution varies over the scene following the iso-range contours.

Several scenario advantages to bistatic SAR operation need to be evaluated. For monostatic operation at long standoff ranges and low grazing angles, the two-way foliage loss is unavoidable for fixed and moving targets in tree lines or narrow roads. As we have seen in the bistatic advantage calculation, this mode will also provide the lowest-target SNR.

For pseudo-monostatic SAR operation the SNR will be inherently higher due to the shorter receiver range. In addition, there is the possibility of both lower foliage loss and lower clutter return depending on the geometry and resolution parameters.

For a 90 degree bistatic geometry, it may be possible to fly the receiver in a manner such that there is little or no foliage loss on that leg of the image formation. If the target is in the open the background clutter will also be less, providing an advantage in clutter to noise ration (CNR). However, the image resolution will be lower due to the $\cos^{-1}(\beta/2)$ dilution of range and cross-range resolution. These factors are best evaluated with a high-fidelity simulation.

9.4 Bistatic SAR Modeling

The resolution of bistatic SAR is limited by the integration time, the receive antenna beamwidth, and the bandwidth of the transmitted waveform. However, there is not a good analytic relationship for the resolution of the bistatic SAR resolution, as there is for monostatic SAR. Thus, it is more reasonable to simulate the scenario and determine the separability of point scatters in the scene.

To explore these advantages, a mixed clutter scene was modeled in RL-STAP. Eight collections of targets were modeled with some of the targets being in the foliage, some clearly in the open and some near tree lines. In addition, the orientation of the targets was varied to evaluate the ability of the SAR to resolve multiple targets at spacings from 2 meters up to 32 meters. The area was assumed to be reasonable flat with the exception of the height of the foliage. Figure 9–13 shows the RLSTAP land-use, land-cover model of the region simulated, along with a picture of the area from Google Earth. All of the groups of targets were assumed to be stationary. The scene was chosen to provide open and forested regions for evaluating effects of bistatic loss parameters from (9.24) for waveform and foliage loss considerations.

The expected advantage of bistatic operation is summarized in Figure 9–14. For the monostatic SAR, the integration times need to be longer than the desired 10 seconds to obtain cross-range resolutions of less than 5 meters, and the foliage loss is higher due to the low grazing angle. Due to the longer operating range, the clutter cells are larger, presenting high clutter scattering returns to the receiver. For the pseudo-monostatic case, the receiver will generally have a higher grazing angle and less foliage loss. Moreover, there is a SNR enhancement due to the shorter range from the receiver to the targets.

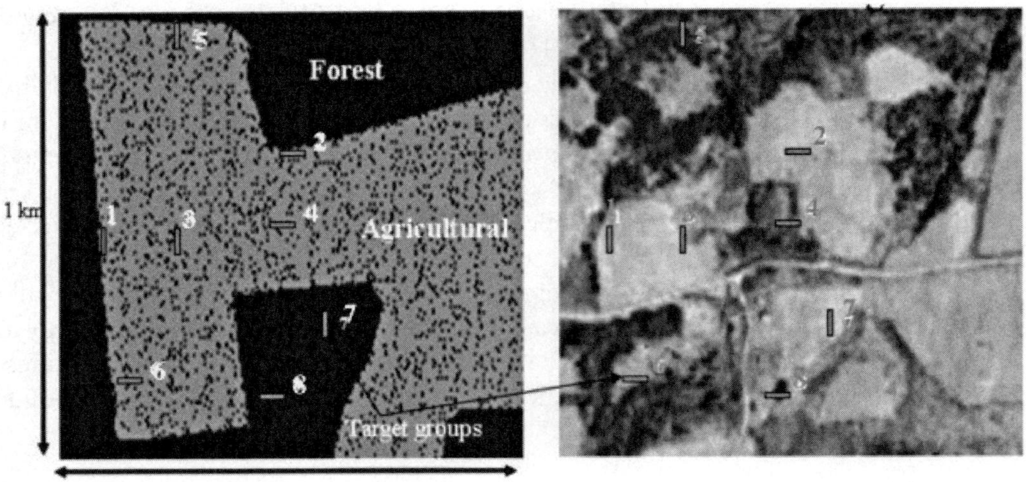

a. Land Use Land Cover Model

b. Photograph of Region

FIGURE 9–13

RLSTAP model for bistatic SAR evaluation © 2009 IEEE [7]

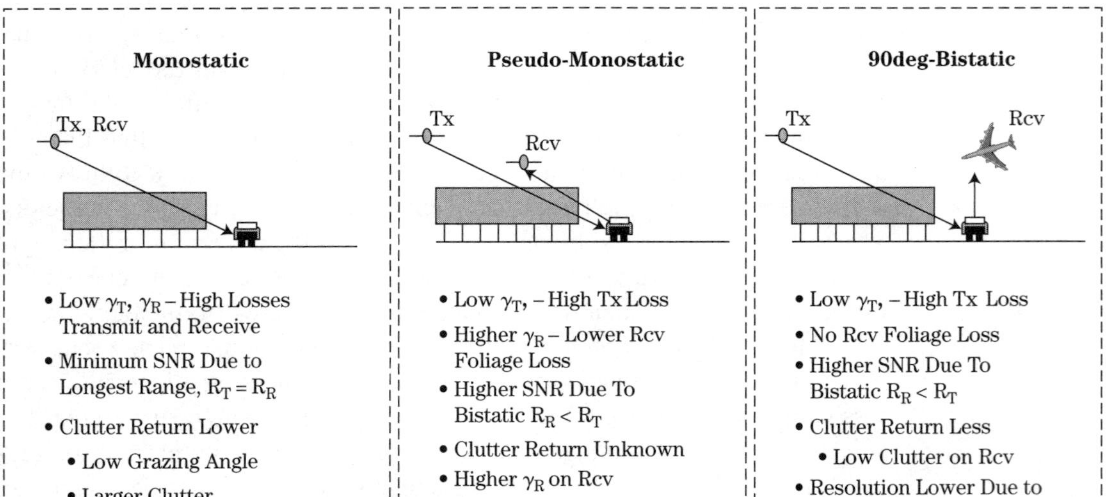

FIGURE 9–14

Scenario conditions for foliage loss and SNR improvement © 2009 IEEE [7]

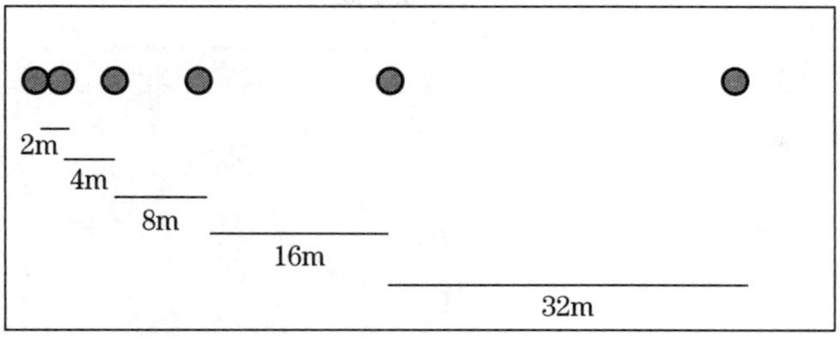

FIGURE 9–15

Point target arrangement for resolution evaluation © 2009 IEEE [7]

The impact of clutter is not immediately obvious since the higher grazing angle and smaller cross-range resolution might be offset by an increased clutter return. Finally, the 90 degree bistatic operation could have lower receive loss and potentially no foliage loss if the receiver is sighted down know trails or open areas. The SNR and SCR will be enhanced due to the shorter range. But the resolution will be directly impacted due to the dilution of resolution with the bistatic angle β. The detailed simulation was carried out to quantify these effects for at least one scenario geometry.

Point scatterers are placed such that there are variations in orientation and foliage attenuation when viewed by the transmitter and receiver. Each point scatterer is a 10 dBsm target, separated by varying spacing to assess the synthetic resolution of the scenario, which is 2, 4, 8, 16, and 32 meters as illustrated in Figure 9–15. The attenuation of each scatterer is dependent on the single versus double path and the grazing angles to the target from transmitter and receiver.

Figure 9–13 showed the scenario used for this evaluation, including the effects of terrain elevation as well as the variation in land cover. A 1 kilometer square region, composed of forest and farmland, was chosen to illustrate the bistatic target and clutter variations. This region provides the diversity in background clutter between the foliage covered and opened regions, where the modeled backscattering characteristics are shown in Figure 9–16.

The scenario has been modeled using RLSTAP, a detailed pulse-by-pulse simulation of the transmitter and receiver characteristics, previously described in Section 8.4.3. This simulation detail provides the clutter and propagation characteristics and target models needed to quantitatively assess the bistatic resolution and SCR as a function of geometries [16]. Because the RLSTAP modeling tool represents the position and velocity of both transmit and receive platforms at each pulse, the effects of navigation errors, range and Doppler walk, and waveform and beam-shape contributions can be evaluated

Table 9–3 summarizes the parameters in the scenario. The orientation angle is with respect to East. The transmitter was due West of the scene at 20 Km slant range and 7 Km altitude. The pseudo-monostatic receiver

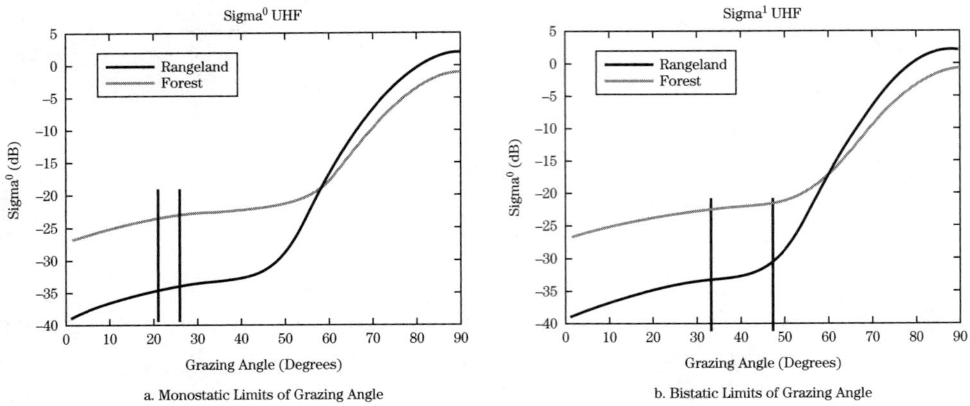

FIGURE 9–16

Clutter models for bistatic SAR evaluation © 2009 IEEE [7]

location is at 180 degrees from East, and the 90 degree bistatic location is −90 (or 270) degrees from East. A modest resolution GMTI RADAR waveform bandwidth of 30 MHz was used so that the slant-range resolution was 5 meters. The rest of the parameters have been discussed earlier in the trade-off analyses. The objective of this simulation is to verify the bistatic advantage claims in terms of signal strength, clutter interference and image resolution.

RLSTAP determines the clutter characteristics versus grazing angle as shown in Figure 9–16. For a bistatic scenario, it employs the bistatic equivalency principal by using the grazing angle along the bistatic angle bisector. The two vertical lines show the span of clutter scattering RCS from near to

Table 9–3 Bistatic scenario loss and target characteristics © 2009 IEEE [7]

Target Group Number/ Orientation	Rx Foliage Obscure	Tx Foliage Obscure	Loss Tx R-20 km	Loss Rx R = 2 km	Target RCS m^2 Monostatic	Target RCS m^2 pseudo Monostatic	Target RCS m^2 90° Bistatic
1-Vertical	No	Yes	6.5 dB	0	0.5	0.708	2.2
2-Horizontal	Yes	No	0	5 dB	10	10	3.2
3-Vertical	No	No	0	0	10	10	10
4-Horizontal	No	No	0	0	10	10	10
5-Vertical	Yes	No	0	5	10	10	3.2
6-Horizontal	No	Yes	6.5	0	0.5	0.7	2.2
7-Vertical	Yes	Yes	6.5	5	0.5	0.7	0.7
8-Horizontal	Yes	Yes	6.5	5	0.5	0.7	0.7

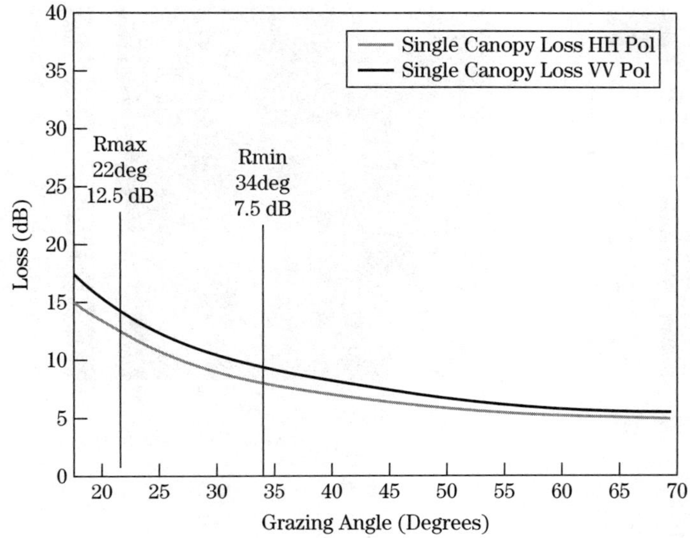

FIGURE 9–17

Foliage 2-way attenuation for bistatic analysis © 2009 IEEE [7]

far range in the 1 Km square image. The clutter return varies more in the bistatic scene than in the monostatic scene. This is due to the higher grazing angle in the short-range bistatic receiver geometry, with a larger variation in backscatter versus range within the scene. This is an important distinction in evaluating the clutter background.

The foliage penetration loss is determined by the FOPEN SAR measurement campaigns presented in Chapter 2. The scenario was modeled as a single-canopy forested region, with a horizontal polarization GMTI transmit characteristics. The curves in Figure 9–17 indicate a two-way path loss through the foliage. In evaluation of the bistatic cases, the one-way loss is taken as half of the two-way loss (in dB). It should be noted that the minimum range to the 1 Km square scene is at a 33 degree grazing angle, which has a lower loss by 5 dB (two-way) than the far range.

Table 9–3 also summarizes the eight target groups and their situation for obscuration from the transmitter and receiver. The losses are evaluated in the middle of the scene, so that there is higher loss for shorter-range target groups than for longer range. The equivalent target RCS is summarized in the columns for a 10 square meter target in the open. The majority of the resolution, SNR, and SCR analyses have been conducted for target groups 3, 4, and 5.

9.4.1 Monostatic SAR

A representative monostatic 2 meter cross-range resolution SAR image has been created from the detailed signal return, using the backprojection algorithm [17]. The SAR image generated from the baseline monostatic scenario,

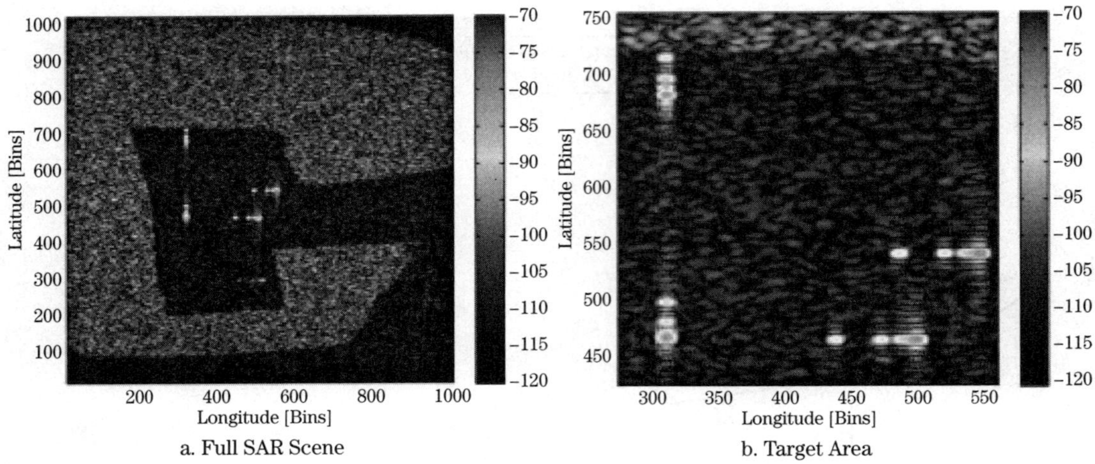

FIGURE 9–18

Simulated monostatic SAR image © 2009 IEEE [7]

with 5 meter range resolution waveform, is illustrated in Figure 9–18. The monostatic RADAR platform is flying "north" at 75 m/sec, "west" of scene center at a 20 Km slant range, and for a 10 second integration time.

The variation in clutter characteristics within the region is clearly indicated, along with the location of each group of target scatterers. Besides the difference in the clutter scatterer types, the power spectral density of the scene is created by the variation in the slant range and transmitter pattern gain.

To assess the signal return and the resolution of these scatterers, slices of the image through three of the target groups are provided. In Figure 9–19a, the slice in latitude is through targets 3 and 5 at longitude bin 251. For a monostatic SAR it is clearly evident that for the integration time of 10 seconds and the cross-range resolution of 10 meters, insufficient resolution exists to separate any targets with less than a 16 meter separation. The longitude slice through target 4 is shown in Figure 9–19b, where an inherent meter range resolution shows separation of each individual scatterer, except for the scatterers separated by 2 and 4 meters. This 1 square kilometer SAR image will be used to evaluate the bistatic advantage for two additional collection geometries.

9.4.2 Pseudo-Monostatic SAR

The first bistatic scene evaluation is for a bistatic receiver flying "north' 2 kilometers "west" the scene center at 75 m/sec, with a stationary transmitter at 20 kilometers. The geometry is such that the middle of the SAR image is aligned with the illumination azimuth from the transmitter to the scene

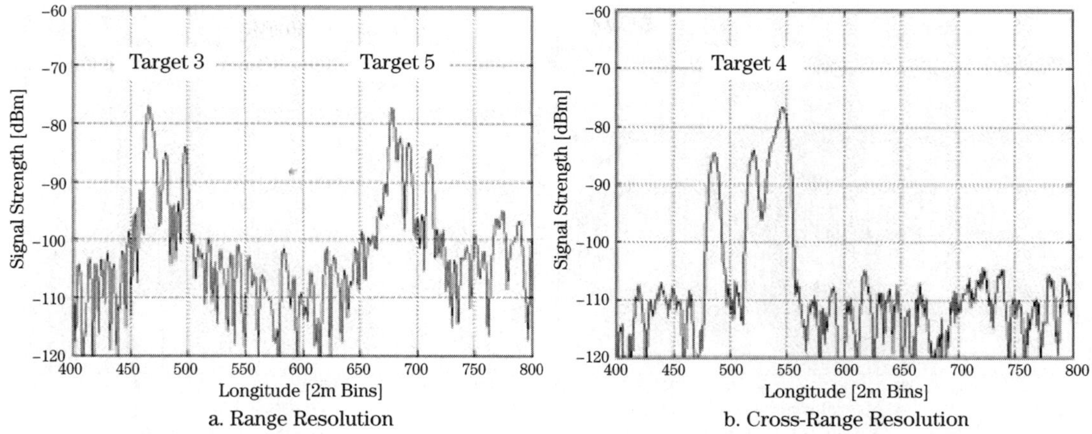

FIGURE 9–19

Cuts through target groups to determine monostatic resolution © 2009 IEEE [7]

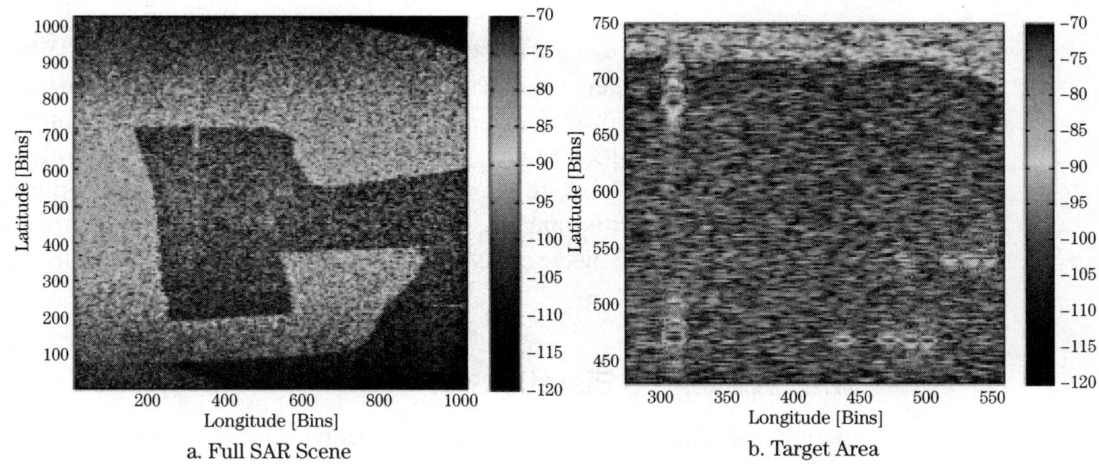

FIGURE 9–20

Pseudo-monostatic simulated SAR scene © 2009 IEEE [7]

center – a pseudo-monostatic geometry. Figure 9–20 shows the 2 meter SAR image constructed from the simulated data for this case. Note that the clutter returns are warmer than the monostatic clutter in Figure 9–18 This is primarily due to the improvement in SNR or SCR from the bistatic geometry. However, there are minor differences in grazing angles and a reduced clutter cell size. In addition, the signal strength is scaled by the RADAR range squared.

Figure 9–21 shows the longitude and latitude slices through the same three targets as shown in the monostatic case. The bistatic enhancement of the

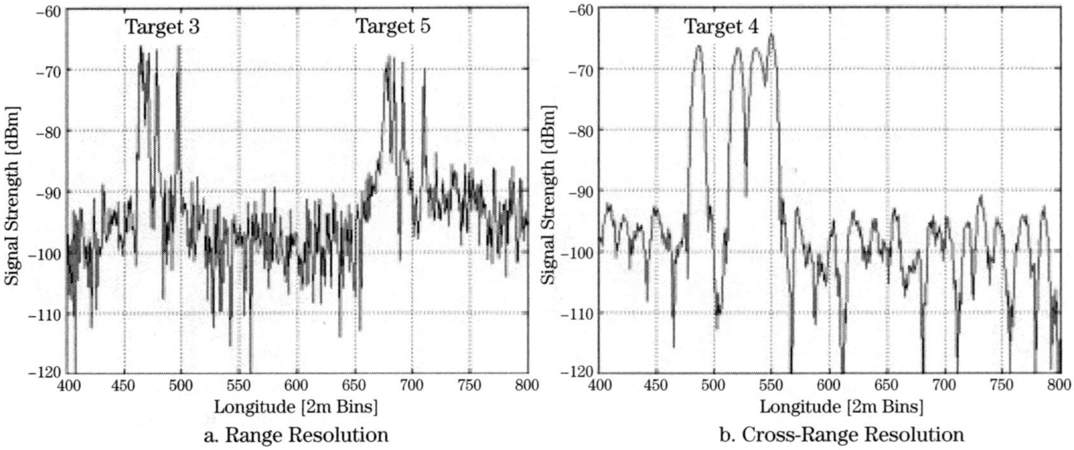

FIGURE 9–21

Cuts through pseudo-monostatic SAR for resolution © 2009 IEEE [7]

target characteristics is evident at the scene center due to their high SCRs when compared with the targets obscured by foliage. The 13 dB improvement in SNR is apparent in the pseudo-monostatic case due to the 10-to-1 reduction in range ratio. The 5-to-1 pseudo-monostatic improvement in cross-range resolution is also noted, where the closer spacing of the targets falls entirely within one 2 meter cross-range resolution pixel of the SAR image. In the monostatic case the 10 meter cross-range resolution fails to fully resolve all the targets.

9.4.3 90 Degree Bistatic SAR

The final case examined is for the bistatic receiver flying on a velocity vector of 90 degrees with respect to the scene illumination. This geometry equates to approximately an 80 degree bistatic angle (β). The receiver is flying "east" at a slant range of 2 Km "north" scene center. The transmitter is located 20 Km "west" of the scene center, as in the two previous cases. Figure 9–22 shows the SAR image for this geometry, with similar characteristics to the pseudo-monostatic image in Figure 9–20. However, the detailed target characteristics shown in Figure 9–23 indicate the advantage and disadvantage of the wide bistatic angle operation. The two-way attenuation of targets 3 and 5 are reduced to one way since the receiver has clear line-of-sight to the targets.

However, the SAR "improvement" in range resolution is lost, since this is in the same dimension of the transmit range waveform. This loss in SAR

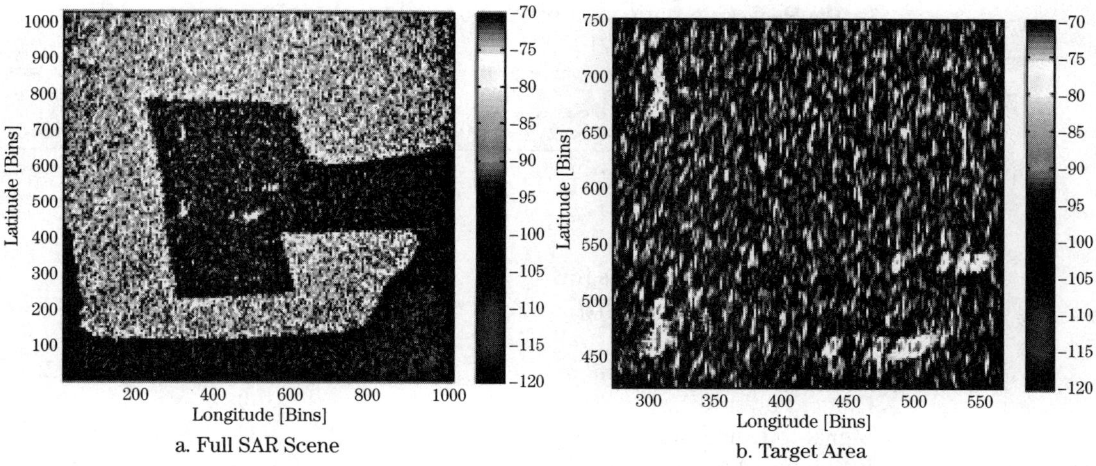

FIGURE 9–22

Simulated SAR scene for 90-degree bistatic resolution evaluation © 2009 IEEE [7]

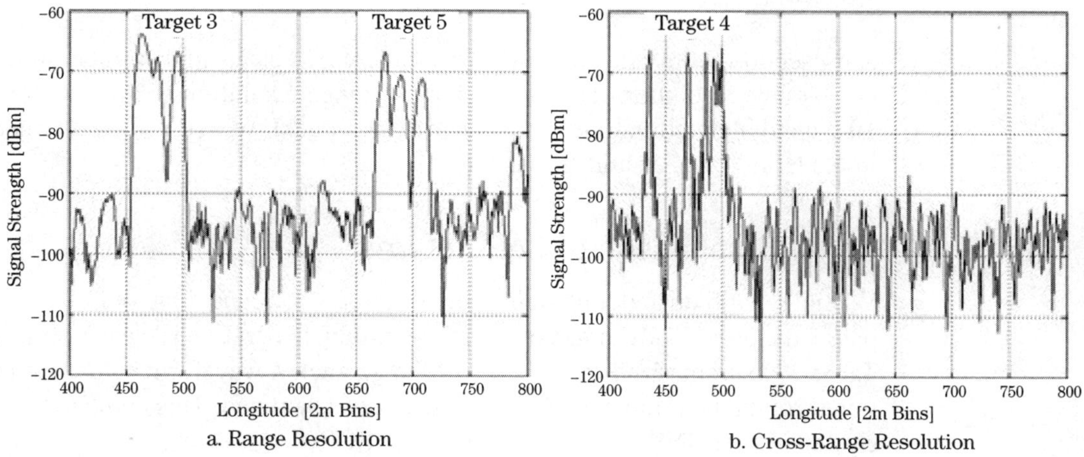

FIGURE 9–23

Cuts through 90-degree bistatic SAR to determine resolution © 2009 IEEE [7]

cross-range resolution can be appreciated by looking at the iso-Doppler contours as shown in Figure 9–9. There is a significant reduction in the density of Doppler contours in the east–west dimension, indicating a reduction in the ability to spatially discriminate scatterers by separation of the Doppler frequencies of targets or ground clutter.

For the 90 degree bistatic case the resolution of the targets along the longitude dimension is similar to the latitude resolution in the monostatic

Table 9–4 Comparison of bistatic SAR characteristics for three SAR scenarios © 2009 IEEE [7]

	Monostatic	Pseudo-monostatic	90-degree Bistatic
Peak Target Signal [dBm]	−78	−64	−64/ − 67*
Longitude Resolution [m]	5.3	2	7.5
Latitude Resolution [m]	10	5.8	2
Target 3 & 4 Foliage Loss [dB]	0	0	0
*Target 5 Foliage Loss [dB]	0	0	4.3*
Clutter Area [dBsm]	17.2	10.6	11.7
Clutter Reflectivity [dBsm/m^2]	−33	−32	−32
Clutter Power [dBm]	−110	−98	−94
Signal-to-Clutter Ratio [dB]	26	31	28/25*
Resolution Advantage [dB]	n/a	6.6	5.5
Bistatic SNR Advantage [dB]	n/a	14	14/11*
SCR Advantage [dB]	n/a	5	2/ − 1*

*Targets under foliage

and pseudo-monostatic cases. However, there is a clear improvement in the cross-range resolution. The expected cross-range resolution is 2 meters for a 10 second integration time of the 2 Km slant range. Five of the six test targets in the latitude dimension are visible.

9.4.4 Comparison of SAR Returns

When the latitude and longitude slices of targets 3, 5, and 4 are closely compared, the bistatic advantage of the range–product operation is clearly shown. Table 9–4 summarized the benefits of the geometry, waveform, and attenuation through the foliage for the three scenarios examined. This comparison is carried out for SNR, image resolution, and SCR.

The range–product of the receiver, being close to the targets, provides the bistatic advantage on SNR as given analytically in Figure 9–10. By observing the peak return from the targets, a direct measure of the signal level is obtained. It should be noted that targets 4 and 5 are in the open for the bistatic receiver. Target 3 is observed through the trees in the 90 degree bistatic case. This explains the 3 dB difference in foliage attenuation.

The resolution of the targets is clearly illustrated in the latitude and longitude slices through the targets. For the monostatic 20 Km slant range SAR image, the resolution cell area is significantly larger than for either bistatic SAR case. This resolution affects both the separation of the targets and the

SCR in the scene. Input data on the geometry and mode parameters was previously given for these cases in Table 9–3.

The CNR for the groupings of targets is a more complicated consideration because of the variation in the grazing angle from the transmitter and receiver platforms and the clutter cell size around the targets. The clutter power was obtained from the SAR images from Figure 9–20 and Figure 9–22, and then was compared with the monostatic SAR in Figure 9–18.

Finally, the image resolution is a very nonlinear combination of the waveform bandwidth from (9.22) and the cross-range geometry from (9.23). For pseudo-monostatic operation, the resolution is primarily set by the time bandwidth product in the range direction and by the integration angle in the latitude dimension. However, for larger bistatic angles, there is a strong interplay between the projection of the two orthogonal components of the transmit waveform and the receiver geometry on the two image axes. The final three lines of the table are measures of the bistatic advantage in resolution, SNR, and CNR. The two entries on the 90 degree bistatic case accounts for the added 3 dB of attenuation loss in target 3.

9.5 Summary

This chapter has illustrated the advantages of a bistatic FOPEN SAR receiver working in concert with a tethered airborne GMTI RADAR illuminator. The potential for moderate resolution SAR operation is made possible by the short-range, higher angular rate collection of a bistatic adjunct receiver. Issues of platform motion are evaluated based on a detailed pulse-by-pulse simulation of a large scene, with impact of foliage scattering and loss and target separation. By placing patterns of targets in the scene, the ability to resolve targets in two cardinal planes is provided. Clearly, the collection geometry of a SAR receiver must be fully understood to provide for desired signal to clutter and resolution.

In summary, a bistatic adjunct receiver can provide detection of fixed targets by using the GMTI RADAR waveform. The operation closer to the target area on a small UAV will give increase in system sensitivity and resolution over the monostatic platform. More importantly, both the GMTI RADAR and fixed target surveillance can be obtained with a single waveform.

Future analysis is required on the synchronization and direct path radiation from the GMTI RADAR platform to the UAV and the effect of volumetric clutter on the fixed target detection. This is especially true in a dense radiofrequency interference (RFI) environment. Bistatic operation can definitely provide improvement in system operational capability.

9.6 References

[1] Bessette, L. A., Crooks, S. M., and Ayasli, S., "P-3 Ultra Wideband SAR Grayling Michigan Target and Clutter Phenomenology," Proc. 1999 IEEE *RADAR* Conference, Boston, MA, May 1999, p. 125.

[2] Parsch, A., "Tethered Aerostats," Directory of U.S. Military Rockets and Missiles" *Directory of U.S. Military Rockets and Missiles* Appendix 4: Undesignated Vehicles, 2004, http://www.designationsystems.net/dusrm/app4/aerostats.html

[3] Tethered Aerostat RADAR System, Air Force Air Combat Command Fact Sheet, http://www.af.mil/factsheets/factsheet.asp?id=3507

[4] Robinson, C. A., "RADAR Counters Camouflage," *Signal*, June 2007.

[5] Sanyal, P. K., Brown, R. D., Little, M. O., Schneible, R. A., and Wicks, M. C., "Space-Time Adaptive Processing Bistatic Airborne RADAR," *Proc 1999 IEEE RADAR Conference,* Boston, MA , May 1999, pp. 114–118.

[6] Ulander, L. M. H. and Martin, T., "Bistatic Ultra-Wideband SAR For Imaging Of Ground Targets Under Foliage," *Proc 2005 IEEE International RADAR Conference,* Washington, DC, May 2005.

[7] Davis, M. E. and Kapfer, R. M., "Bistatic SAR Using A Tethered Ground Moving Target Indication RADAR Illumination," *Proc. 2009 IEEE RADAR Conference,* Rome, Italy, May 2009.

[8] Willis, N.J., *Bistatic RADAR*, Artech House, Norwood, MA, 1991, p. 246.

[9] Rigling, B. D., "Spotlight Synthetic Aperture RADAR," in *Advances in Bistatic RADAR*, Ed. Willis, N. J. and .,Griffiths, H. D., SciTech Publishing, Raleigh, NC, 2007, Chapt. 10.

[10] Rigling, B. D. and Moses, R. L., "Polar Format Algorithm for Bistatic SAR," *IEEE Trans on Aerospace and Electronic Systems* Vol. 40, No. 4, October 2004, pp. 1147–1159.

[11] Munson, D. C., O'Brien, J. D., and Jenkins, W. K., "A Tomographic Formulation of Spotlight-Mode Synthetic Aperture RADAR," *Proc of the IEEE* Vol. 71, No. 8, August 1983, pp. 917–25.

[12] Li, Z., Zeng, D., Long, T., Wang, W., and Hu, C., "Analysis of Time Synchronization Errors in Bistatic SAR," *Proc. 2008 International Conference on RADAR,* Adelaide, Australia, October 2008.

[13] Yang, Y., Ping, Y., and Li, R., "Back Projection Algorithm for Spotlight Bistatic SAR Imaging," *Proc 2006 International Conference on RADAR,* Shanghai, China, October 2008.

[14] Cherniakov, M, Zeng, T., and Plakidis, E., "Ambiguity Function for Bistatic SAR and Its Application in SS-BSAR Performance Analysis," *Proc 2003 IEEE RADAR Conference,* Huntsville, AL, May 2003.

[15] Hartnett, M. and Davis, M. E., "Bistatic Surveillance Concept of Operations,"", *Proc. 2001 IEEE RADAR Conference,* Atlanta, GA, April 2001.

[16] Davis, M. .E., Maher, J., Hancock, R., and Theis, S., "High Fidelity Modeling of Space-Based RADAR," *Proc. of the 2003 IEEE RADAR Conference,* Huntsville, AL, May 2003.

[17] Kapfer, R., Davis, M.E., and Krumme, M., "Sparse Array Performance with Subarray and Timing Errors," *Proc 2006 IEEE RADAR Conference,* Verona, NY, April 2006.

Glossary

Acronym	Meaning	Usage
ACR	Area Coverage Rate	Measure of RADAR operation for rate of collecting target returns over unit area
ADC	Analog-to-Digital Converter or Conversion	Device or technique that converts an analog signal to digital words for processing
AFCRL	Air Force Cambridge Research Laboratory	Research Laboratory that developed early FOPEN SAR system, Hanscom AFB MA
AFRL	Air Force Research Laboratory	Laboratory responsible for research and development of RADAR technology
AIRSAR	Airborne synthetic aperture RADAR	NASA JPL airborne system
AM	Amplitude Modulation	Modulation technique on interfering signals using time varying amplitude
AMBR	Ambiguity Ratio	Level of unwanted signal in detection process due to either antenna or waveform range and Doppler ambiguities
ARL	Army Research Laboratory	US development organization for BoomSAR experimental UWB system
ATD	Advanced Technology Demonstrator	Prototype system with form and function approaching operational unit
ATD/C	Automatic Target Detection/Classification	Processing procedure to differentiate between targets and clutter and provide probabilistic estimate of target class
ATI	Along Track Interferometry	Technique for discriminating moving targets in dual channel SAR

Acronym	Meaning	Usage
BNN	Bayesian Neural Network	Processing technique for class discrimination using neural networks and Bayesian probability
BoomSAR	Boom Synthetic Aperture RADAR	Experimental UWB system integrated on controlled ground moving platform by ARL, Adelphi MD
BPA	Back Projection Algorithm	FOPEN SAR image formation algorithm
CARABAS	Coherent All RAdio BAnd Sensing	VHF FOPEN RADAR developed by FOA (now FOI), Sweden
CERDEC	Communications Electronics Research and Development Command	US Army engineering unit responsible for procurement and engineering of TUAV FOPEN systems
CFAR	Constant False Alarm Rate	Receiver technique that sets a detection threshold to limit false alarm rate as a function of the SNR
CLS	Coherent Least Squares	Adaptive Processing Technique that Evaluates Least Squares of Signal Error from Ideal Response
CLSC	Coherent Least Squares with target Clipping	RFI removal algorithm that reduces interference from received signal using CLS technique, along with strong target removal
CONOPS	Concept of Operations	Military term for how a system will be employed and tasked
COTS	Commercial Off The Shelf	Hardware or signal processing elements that can be configured from commercially available products
DARPA	Defense Advanced Research Projects Agency	US research organization
DEM	Digital Elevation Model	Description of local terrain height
DTED	Digital Terrain Elevation Data	Standard for representing terrain elevation of Earth's surface
EIRP	Effective Isotropically radiated power	Power radiated by a device in all directions, i.e. with 0 dBi gain

Acronym	Meaning	Usage
ERIM	Environmental Research Institute Michigan	Research organization in Ann Arbor MI active in early SAR systems
FCC	Federal Communications Commission	Governing body in United States for communications regulations
FDR	Frequency Dependent Rejection	NTIA technique for assessing total receiver rejection of UWB signal
FEBA	Forward Edge of Battle Area	Military term for line between conflicting armies
FFT	Fast Fourier Transform	Technique to filter signals using an efficient application of the Fourier Coefficients
FIR	Finite Impulse Response	Digital signal processing technique for filter construction
FJB	Frequency Jumped Burst	Technique for transmitting UWB signal by use of a timed sequence of multiple frequency-diverse pulses
FM	Frequency Modulation	Modulation technique on interfering signals using temporal variation of frequency
FOA/FOI	Swedish Defense Research Agency	CARABAS RADAR developer in Linkoping Sweden
FOLPEN	FOLiage PENetration	FOPEN SAR system developed by SRI
FOPEN	FOliage PENetration	RADAR mode to detect objects under tree cover
FORESTER	FOPEN Reconnaissance, Surveillance Tracking and Engagement RADAR	FOPEN GMTI system flown on A-160 UAS
GCDS	Ground Communications and Distribution System	Remote unit for data link connection to UAS and distribution of image products
GeoSAR	Geospatial Synthetic Aperture RADAR	Dual band and polarization mapping RADAR developed by NASA JPL
GHz	Gigahertz	Billion cycles per second

Acronym	Meaning	Usage
GMTI	Ground moving target indication	Technique for detecting targets moving in a ground environment
GPS	Global Positioning System	Satellite based navigation system
HALE	High Altitude Long Endurance	Class of UAS that flies at very high altitude and long mission timelines
HF	High Frequency	
ICM	Internal Clutter Motion	Modulation of clutter by environment to present small Doppler frequency effects
IF	Intermediate Frequency	Receiver down-conversion technique with a frequency intermediate between transmit and baseband
IFFT	Inverse fast Fourier transform	Signal processing technique to determine time samples from a signal's Fourier coefficients
IFSARE	Interferometric SAR Elevation	Terrain height mapping RADAR developed by ERIM
IMU	Inertial Measurement Unit	Subsystem that measures platform velocity, acceleration and orientation
INS	Inertial Navigation System	Subsystem that determines platform position, velocity, time and orientation over flight path
InSAR	Interferometric Synthetic Aperture RADAR	RADAR mode for terrain characterization
IPR	Impulse Response	Metric of SAR image response to a waveform of specified bandwidth
ISLR	Integrated Sidelobe Ratio	Ratio of waveform or antenna sidelobes power to power in mainlobe
JPL	Jet Propulsion Laboratory	NASA research organization in Pasadena CA, developed AIRSAR and GeoSAR
JSTARS	Joint Surveillance and Target Acquisition RADAR System	X-band GMTI RADAR system developed by US Air Force

Acronym	Meaning	Usage
LMS	Least Means Squared	Adaptive processing technique that minimizes the square of a relevant signal processing metric
LNA	Low Noise Amplifier	Circuit that amplifies signal with minimum added noise and sufficient gain to establish receiver sensitivity
LWL	Land Warfare Laboratory	US Army Lab developed early FOPEN GMTI system
M-FOPEN	Multipurpose FOliage PENetration	Early man portable FOPEN GMTI system by SURC, Syracuse NY
MAP	Maximum A Posterior	Probability after a measurement sequence
Mbps	Mega bits per second	Signal conversion or transfer rate in terms of millions of digital bits
MCE	Mission Control Element	Remote control center for flight and mission control of UAS
MDV	Minimum Discernable Velocity	GMTI characteristic of slow target detection
MHz	Megahertz	Million cycles per second
ML	Maximum Likelihood	Signal condition measure using maximum SNR measure
MNR	Multiplicative Noise Ratio	SAR Image quality metric
MOCOMP	Motion Compensation	SAR processing step to remove platform motion from the synthetic aperture RADAR collection
MOP	Measures of Performance	Critical requirements for RADAR system to meet operational requirements
MSSL	Mean Squared Sidelobe Level	Measurement of the square of the mean of the waveform or antenna sidelobes
NADC	Naval Air Development Center	US development organization for P-3 UWB RADAR
NASA	National Air and Space Administration	

Acronym	Meaning	Usage
NBFM	Narrow Band Frequency Modulation	Modulation technique on interfering signals using narrow band FM
NTIA	National Telecommunications and Information Administration	US regulatory body for frequency use and licensing
OFR	Off-frequency Rejection	NTIA technique for assessing rejection of UWB signal outside receiver passband
OTR	On-tune Rejection	NTIA technique for assessing rejection of UWB signal within receiver passband
PFA	Polar Format Algorithm	SAR processing algorithm using quadratic phase as in a polar flight path
PML	Parametric Maximum Likelihood	Adaptive Processing Technique that uses parametric model to measure minimum power from signal.
PRF	Pulse Repetition Frequency	Rate of RADAR pulse Transmittion
PRI	Pulse Repetition Interval	Time interval between RADAR transmit pulses
PWF	Polarization Whitening Filter	Technique for minimizing clutter speckle using multiple polarization SAR
QNR	Quantization Noise Ratio	Additive noise contributed by the quantization of the ADC or signal digital control circuits
RADAR	RAdio Detection And Ranging	
RADCON	RADAR Detection of Concealed Targets	AFRL program to develop real-time processing for target ATD/C
REFORGER	REturn of FOrces to GERmany	NATO exercises in Germany
RF	Radio Frequency	General class of emitters operating in bands set aside for radio operation

Acronym	Meaning	Usage
RFI	Radio Frequency Interference	
RLSTAP	Rome Laboratory Space-Time Adaptive Processing	Modeling and simulation environment developed by AFRL, TSC and CAESoft to simulate multiple channel RADARs
RMA	Range Migration Algorithm	FOPEN SAR image formation algorithm
ROC	Receiver Operating Characteristic	Technique for measuring target detection and false alarms
ROI	Region of Interest	Area in SAR image derived from CFAR to enable further characterization of target versus clutter
RPV	Remotely piloted vehicle	
RVP	Residual Video Phase	Error in SAR signal when the modulation phase has not been completely removed.
SADFRAD	Single Aperture Dual Frequency RADAR	Early FOPEN SAR experimental system
SAR	Synthetic Aperture RADAR	
SCR	Signal-to-Clutter Ratio	
SEACORE	South East Asia Communications Research	Early development of effects of foliage on communications in tropical jungles.
SINR	Signal-to-Interference-plus-Noise Ratio	Measure of RADAR signal power to the sum of background noise and all interference power
SIR	Signal-to-Interference Ratio	Ratio of the power in the signal to the total noise plus interference power
SLAR	Side Looking Array Radar	RADAR mode for battlefield characterization
SMI	Sample Matrix Inversion	Adaptive processing technique to calculate weights from the covariance matrix of a signal vector
SOTAS	Stand Off Target Acquisition System	Early X-band GMTI system

Acronym	Meaning	Usage
SRI	Stanford Research Institute	FOLPEN developer, Menlo Park CA
STAP	Space Time Adaptive Processing	Method to adaptively cancel background interference by using both spatial and temporal channel degrees of freedom
SURC	Syracuse University Research Corporation	Early FOPEN RADAR developer for Army LWL
TARS	Tethered Aerostat Radar System	Surveillance RADAR installed in an aerostat platform that is tethered to a ground station
TCR	Target-to-Clutter Ratio	Ratio of power in target return to clutter in the same resolution cell
TEM	Transverse Electro-Magnetic	RF power transmission where electric and magnetic fields are orthogonal
TUAV	Tactical Unmanned Air Vehicle	Class of UAS that flies at middle altitudes and moderate ranges
TV	Television	
UAS	Unmanned aircraft system	
UHF	Ultra High Frequency	300–1000 MHz frequency band
UWB	Ultra Wide Bandwidth	RADAR operation with greater than 25 percent bandwidth
VHF	Very High Frequency	30–300 MHz frequency band
VSWR	Voltage Standing Wave Ratio	Measurement of the match and reflection of microwave power into a load
WiFi	Wireless Fidelity	Local UWB communications techniques

Index

A

B

J

K

L

M

N

O

P

R

S

₹85·00